Climate Change and Sustainable Development

From rising sea levels and extreme weather events to food insecurity and forced migration, the effects of climate change are real and accelerating.

Climate Change and Sustainable Development offers a comprehensive compilation of contemporary research, covering both fundamental and advanced aspects of climate science. It delves into a wide range of critical topics, including the greenhouse effect, food security, social and environmental issues, threats to marine ecosystems, the role of green energy in reducing carbon emissions, life cycle assessments, cost-effective sustainable strategies, public health implications, carbon capture technologies, and climate adaptation strategies.

Features:

- Explores both fundamental and advanced topics such as the greenhouse effect, carbon footprint reduction, public health responses, and adaptation strategies.
- Integrates insights from environmental science, economics, agriculture, sociology, and earth sciences to provide a holistic understanding.
- Includes the latest findings and case studies, ensuring up-to-date knowledge aligned with global sustainability goals.
- Structured for easy understanding, making it suitable for students, researchers, educators, and professionals alike.
- Emphasizes actionable strategies for mitigation, adaptation, and innovation in sustainable development.

This volume is especially valuable for scientists, academics, students, educators, and professionals across diverse disciplines such as environmental science, agriculture, earth science, economics, sociology, and geology. By addressing every key element related to climate change, this book equips readers with a clear understanding of its causes, consequences, and the broader implications for human society.

The aim of this book is to inform, inspire, and empower readers—from policymakers and researchers to students and concerned citizens. The challenges ahead are formidable, but they are not unmanageable. Through knowledge, collective action, and unwavering commitment, we can confront climate change and build a more sustainable world for future generations.

Climate Change and Sustainable Development

Edited by
Aasawari Jadhav, Prakash Raut,
and Sukanta Mondal

CRC Press
Taylor & Francis Group
Boca Raton London New York

CRC Press is an imprint of the
Taylor & Francis Group, an **informa** business

Designed cover image: Shutterstock

First edition published 2026
by CRC Press
2385 NW Executive Center Drive, Suite 320, Boca Raton FL 33431

and by CRC Press
4 Park Square, Milton Park, Abingdon, Oxon, OX14 4RN

CRC Press is an imprint of Taylor & Francis Group, LLC

ISBN: 9781032795614 (hbk)
ISBN: 9781032795621 (pbk)
ISBN: 9781003492733 (ebk)

DOI: 10.1201/9781003492733

Typeset in Times
by codeMantra

Contents

Preface

Climate change is one of the inevitable environmental realities, which has had enormous effects on living earth. It has become a global concern due to its severe and possibly permanent negative alterations to our planet's geological, biological, environmental, and ecological systems. These changes have led to the emergence of large-scale environmental hazards affecting biota, including human health. The world is experiencing extreme weather patterns, including increases in heatwaves, hurricanes, heavy flooding, insect-borne diseases, and various microbial infections. Climate change is not only affecting the health sector but is also placing a significant burden on agriculture, wildlife, biodiversity, and marine resources. Both the direct and indirect effects of climate change are alarming the country's financial segment. The long-term shift in temperature is accountable for focusing the carbon footprint of the corporate world. Carbon emission calculation load has become the necessary requirement for industrial domain. The technocrats are inventing green technologies to reduce the carbon effects of conventional practices. Carbon-neutral campuses in every organization have become the need of the hour. Climate change has become one of the integral parts of every government policy framework. Therefore, sensitization of general population about climate change and its effects is also an essential aspect.

Climate Change and Sustainable Development is an outstanding collection of current research that integrates basic and advanced concepts of climate change which includes greenhouse effect, food security, social issues, threat to marine biota, green energy for lesser carbon footprint, impact on economy, life cycle assessment, low-cost sustainable practices for climate change, climate change and human health interventions, carbon capture mechanisms for industries, and adaptation to climate change. This book is particularly attractive for scientists, researchers, students, educators, and professionals in environmental science, agriculture, earth sciences, economics, sociology, and geology. This book will make the readers aware about every single concept associated with climate change and enable them to understand the phenomenon of climate change with its impacts on human society.

This book consists of 15 chapters that focus on current approaches and strategies for sustainable development under climate change scenario. Chapter 1 demonstrates the relationship between climate change and greenhouse effect. Chapter 2 enlightens the aspects related to food security due to changing patterns of climate. As severe climate will deplete the agricultural yield, preserving global food security requires policies and legislative actions that support climate mitigation and resilience initiatives, improve readiness for unforeseen events, and promote global cooperation. Chapter 3 deals with the threats to the marine biota which explains the effects of global warming, ocean acidification, deoxygenation, and rising sea levels on marine organisms, emphasizing both general consequences and species-specific susceptibilities. The diminishing cryosphere results in habitat degradation in polar areas and exacerbates warming cycles. Chapter 4 explains the complex interrelationship between biodiversity and climate change, highlighting the vital role that renewable energy sources play in mitigating carbon emissions, i.e. carbon footprint. It also discusses how human-caused climate change affects biodiversity by changing species distributions, upsetting ecosystems, and raising the risk of extinction. Chapter 5 examines the significance of using sources of renewable energy to lower emissions of greenhouse gases, which in turn aids to curb climate changes. It begins by outlining several technologies of green energy such as geothermal, hydro, wind, and solar energy while underscoring their environmental advantages over traditional fossil fuels. This chapter examines recent advancements in energy storage solutions and grid integration, which are essential for overcoming intermittency issues and ensuring a stable energy supply.

Forests play a critical role in mitigating climate change by sequestering carbon and maintaining ecological stability. Therefore, Chapter 6 explores the forest carbon cycle and the mechanisms through which forests capture and store atmospheric CO_2, highlighting their importance in climate change mitigation. It examines the adverse effects of deforestation, driven by factors such as agricultural expansion, logging, mining, urbanization, and natural disturbances and outlines its impacts on greenhouse gas emissions, biodiversity loss, and water cycles. This chapter also discusses sustainable forest management (SFM), focusing on its challenges and the importance of preserving old-growth forests, maintaining biodiversity, and addressing biotic stress factors. Chapter 7 deals with the life cycle assessment as a tool for climate change mitigation. This chapter elaborates the monocrystalline silicon solar technology, which remains a dominant choice in the photovoltaic industry for power generation. A life cycle assessment was carried out to assess its environmental impact on India, focusing on the entire production process from raw material extraction to the final module. This cradle-to-gate study accounts for all energy and material inputs, beginning with silica extraction and ending with the manufacturing of a one-square-meter solar panel. This chapter shows key environmental aspects, including global warming potential, human toxicity, and critical environmental concerns in module production.

Chapter 8 examines accessible, low-cost, sustainable practices as viable tools for mitigating climate change, focusing on practical and adaptable strategies. Key initiatives include agro-ecology and waste management, which support sustainable farming and recycling while offering economic and environmental advantages like cost savings, job creation, environmental protection, and resource conservation. This chapter also highlights successful examples worldwide. Extreme weather events and environmental degradation have been found to be associated with mental health issues such as anxiety, sadness, and post-traumatic stress disorder (PTSD). Climate change disproportionately affects vulnerable populations, such as children, the elderly, and marginalized communities. Socioeconomic factors influencing health contribute to variations in the impact experienced by different demographic groups. Therefore, Chapter 9 emphasizes the immediate requirement for interdisciplinary approaches and new solutions to safeguard public health in response to climate change. The text discusses the future problems and opportunities in addressing the connection between climate change and human health.

Chapter 10 deals with the carbon capture mechanisms for industries. The emission-intensive industries are under increasing pressure to reduce their carbon dioxide emissions. Despite the industry sector's limited adoption of carbon capture and storage (CCS), industry and academic evaluations have raised hopeful hopes for CCS's potential to lower emissions. The efforts done globally by many countries with their interest in carbon capture, utilization, and storage (CCUS) are narrated in this chapter. Chapter 11 emphasizes climate adaptation changes and international methods for climate change adaptation, including the United Nations Framework Convention on Climate Change (UNFCCC). It also highlights the importance of a clean development framework and essential international strategies for ensuring sufficient adaptability. Chapter 12 explains the causes of high carbon emissions in urban cities. Various factors described in this chapter such as usage of cement, machinery, changing lifestyles, industrialization, heating and cooling instruments are responsible for increasing urban heat. Flood disasters are among the most frequent and devastating natural hazards, impacting millions globally each year. Chapter 13 explains the need for effective flood risk management, policy interventions, and resilience-building measures to mitigate the socioeconomic impacts of these recurrent events. This chapter analyzes flood disaster trends in the Indian scenario using data from the EM-DAT database, focusing on key aspects such as affected populations, economic losses, and associated inflationary pressures.

Chapter 14 deals with the entrepreneurship ideas and climate change mitigation mechanisms. Although climate change is a worldwide issue that requires attention, numerous entrepreneurial innovations have emerged to tackle the problem, resulting in the establishment of prominent firms. These concepts are crucial in climate change mitigation and adaptation. Innovations in renewable energy represent one of the most effective entrepreneurial ventures. This chapter summarizes the innovations and job opportunities in the areas of green energy, carbon footprint reduction, etc. Chapter 15 explores the transformative potential of environmental education as a strategic solution to combat climate change. By integrating climate literacy into educational systems, it underscores the pivotal role of empowering individuals with knowledge, skills, and values essential for sustainable living. The discussion aligns environmental education with Sustainable Development Goals (SDGs), highlighting its capacity to foster critical thinking, behavior change, and collective action. This chapter advocates for a collaborative, data-driven, and inclusive approach to climate education for a sustainable future.

Sustainable development is not merely an aspiration; it is a necessity. By balancing economic growth with environmental stewardship and social equity, we can create resilient societies that thrive within planetary boundaries. This book explores various aspects of innovations that are shaping a greener future. The aim of this book is to inform, inspire, and empower readers—from policymakers and researchers to students and concerned citizens. The challenges ahead are formidable, but they are not unmanageable. Through knowledge, collective action, and unwavering commitment, we can confront climate change and build a more sustainable world for future generations.

It is a pleasure to publicly acknowledge our enormous debt to contributors who assisted materially in the preparation of this book. We are grateful to our families who cheerfully tolerated and supported many hours of absence for finishing the book project. Thanks are also due to Taylor & Francis publishing team for their extra care and patience in publishing this book.

Aasawari Jadhav
Prakash Raut
Sukanta Mondal

About the Editors

Aasawari Jadhav is currently working as Assistant Professor and I/c Head, Department of Environmental Science, Shivaji University, Maharashtra, India. Her major research interests involve environmental monitoring in air, water, wastewater, noise, soil, environmental microbiology, and climate change. She has been awarded with the UGC NET Junior Research Fellowship in 2006. She is working as Principal Investigator for National Ambient Air Quality Monitoring Project for the Kolhapur region. She is a Lead Auditor for Environmental Management System (ISO 14001: 2015). She is involved in the preparation of Green audit reports of many educational institutes and industries. She has published more than 47 research papers in national/ international journals and presented more than 30 papers at national/international conferences. She is the Editor of Environmental Studies Textbook for undergraduate students of Distance mode and Co-Editor of Environmental Studies Textbook for undergraduate students. Being the Chairman of Board of Studies in Environmental Science, she has designed many environmental syllabi.

Prakash Raut has an M.Sc. and Ph.D. in Zoology with various degrees in education such as D.H.Ed., B.Ed., M.Ed., and D.E.M. He is an Ex-Professor and Head of the Department of Environmental Science, former I/C Dean of Interdisciplinary Studies, former I/C Head of Department of Geography and Zoology at Shivaji University, Kolhapur. Presently, he is the Chief Executive Officer of Shivaji University Research and Development Foundation, a Section 8 Company of Shivaji University. He has received many other educational degrees and has 39 years of teaching experience. He specializes in Environmental Toxicology, Ecology, and Biodiversity. He has 97 research publications to his credit, published in national and international journals of repute on Invertebrate Endocrinology, Toxicology, and Ecology. Under his guidance, 17 students have completed Ph.D., 6 students M.Phil., and 22 students completed their M.Tech. degree. He has attended 86 national and international conferences, seminars, and workshops in India and abroad to present papers and deliver lectures. He has received many honors such as the Gold Medal and FAEB Fellowship of Academy of Environmental Biology, Lucknow. He was honored by Shivaji University with the "University Best Teachers Award" in 2017. He was also felicitated with the "Vasundhara Gourav Award 2018" by Kirloskar Vasundhara Festival. He is working and has worked on various bodies, committees, and establishments such as Senate, Management Committee, BOS, BOE, Coordinator, Chief Rector, etc. He is a member of various national and international organizations like BNHS, IBCN, WWF, and various other research societies. He has completed research projects of MoEFCC, DAE-BRNS, BARC, and UGC with funding of around 2.22 Cr with themes on GIS study of Wildlife Sanctuary, Flora and Fauna, Demography, and Geo-medical study of groundwater.

Sukanta Mondal is currently working as Principal Scientist in the Physiology Division at ICAR-National Institute of Animal Nutrition and Physiology, Bangalore, India. His major research interests involve the impact of stress on maternal recognition of pregnancy, cloning, characterization and expression of genes regulating early embryonic loss, and molecular characterization and expression of hormone receptors. He is a noted environmental biologist who has deciphered the molecular basis of climate change impact on maternal recognition of pregnancy and embryo survival in sheep. He has published over 200 research publications in various national/international journals and presented more than 30 papers at various national/international conferences. His pioneering and path-breaking contribution in hormone receptor as well as sheep and buffalo genomics has been recognized nationally, and he has received several prestigious awards like the Prof. G. P. Talwar Midcareer Scientist award (2014), Rashtriya Gaurav award (2014), Prof. G. K. Pal award (2013), Fellow of Indian Chemical Society (2013), Fellow of Society for Applied Biotechnology (2012), Fellow of Indian Association of BioMedical Scientists (2010), Dr. K. Anji Reddy award (2011), Prof. P. B. Sen Memorial Oration award (2008), Biotechnology Overseas Associateship award (long term, 2007–2008), and ISSRF Young Scientist Award (2006). He has been bestowed with several fellowships including National Scholarship (Govt. of India), S J Jindal Trust fellowship, Junior Research Fellowship (NDRI), Senior Research Fellowship (IVRI), and CRBR Postdoctoral fellowship (Canada). He was serving as Joint Secretary of the National Environmental Science Academy, Executive member of the Physiological Society of India, and Editorial board member of the Indian Association for Animal Production. He was Associate Editor of journals including the *American Journal of Biochemistry and Molecular Biology*, *Asian Journal of Biotechnology*, *Asian Journal of Cell Biology*, *Current Research in Poultry Science*, and *Biotechnology.*

Contributors

Amanullah Adeel
Department of Environmental Science
Shivaji University
Kolhapur, Maharashtra, India

Abdulahad Ahadi
Department of Geoscience
Kabul University
Kabul, Afghanistan

Jovine Bachwenkizi
Department of Environmental and Occupational Health
Muhimbili University of Health and Allied Science (MUHAS)
Dar es Salaam, Tanzania

D. Bhagawan
Department of Environmental Science
School of Earth Sciences
Central University of Rajasthan
Ajmer, India

Amelia Chebyala
Department of Environmental Science
Shivaji University
Kolhapur, Maharashtra, India
and
Department of Environmental and Occupational Health
Muhimbili University of Health and Allied Science (MUHAS)
Dar es Salaam, Tanzania

Tanvi Choudhary
Department of Environmental Science
School of Earth Sciences
Central University of Rajasthan
Ajmer, India

Sanjivani T. Chougale
Department of Environmental Science
Shivaji University
Kolhapur, Maharashtra, India

Pooja Ramchandra Dharmoji
Department of Environmental Science
Shivaji University
Kolhapur, Maharashtra, India

Vaishnavi A. Falke
M.Sc. Student of School of Environmental Science and Management
CSIBER
Kolhapur, India

Alaa Ayman Hasan
Department of Botany
Latakia University
Latakia, Syria

Mustafa Hasani
Daira Sabz Environmental Consulting Services Company
Kabul, Afghanistan

A.S. Jadhav
Department of Environmental Science
Shivaji University
Kolhapur, Maharashtra, India

Pramod N. Kamble
Department of Environmental Science
School of Earth Sciences
Central University of Rajasthan
Ajmer, India

Fatima Sami Khalifa
Department of Environmental Science
Shivaji University
Kolhapur, Maharashtra, India

V. S. Khandagale
Department of Education
Shivaji University
Maharashtra, India

Rajesh Kumar
Department of Environmental Science
School of Earth Sciences
Central University of Rajasthan
Ajmer, India

Iddi Mapande
Department of Environmental and Occupational Health
Muhimbili University of Health and Allied Science (MUHAS)
Dar es Salaam, Tanzania

Hassan Mhanna
Department of Environmental Science
Shivaji University
Kolhapur, Maharashtra, India

Rasiya C. Padalkar
Department of Environmental Science
Shivaji University
Kolhapur, Maharashtra, India

Sachin S. Panhalkar
Department of Geography and Coordinator
Centre for Climate Change and Sustainability Studies
Shivaji University
Kolhapur, Maharashtra, India

Abhinandan Patil
D. Y. Patil Education Society (Deemed to be University)
Kolhapur, India

Pooja M. Patil
School of Environmental Science and Management
CSIBER
Kolhapur, India

Sushama Tanaji Patil
Department of Environmental Science
Shivaji University
Kolhapur, Maharashtra, India

S. H. Pawar
Centre for Research and Technology Developments (CRTD), Sinhgad Institutes Solapur
Solapur, India

Somanath D. Pawar
Department of Statistics
Shivaji University
Kolhapur, Maharashtra, India

Satvasheel Powar
School of Mechanical and Materials Engineering
Indian Institute of Technology Mandi
Mandi, Himachal Pradesh, India

Oscar Pungut
Department of Medical Sciences and Technology, Mbeya University of Science and Technology (MUST) Must Road
Iyunga Industrial Area Mbeya, Tanzania

Reem Al-Shikh Rasheed
Department of Zoology, Faculty of Science
Latakia University
Latakia, Syria

P. D. Raut
Department of Environmental Science
Shivaji University
Kolhapur, Maharashtra, India

Sajeed S. Shaikh
Independent researcher
Kolhapur, India

Saurav Shekhar
Department of Environmental Science
School of Earth Sciences
Central University of Rajasthan
Ajmer, India

Pawar Pratiksha Shivaji
Department of Environmental Science
Shivaji University
Kolhapur, Maharashtra, India

Shweta Singh
Bureau of Economic Geology, Jackson School of Geosciences
The University of Texas at Austin
Austin, Texas, USA

Jeetendra Kumar Vaishya
National Medicinal Plants Board
Ministry of Ayush
New Delhi, India

P. R. Vasagadekar
Department of Environmental Science
Shivaji University
Kolhapur, Maharashtra, India

Prakash T. Waghmare
Department of Geography
Shri Shiv Shahu Mahavidyalaya
Kolhapur, Maharashtra, India

Tushar T. Waghmare
Department of Geography
Yashwantrao Chavan Warana Mahavidyalaya, Warananagar
Warananagar, Maharashtra, India

Madhuri Walvekar
Department of Zoology
Shivaji University
Kolhapur, India

1 Climate Change and Greenhouse Effect
A Global Problem

A.S. Jadhav, Sanjivani T. Chougale, and Amanullah Adeel

1.1 INTRODUCTION

Global climate change appears to be complex and interconnected. Certain individuals assert that this issue is attributable to human actions. Despite being a natural phenomenon on Earth, some individuals argue that climate change is inherently natural. However, there are individuals who reject the concept of climate change. Consequently, which option is it? Is climate change a reality? Is anthropogenic global warming occurring? What effects has the evolving global climate exerted on ecosystems? What impact has the shifting global climate had on the fauna and humans residing in such ecosystems? Human activity is a contributing factor to global climate change, as evidenced by facts and statistics from organizations and websites dedicated to raising awareness of the subject. The rise in greenhouse gas (GHG) emissions from fossil fuel usage adversely affects functional ecosystems. Consequently, we will present evidence to address the aforementioned questions from two contrasting perspectives. Is global climate change authentic? Researchers and climatologists have determined that global climate change is a highly concerning issue. The formation of various international organizations, including the United Nations Compensation Commission (UNCC), the United Nations Framework Convention on Climate Change (UNFCCC), the Intergovernmental Panel on Climate Change (IPCC), and the United Nations Environment Programme (UNEP), demonstrates that humanity has addressed global climate change with gravity. Nonetheless, counterarguments against climate change frequently arise from sceptics. The dominant counterargument posits that climate change is a natural occurrence and warrants no concern. This is substantiated by the historical occurrence of warming and cooling cycles on Earth. The volume of GHGs naturally rose and subsequently fell. Some contend that we might currently be seeing another cycle of these natural transformations on our planet. A further counterargument is that "winter remains genuinely cold" and that they perceive no alteration in the climate. Consequently, they deduce that the conclusions on climate change may be inaccurate.

Earth has seen warming and cooling cycles, which have resulted in adverse consequences. The Rainforest Alliance asserts that the cycle will become increasingly severe and frequent due to the accelerated rate of human-induced GHG emissions

DOI: 10.1201/9781003492733-1

compared to historical levels. Each cycle was succeeded by significant environmental upheavals, including widespread extinctions. This global organization, the Rainforest Alliance, is dedicated to safeguarding rainforests and preserving ecosystems. While exceptionally low temperatures still occasionally manifest throughout the winter, their frequency is diminishing. Furthermore, 97% of climatologists and specialists concur that human activities have contributed to global warming. Despite the denial of global climate change, nearly all climatologists agree that it has escalated to a concerning degree, warranting greater attention. Is global climate change anthropogenic? Human activities, particularly the combustion of fossil fuels, are responsible for climate change due to the emission of GHGs. Gases, including CO_2, CH_4, N_2O, and water vapour, elevate the planet's temperature through the absorption and retention of heat. Experts at the National Aeronautics and Space Administration (NASA) and the Jet Propulsion Laboratory assert that GHG emissions from human activities elevate Earth's average temperature, resulting in various repercussions for our ecosystems. The velocity of alterations in Earth's global climate is exceedingly swift; were natural processes the cause, the rate of change would be significantly diminished.

A significant quantity of GHGs has been generated over the years of persistent fossil fuel consumption and combustion. The combustion of fossil fuels has persisted since the inception of the Industrial Revolution. The quantity of GHGs has significantly escalated in the modern age due to the volume of carbon emissions generated by human activities. Vehicle fuel, including that for cars, boats, and airplanes, is sourced from fossil fuels. Individuals utilize fossil fuels to consistently generate GHGs on a global scale.

1.1.1 What Impact Has Climate Change Had on Ecosystems?

The repercussions of climate change continue to be ignored by individuals who have not personally encountered its effects in their locality. Consequently, this does not imply the absence of repercussions. Significant effects of climate change encompass the melting of ice and glaciers, rising sea levels, droughts, flooding, and alterations in climate patterns. Inhabitants residing on land or near water may have problems due to increasing water levels. Consequently, coral reefs may be affected. Coral reefs are essential to the oceanic ecology since they provide habitats for marine organisms and serve as a buffer against storms and waves for coastal land. Weather can affect ecosystem water availability and temperature. Nonetheless, the meteorological patterns of certain ecosystems are significantly influenced by global climate change. Variations in climate can result in variations in weather. A scientific study indicates that anthropogenic activities have exacerbated the severity and frequency of anomalous weather and heat waves, leading to the occurrence of significant storms. Recent years have seen a preponderance of these storms. The annual frequency of storms, including cyclones, hurricanes, typhoons, wildfires, floods, and droughts, is rising. From a human perspective, one of the most significant effects is the alteration of rainfall patterns. These alterations have resulted in flooding in certain areas; nevertheless, inadequate precipitation in other ecosystems may cause droughts. What impacts has the shifting global climate exerted on the fauna and human populations residing inside those ecosystems? Contrary to common assumption, humans are the most susceptible beings to the repercussions of climate change. An illustration of this is the sustenance we consume. An illustrative instance is the food ecosystem. Insufficient rainfall can lead to droughts that adversely affect the seeds and plants essential for cultivating fruits and vegetables. Climate change may accelerate the proliferation of disease. Intensifying and more frequent extreme heat waves can adversely affect certain crops and elevate the risk of illness and mortality, especially among at-risk populations. Elevated temperatures and heightened humidity facilitate the proliferation of disease-carrying insects, including ticks and mosquitoes. For example, Lyme disease and malaria have the potential to kill multiple individuals and spread more easily. Other organisms, besides humans, are also endangered by climate change. Numerous living species will encounter various life-threatening dangers due to climate change. Due to their reliance on ice for existence, several Arctic animal species, including penguins and polar bears, are presently endangered. However, the ice that sustains their buoyancy and enables their hunting activities diminishes as temperatures increase. The changing ocean currents have fundamentally transformed the ecology of various species, including the North Atlantic cod. Over time, numerous species have encountered various challenges associated with climate change. While several groups face permanent extinction, others are merely commencing their fall. To benefit both ourselves and other living organisms in our world, we must initiate progress on the appropriate path. Which option is it, then? The response to that inquiry is affirmative; both natural and anthropogenic global climate change exist. It impacts numerous ecosystems on Earth. It affects the various animal species residing in these settings. Despite ongoing disputes, climate change is fundamentally anthropogenic; thus, stringent measures are imperative. However, the question remains as to how we can prevent or lessen its impact. We can start by promoting a greater awareness of environmental issues. We can diminish resource use, enhance recycling efforts, and augment the utilization of renewable energy sources. Nonetheless, increasing knowledge of climate change is the most effective route to action. Increased awareness of this issue among the populace can expedite the prevention of future harm. Instead of attempting to demonstrate that climate change is a natural phenomenon, it is imperative to recognize that it is a global issue, necessitating the exploration of strategies to mitigate hazards.

1.1.2 General Background

Climate change has resulted in an increase in global temperatures (Battisti and Naylor 2009). The Earth's climate issue intensified with the onset of the Industrial Revolution (Adeel and Jadhav 2024). Prompt diagnosis and adequate care may enhance the likelihood of mitigating its detrimental effects. Evaluating the effects of climate change on individual sectors is impractical (Jurgilevich et al. 2017); nonetheless, enhanced comprehension of climatic uncertainties and their integration into local and national policy may lead to the identification of key drivers. Natural sources of GHGs encompass volcanoes, wildfires, and seismic events (Murshed 2020, 2022; Usman and Balsalobre-Lorente 2022). An extraordinary international accord was established during COP21 in 2015 to combat climate change. Each nation pledged to reduce GHG emissions by all feasible means. The implementation of these strategies necessitates financial resources; hence, developing nations may need assistance from rich countries. The primary objective is to maintain the average global temperature at 1.5°C until mid-century and below 2°C by the century's conclusion. Achieving these targets would constitute the inaugural human endeavour to regulate climate change (Sharif et al. 2020; Sharma et al. 2020). This section highlights that climate change is a significant environmental issue, primarily driven by human activity. This chapter examines how climate change is a global pressing matter.

1.2 GREENHOUSE EFFECT

In the late 19th century, Swedish physicist Arrhenius was the first to explore the correlation between CO_2 levels and rising temperatures. He utilized a single-layer model alongside a thermal equilibrium model of the Earth's surface and atmosphere. In assessing the climate's sensitivity to thermal forcing, he evaluated two primary positive feedback mechanisms: the interactions of water and snow albedo, given the limited data available at that time.

1.2.1 Greenhouse Gas Effects

GHGs are the atmospheric gases that absorb and emit infrared radiation, hence regulating heat. Both naturally occurring and anthropogenic GHGs are present in the troposphere. Collectively, GHGs absorb a significant portion of the upper atmospheric radiation emitted by terrestrial objects, constituting a negligible component of the atmosphere. The radiation generated by the atmosphere partially offsets the loss of upper-altitude flux caused by the aforementioned absorption. GHG emissions, along with their storage, offsetting, sequestration, and environmental sinking, might yield varied impacts. The generation of GHGs is significantly affected by land use and land cover (LULC). The surface is defined by LULC if it acts as a source or sink of GHGs. Methane is generated in a region where waste products are present on the surface. Should the region be classified as a wetland, it functions as a carbon sink (Figure 1.1).

1.2.2 Effects of GHGs

Water vapour in the troposphere constitutes one of the most potent, but tiny, concentrations of GHGs, contributing to cloud formation and the vaporization of water (Cassia et al. 2018). More accurate calculations indicate that clouds and water vapour contribute 25% and 49%, respectively, to long wave (thermal) absorption. The quantities of water vapour vary spatially and are not directly influenced by human activity. Water vapour, which simultaneously emits and absorbs energy throughout a substantial portion of long wave radiation, is a key contributor to the GHG effect in the atmosphere (Manabe 2019). In contrast to persistent GHGs like carbon dioxide, the water vapour cycle initiates with evaporation from the Earth's surface (e.g., the ocean) and exits the atmosphere rapidly via condensation and precipitation. Water vapour often persists in the atmosphere for a maximum

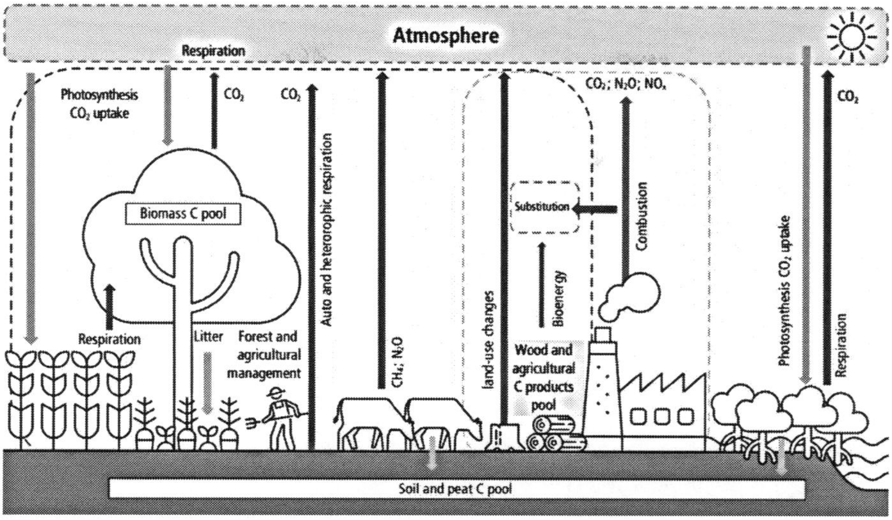

FIGURE 1.1 Greenhouse formation following land management (IPCC 2022).

of several weeks. Methane is another significant gas present in the environment. This gas emission originates from the combustion of fossil fuels, agricultural practices, and livestock farming. The concentration of CH_4 has increased twofold since prior to the Industrial Revolution. In 2014, the global average concentration was 1.8 μmol/mol. Despite constituting merely 0.5% of atmospheric CO_2, the rising emission of CH_4 is a significant cause for concern. CO_2 is 30 times more effective than CO_2 as a GHG (IPCC 2014). Sources of CH_4 emissions to the atmosphere encompass the mining, extraction, and transportation of coal, natural gas, and oil. Methane is emitted during the decomposition of organic waste in landfills and municipal solid waste, as well as in agricultural and animal husbandry practices. The assessment of metabolic methane emissions from livestock husbandry takes into account the kind of animal, its productivity, and its dietary composition. Third, although N_2O is a powerful GHG, it indirectly facilitates the production of O_3. In the last century, global emissions have increased significantly, primarily as a result of human activity (IPCC 2014). N_2O and NO are emitted by the soil. It is essential to distinguish nitrous oxide (N_2O) from nitrogen dioxide (NO_2) and nitric oxide (NO). Nitrogen dioxide and nitric oxide are not designated as GHGs, despite their substantial contribution to the formation of tropospheric ozone, which is a GHG. Nitrous oxide exists in the atmosphere due to various natural and anthropogenic sources, many of which are challenging to quantify. Thus, there is consensus that maintaining equilibrium between the atmospheric sources and sinks of nitrous oxide is difficult.

Ultimately, the combustion of wood products, coal, oil, gas, and solid waste emits carbon dioxide into the atmosphere. Carbon dioxide may also be emitted during some chemical reactions, like those involved in cement production. Plants extract it from the atmosphere as part of the biological carbon cycle, a process referred to

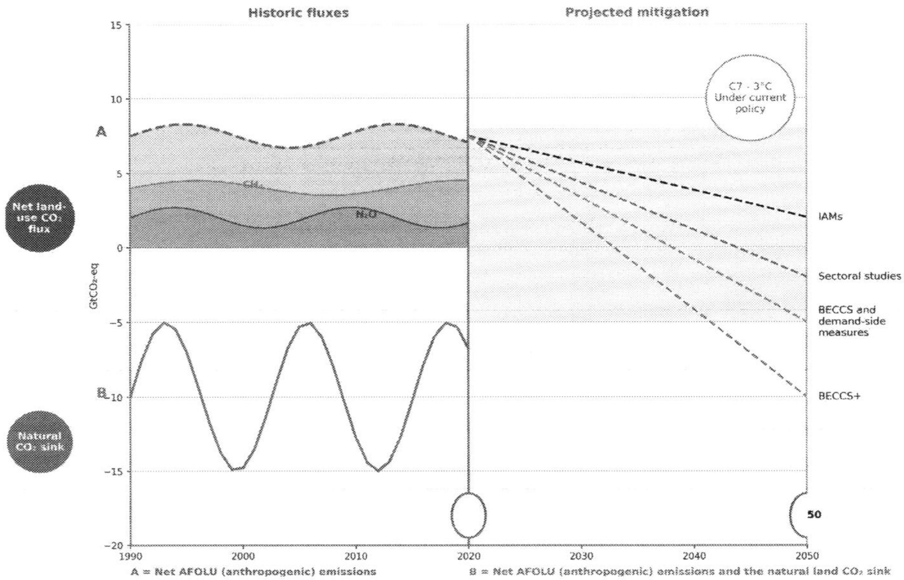

FIGURE 1.2 Historical land sector GHG flux estimates from the IPCC, together with options for mitigation for forestry, agricultural, and other land users (AFOLU) through 2050.

as "sequestration." The substantial accumulation of CO_2 in the atmosphere, rather than its intensity, renders it the most perilous GHG currently. Forecasted mitigation, informed by historical anthropogenic and natural flows, indicates a reduction until 2050 (Figure 1.2).

1.2.3 Major Sectorial Sources of GHGs

Notably, 21% of direct CO_2 emissions originate from industry, whereas 24% stem from land uses excluding agriculture and forestry (IPCC 2014). Three GHGs generated by various sectors predominantly regulate Earth's temperature: CO_2, CH_4, and N_2O. The current topsoil temperature is 14°C, rising to 19°C upon the removal of GHGs. According to data from the IPCC (2014), anthropogenic GHG emissions are consistently elevated. Over the past 30 years, the global temperature has risen by around 0.8°C. This escalation transpired due to GHG emissions in the 1970s. The IPCC (2022) reported the direct and indirect emissions of various sectors as follows: direct emissions are 5.6% for buildings, 15% for transport, 22% for agriculture, 24% for industries, 10% for other energy, and 23% for electricity plus heat. The direct plus indirect emissions are 16% for buildings, 15% for transport, 22% for agriculture, 34% for industries, and 12% for other energy (Figure 1.3a).

Each sector contributes variably to GHG emissions. The quantity of GHG emissions is contingent upon the nature of the activity, its duration, chemical usage, geographical factors, and regulatory frameworks. This phrase fundamentally represents the issue of "energy." Energy consumption across all sectors contributes to GHG

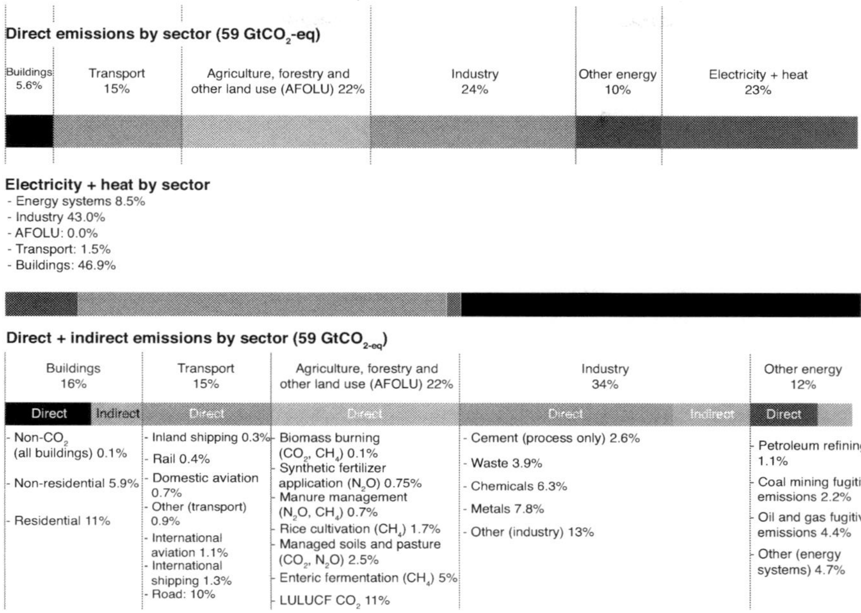

FIGURE 1.3a Total annual human-caused greenhouse gas emissions, direct and indirect, broken down by sector and subsector (in $GtCO_2$-eq). Emissions are attributed to the sector from which they originate in direct emissions estimates (scope 1 reporting). The transfer of emissions from the generation of heat and electricity to the final use sector is referred to in this context as "indirect emissions" (scope 2 reporting).

emissions. Energy is an essential component throughout agriculture, transportation, industry, and urbanization, thereby serving as a significant source of GHG emissions. The primary constituents of fossil fuels are hydrogen and carbon. Fossil fuels are combusted when oxygen and hydrogen react to produce carbon dioxide (CO_2) and water (H_2O). During these processes, heat is emitted, which we subsequently harness as energy. The quantity of heat produced is contingent upon the proportions of carbon and hydrogen in the fuel, as is the volume of CO_2 emitted. Natural gas, mostly consisting of methane (CH_4), possesses a high hydrogen concentration, which leads to reduced CO_2 emissions when combusted compared to other fossil fuels for equivalent heat output. For instance, combusting an equivalent amount of energy from natural gas generates roughly 50% of the CO_2 emissions associated with coal. The IPCC (2022) documented global sector-wide energy production for the year 2019 (Figure 1.3b).

The protocol establishes national targets for the reduction of GHG emissions. The protocol's signatories must develop and submit an annual national inventory of their human GHG emissions. The IPCC (2014) created protocols for the development of inventories. Emission statistics for a country may be considered; however, this necessitates accurate inventory data obtained from precise GHG measurements. The IPCC (2022) report illustrates the transportation sector on a continental scale, revealing that North America exhibited the largest emissions from 1990 to 2019 (Figure 1.3c).

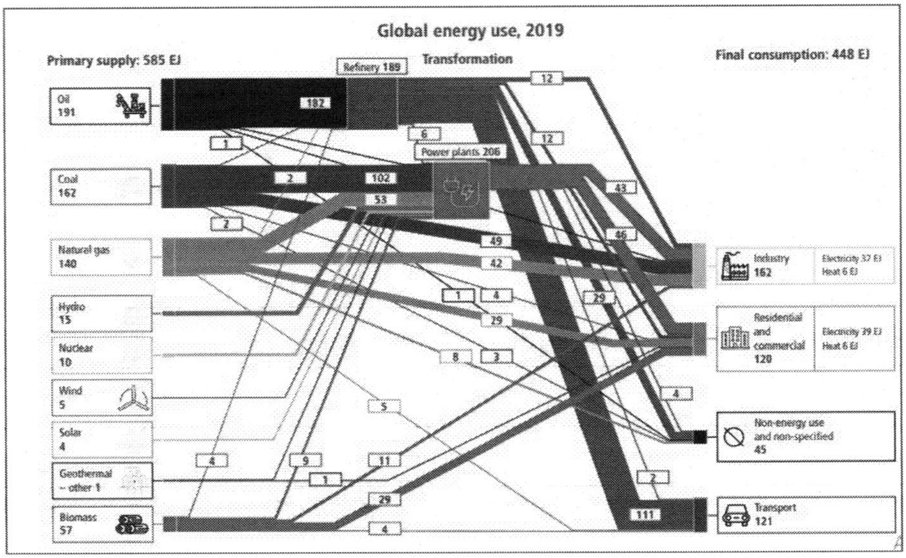

FIGURE 1.3b Global energy type production.

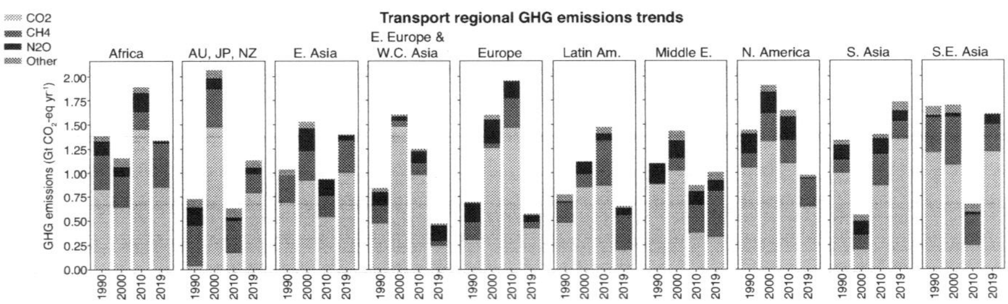

FIGURE 1.3c Global transport greenhouse gas emission trends.

1.3 GLOBAL IMPACT OF CLIMATE CHANGE

Climate change and environmental degradation are interrelated issues threatening the future of our world (Piguet 2022). The prioritization of global development, often to the detriment of natural systems and their consequences, is the primary cause of these issues (Wang et al. 2023). Since the Industrial Revolution, the utilization of natural resources and the combustion of fossil fuels have driven global economic systems. Consequently, the environment has experienced substantial alterations, including urbanization and modifications in land use, which prioritize development over forestry (Wang et al. 2023). Fossil fuel utilization has significantly increased GHG emissions for various reasons, thereby altering global temperature and precipitation patterns (Tol 2020). Wang et al. (2023) report that the average global temperature increased by 0.86°C by 2022 compared to the 20th-century average, indicating 46 consecutive years of rises over this baseline. Examples of climate change's impact on the world encompass severe weather phenomena, including wildfires in Australia and the United States, which have expedited polar ice melting, sea-level elevation, modified river flows, extreme precipitation, droughts, species extinction, and the proliferation of infectious diseases (Cai et al. 2024; Dong et al. 2024; Kumar et al. 2024; Liu et al. 2024). In light of these dangers to the environment and humanity, it is imperative to create adaptation and mitigation strategies prior to the onset of irreversible impacts.

Scientists increasingly concur that mitigation and adaptation strategies for climate change are vital for addressing the issues arising from global warming. Even if all anthropogenic emissions ceased suddenly, climate change would persist. Decades will pass before climate change mitigation actions have observable effects on rising temperatures; therefore, global systems must be reformed to accommodate ongoing changes that will persist in the foreseeable future (Wang et al. 2023). Developing and executing solutions across all global development systems may be essential to enhancing resilience and adapting to climate change. Two instances of these strategies involve cultivating drought-resistant crops to combat water constraints and constructing coastal barriers to safeguard towns from rising sea levels. To prepare for extreme weather events, one can utilize artificial intelligence (AI), advanced monitoring and modelling technologies, and satellite-based remote sensing (Heavens et al. 2013; Voosen 2018). These technologies utilize both historical and contemporary data to facilitate the development of Earth systems and climate models. These models, which illustrate the frequency and intensity of extreme weather events, can be utilized to predict their occurrence and impacts. Establishing early warning systems enables the preservation of human populations and ecosystems by applying predictive capabilities to vulnerable areas (Heavens et al. 2013). By 2050, when carbon neutrality is anticipated to be achieved, this objective must be fulfilled. This indicates that the transition from carbon-based energy to low-carbon alternatives is both urgent and unprecedented, notwithstanding the swift expansion of clean and renewable energy sources (Fuhr 2021). To achieve the climatic objectives established in the Paris Agreement, the present level of effort is inadequate. Marquardt et al. (2023) asserted that, despite enduring commitments to significantly diminish GHG emissions, developing societies are increasingly challenged to achieve carbon neutrality while fostering economic growth. Strategies for alleviating climate change have been founded on enduring sustainable development goals (SDGs). This encompasses augmenting the capture, utilization, and storage of carbon dioxide (CO_2); safeguarding and enhancing carbon sinks; and modernizing and integrating food, water, and energy systems. Furthermore, modifications in nutrition, behaviour, and lifestyle are essential to mitigate the impacts of climate change. Climate change endangers all facets of existence; thus, the global society must collaborate and expedite efforts to mitigate these perilous transformations (Ratwatte et al. 2023).

1.3.1 Effects of Climate Change

1.3.1.1 Elevation of Sea Levels

Land-based ice melting considerably influences sea-level rise. Two factors are responsible for this: the melting of mountain glaciers and ice caps, along with the retreat and melting of the Antarctic and Greenland ice sheets. The World Glacier Monitoring Service has recorded almost 130,000 glaciers in its World Glacier Inventory, reflecting the extensive quantity of glaciers on Earth. MIMURA (2013) posits that the volume of water contained in glaciers is estimated to correspond to a global sea-level rise of 0.5 m. In regions like the Himalayas, glacial lakes have emerged, and glaciers are receding due to rapid melting and runoff, driven by the increased vulnerability of high-altitude glaciers and ice caps to warming temperatures.

Research on sea-level rise has progressed markedly in multiple domains since the publication of IPCC AR4. Advancements in processing techniques for tide gauge and satellite altimeter data have rendered sea-level change data more exact and comprehensive, as observed by Cazenave and Llovel (2010) and Ray and Douglas (2011). The increase in ocean heat content has emerged as a substantial factor in the rise of mean sea levels as seawater undergoes thermal expansion due to global warming. To get more accurate estimates of changes in the ocean's heat content, ways have been found to change historical data from disposable bathythermographs (XBTs). Comprehension of the mechanisms affecting alterations in ice sheet mass in Greenland and Antarctica has progressed, despite the challenges associated with conducting accurate quantitative evaluations of these changes. Recent projections of sea-level rise have been developed using process-based models, including atmospheric and oceanic general circulation models (AOGCMs), for the 21st century and beyond. Moreover, by analysing historical data on the relationship between mean sea level and global mean temperature, semi-empirical models—lacking a robust scientific foundation—have been created to forecast sea-level rise. Numerous studies in the impact assessment sector have been done to comprehend the potential repercussions of sea-level rise (Mimura 2013).

1.3.1.2 Effects of Climate Change on Precipitation Patterns

Numerous studies have shown that Indian precipitation in Northeast China and South China generally exceeds that in West China and Northwest China (Goswami et al. 2006; Krishnan et al. 2020; Nayak et al. 2021; Prakash et al. 2015). The phenomena of El Niño and La Niña alter the Indian Ocean Dipole, hence affecting rainfall patterns and augmenting the frequency of this phenomenon. The Madden-Julian Oscillation (MJO) influences the eastward movement of clouds and rainstorms over the tropics. The Himalayan mountain range's capacity to obstruct winds from transporting moisture during the monsoon season influences rainfall distribution. The region's topography and anthropogenic activities, including deforestation and urbanization, influence variations in precipitation patterns. The southwest region experiences the highest precipitation during the monsoon season, which occurs from June to September, while the northwest region receives the least (Goswami et al. 2006). Temporal variations in precipitation distribution have been observed. A reduction in monsoon precipitation in some areas has been linked to alterations in land use, atmospheric aerosols, and GHGs (Maity et al. 2021; Nayak et al. 2021; Singh et al. 2019). Rainfall patterns are a significant component of the Earth that can be profoundly affected by climate change resulting from increasing global temperatures. Studies indicate that climate change has already altered rainfall patterns in multiple places (Von Storch et al. 1993). Sambas, located in West Kalimantan, are crucial for food availability, particularly within the province. Precipitation is a vital component of agricultural meteorology that influences food security and maintains the water supply in West Kalimantan. Consequently, to anticipate and mitigate the likelihood of future disasters such as extreme droughts or floods, it is essential to examine the impact of climate change on Sambas' water supply (Sipayung et al. 2018).

1.3.1.3 Climate Change Impacts Flora

Human activities not only generate heat through energy consumption but also alter energy flow and influence the global climate by modifying the composition of atmospheric gases (Karl and Trenberth 2003). The primary immediate consequence of elevated GHG concentrations in plants is an increase in atmospheric CO_2. Increased CO_2 levels prompt plants to promptly enhance net photosynthesis and reduce stomata aperture (Long et al. 2004). Moreover, experts on climate change anticipate an increase in medium- and long-term surface temperatures and CO_2 concentrations, accompanied by erratic rainfall patterns and a rise in the frequency and severity of extreme weather events, including heat waves, droughts, and floods. To a smaller extent, plant absorption of O_3 may potentially hinder photosynthesis and induce oxidative stress (Mittler and Blumwald 2010). These acute events, which typically exceed the adaptive ability of the plants, result in diminished agricultural and arboreal yields (Ciais et al. 2005).

1.4 PRESENT CONDITIONS AND DEVELOPMENTS IN CLIMATE CHANGE

Human activity is the principal catalyst of global warming, with GHG emissions being the primary culprit. Between 2011 and 2020, the global average temperature rose by around 1.1°C. During this period, well-integrated GHGs likely resulted in a warming of 1.0°C–2.0°C, while other anthropogenic factors (mostly aerosols) likely caused a cooling of 0.0°C–0.8°C (IPCC 2022). The global surface temperature was likely influenced by internal variability by 0.2°C to +0.2°C, while fluctuations in solar and volcanic activity were probably responsible for shifts of 0.1°C to +0.1°C. Homogeneously distributed GHG concentrations have been rising since around 1750, and it is unequivocal that anthropogenic emissions are responsible for these increases. From 1850 to 2019, the cumulative historical net CO_2 emissions amounted to $2{,}400 \pm 240$ $GtCO_2$, with 42% (high confidence) occurring between 1990 and 2019, and 58% occurring between 1850 and 1989. In 2019, atmospheric concentrations of methane and nitrous oxide were the highest recorded in at least 800,000 years, with a high degree of certainty, reaching 1,866 parts per billion for CO_2 and 332 parts per billion for nitrous oxide. In 2019, atmospheric CO_2 concentrations reached their greatest levels in at least two million years, with a high degree of confidence. As of 2019, the estimated global net human GHG emissions were 596.6 $GtCO_2$-eq, representing an increase of about 12% (6.5 $GtCO_2$-eq) since 2010 and a rise of 54% (21 $GtCO_2$-eq) since 1990. Carbon dioxide from industrial operations and fossil fuel combustion constitutes the predominant share and increase in total GHG emissions, followed by methane. Since its initial introduction at low levels in 1990, fluorinated gases, or F-gases, have experienced the highest relative growth rate. The highest average annual

GHG emissions were reported from 2010 to 2019; nonetheless, the growth rate during this period (1.3% per year) was lower than that of 2000–2009 (2.1% per year). Total historical CO_2 emissions, CO_2 from fossil fuel combustion, and net CO_2 emissions from land use, land-use change, and forestry demonstrate considerable geographical variation. In 2019, 41% of the global population resided in nations generating less than 3 tCO_2-eq per capita, with a considerable segment of this demographic lacking access to contemporary energy services. Approximately 35% of the global population resides in nations that emit over 9 tCO_2-eq per capita, excluding CO_2-LULUCF. If you take out CO_2-LULUCF, the average per capita emissions of least developed countries (LDCs) and Small Island developing states (SIDS) are 6.9 tCO_2-eq. The LDCs level is 4.7 tCO_2-eq and SIDs is equal to 1.7 tCO_2-eq (Calvin et al. 2023).

1.5 MECHANISM OF CLIMATE CHANGE

1.5.1 Mechanisms of Thermal Variation

1.5.1.1 Financial Allocations for Energy and Water

Previous research has primarily concentrated on the local processes and mechanisms that can initiate changes linked to an increased greenhouse effect. Nonetheless, atmospheric circulation may adjust in a manner that amplifies these fluctuations. For instance, convergent circulation continuously diminishes when precipitation declines due to the aforementioned remote top-down mechanism, which either enhances subsidence or stabilizes the atmosphere. This diminishes moisture convergence and hence lessens precipitation. On the other hand, when evaporation and precipitation get stronger because of the local bottom-up mechanism, there is upward motion linked to near-surface convergence and more precipitation. This tendency persists when models forecast substantial dry or wet changes (Giannini 2010).

1.5.1.2 Mechanisms for Solar-Climate Interconnections

The subsequent coupling processes are comprehensively understood in the second category of mechanisms sometimes referred to as "top-down" mechanisms. These mechanisms also include a robust first phase that entails heating the stratosphere via an increase in UV radiation. Even so, there are still questions that need to be answered about the size of certain geographical effects, the roles and connections between different coupling systems, and the specifics of temperature and ozone indicators in the stratosphere. Before looking at the possible pathways connected to the third mechanism—cloudiness responses to changes in the atmosphere's ionization caused by galactic cosmic rays—more information needs to be gathered. To figure out how the pathway through ion-induced condensation nuclei might affect other ways that droplets form, it is important to fully understand the steps that come after. Additional observational evidence regarding cloud edge features, including charge and droplet sizes, is required for procedures involving the electric circuit.

1.5.1.3 Earth in the Solar System

The geothermal heat sources from the sun and Earth's interior form the planet's closed energy system. At 30°C, the former is deemed rather stable at 0.06 W/m^2, while the potential influence of incoming solar radiation is 6,000 times more significant. The mantle and crust comprise roughly 0.0257% of the total solar energy incident on the Earth's surface. Since 1950, solar irradiance has remained constant, albeit with fluctuations within an 11-year cycle. Approximately 0.0257% of the total solar radiation that reaches the Earth's surface is due to heat transfer from energy produced in the crust and mantle. Solar irradiation varies throughout an 11-year period; however, it has not risen since 1950. A variety of factors, including cloud cover, surface albedo, snow cover, soil moisture, vegetation cover, latitude, thermal conductivity, and latent soil heat, affect the quantity of incoming radiation. The total energy produced globally is estimated to be 0.0077% of the solar radiation that reaches Earth. Kathayat et al. (2017) assert that solar radiation constitutes around 99.95% of the entire energy affecting the global climate.

1.5.1.4 Thermal Extremes

A 2021 report published by the IPCC indicates that human activity has resulted in a 1.1°C rise in Earth's temperature. According to the US EPA (2021), the incidence of heat waves has risen in the USA (Figure 1.4). Furthermore, the IPCC said that within two decades, the temperature will unavoidably rise to 1.5°C above pre-industrial levels, resulting in dire repercussions for human health. If GHG emissions do not rapidly decline, global warming will surpass 1.5°C, potentially resulting in disastrous consequences for humanity (IPCC 2023). A notable example of the multifaceted progression of the climate crisis is the rising incidence of extreme weather phenomena, such as heat waves, wildfires, floods, droughts, tornadoes, and tropical cyclones

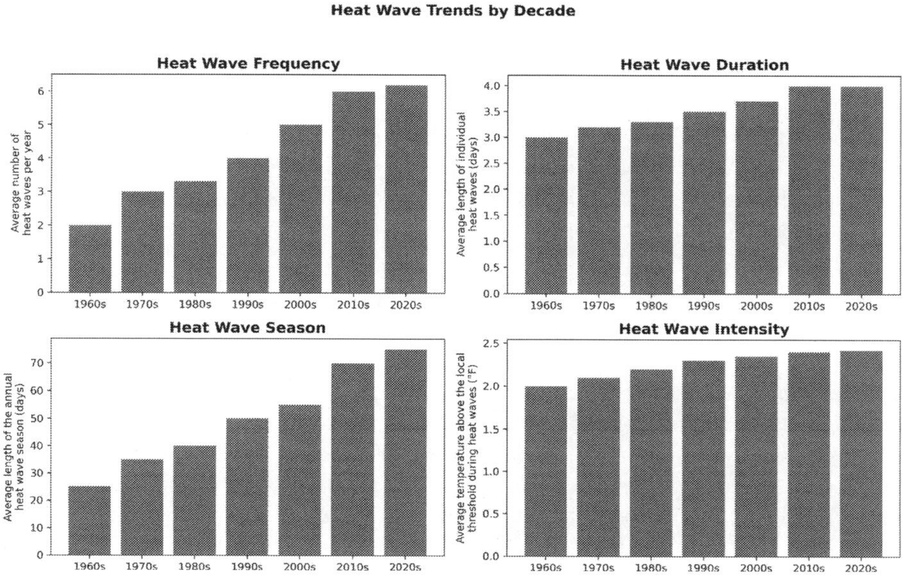

FIGURE 1.4 Heat waves in the United States from 1961 to 2021: An analysis of their characteristics by decade (US EPA 2021).

(Salas and Ebi 2021). Vulnerable people are disproportionately impacted by these occurrences due to poverty, housing or food instability, ageing, racism, and other detrimental social factors affecting health. They adversely affect health by directly causing injuries and illnesses, as well as disrupting the functioning of healthcare institutions (Arcaya et al. 2020).

1.6 CLIMATE MODEL AND PREDICTION

1.6.1 Climate Modelling and Forecasting

Climate models suggest that, notwithstanding substantial endeavours to mitigate GHG emissions, global warming is anticipated to continue until at least the 2050s. The delay stemmed from the time necessary to implement emission reduction measures. The extent of future warming will depend on present actions. Substantial reductions in carbon dioxide emissions could halt global warming by mid-century, leading to a temperature increase of only 1.5°C–2°C. Nonetheless, if emissions persist at elevated levels, temperatures might surpass 4.5°C by century's end. Although forecasts indicate a likely temperature increase of 1.5°C between 2021 and 2040, maintaining temperatures below the thresholds of 1.5°C or 2°C will be difficult without prompt and significant reductions in emissions (IPCC 2023).

Researchers have utilized the intermediate global climate model (IGCM) to elucidate the activities of lipids. This particular model operates at a T21 horizontal resolution, roughly equivalent to 6°, and comprises 22 vertical levels. The name "intermediate" is used because its physical parameterizations (surface flux, radiation, clouds) are akin to those of the most sophisticated models from the 1980s. These parameterizations enable a greater number of computations due to their comparatively reduced processing requirements. A conventional mixed-layer ocean model is employed to compute the variance in ocean surface temperature. The model's seasonal and regional variations in sea surface temperature (SST) are guaranteed to align with the SST recorded in the control run through designated oceanic heat fluxes. The IGCM's response to various climate perturbations exhibits notable similarities to the advanced GCM, ECHAM4 (Joshi et al. 2003), when adjusted for each model's reaction to alterations in CO_2 concentrations.

1.7 POLICY AND GOVERNANCE IN CLIMATE CHANGE: ADDRESSING GLOBAL CHALLENGES

To effectively address the extensive ramifications of climate change, a multifaceted and robust global concern necessitates a comprehensive policy and governance structure. The complex array of factors, such as GHG emissions, deforestation, and industrial pollution, that drive climate change necessitates a unified and multifaceted response at both national and international levels. Effective climate action is enabled by clear objectives, cooperation across several policy domains, and inclusive governance mechanisms. If broadly adopted, regulatory and economic instruments might significantly diminish emissions and enhance climate resilience. The use of diverse information sources is advantageous for development resilient to climate change.

Effective climate governance, inclusive and open decision-making, equitable practices, regulatory stability, and enhanced access to financial resources and technology are essential for both mitigation and adaptation. Effective climate action is facilitated by robust coordination among local, municipal, national, and subnational entities, alongside partnerships with civil society and marginalized populations. The enhanced application and scalability of regulatory and economic mechanisms may result in significant reductions in emissions. Enhancing and strengthening regulatory procedures can optimize mitigation outcomes in sectorial applications according to national conditions.

Carbon price tools have helped low-cost efforts to cut down on emissions, but they haven't been as good at pushing for higher-cost steps that are needed for even more reductions, either when used alone or at the rate that was in place during the evaluation period. Revenue can be utilized, among other purposes, to assist low-income households in mitigating the distributional and equity impacts of carbon pricing mechanisms, including emissions trading and carbon taxes. Enhancing regulatory frameworks and abolishing fossil fuel subsidies can diminish emissions and promote sustainability; yet, the negative impacts on distribution may require careful mitigation. Comprehensive policy packages, encompassing price modifications and public investment pledges, can facilitate the attainment of both immediate economic objectives and carbon reduction targets. Leveraging diverse knowledge systems and employing significant participation and engagement strategies foster the development of socially acceptable solutions and climate-resilient (Figure 1.5) progress (Calvin et al. 2023).

India formulates its climate change strategy based on a benefits framework that prioritizes the optimization of development and climate outcomes (Dubash et al. 2013). These were initially addressed in the 2008 NAPCC and elucidated in India's Five-Year Plan, which included initiatives to foster economic progress while enhancing the nation's climate. The 2008 introduction of the NAPCC aimed to redirect India's developmental trajectory by integrating climate issues with overarching developmental objectives through sustainable development strategies that advantage both the economy and the environment (Byravan and Rajan 2013). The NITI Aayog is responsible for coordinating efforts to achieve the SDGs. NITI Aayog collaborates with the Ministry of Statistics and Program Implementation (MoSPI) and other relevant ministries to develop indicators that represent the SDGs. A preliminary mapping analysis has been done to identify key ministries and relevant agencies for individual programmes and aims in relation to current central government efforts.

1.8 CONCLUSION

Climate change is a worldwide phenomenon fundamentally linked to our activities. The primary source of this energy consumption is coal and fossil fuels utilized for transportation, electricity generation, industrial activities, agriculture, and many everyday services. Climate change is a consequence of GHG emissions, including carbon dioxide, nitrous oxide, and methane. Current climate change trends suggest that elevated levels must be diminished to ensure a secure globe for humans, animals, biodiversity, and ecosystems. This global issue has led to ice melting, which in

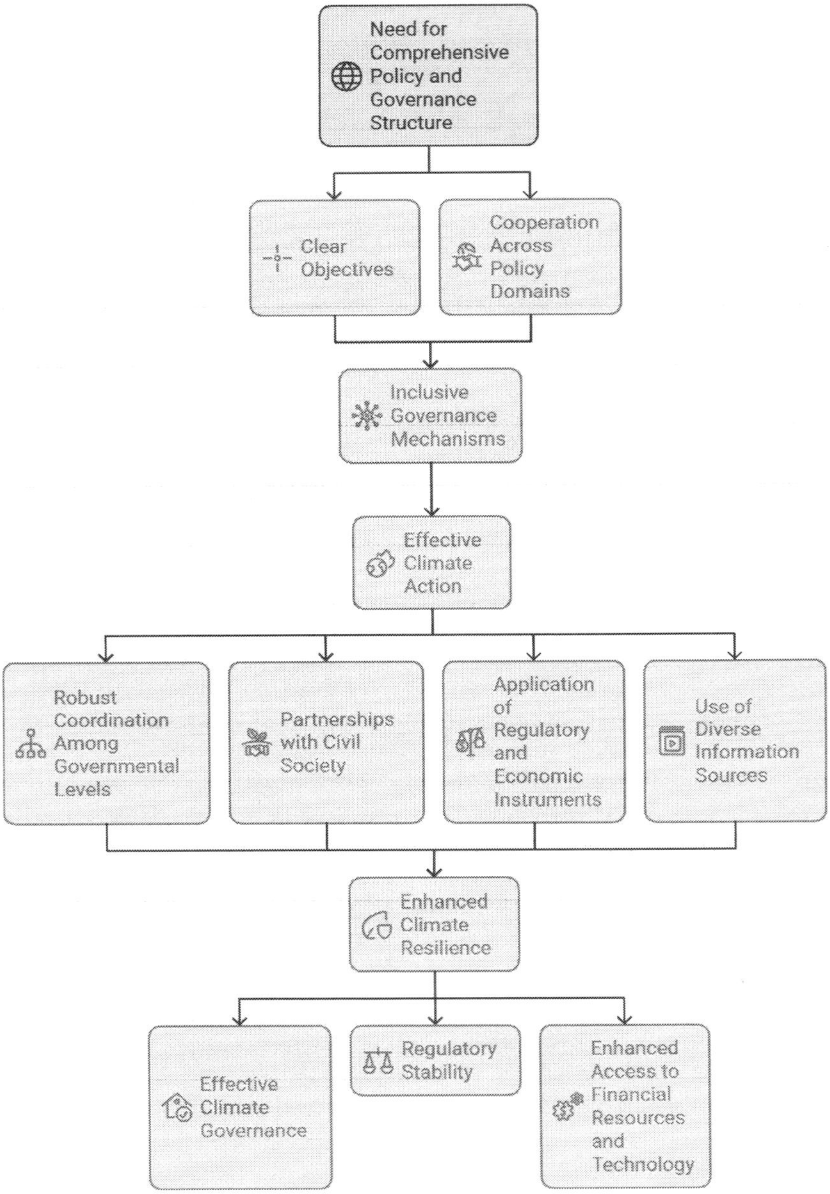

FIGURE 1.5 Policy and governance in climate change.

turn has caused sea levels to rise, posing a significant threat to millions of individuals living in coastal areas. Climate change has dire repercussions for plants, agriculture, and forests. This issue necessitates worldwide collaboration and initiatives through legislation, funding, mitigation, resilience, and adaptation to diminish and prevent further harm.

REFERENCES

Adeel, A., and A. S. Jadhav. 2024. "Identifying the Optimal Sugarcane Residue Percentage for Soil Enhancement: A Statistical and GIS Approach." *Circular Economy and Sustainability*. doi:10.1007/s43615-024-00456-5

Arcaya, M., E. J. Raker, and M. C. Waters. 2020. "The Social Consequences of Disasters: Individual and Community Change." *Annual Review of Sociology,* 46(1): 671–691. doi:10.1146/annurev-soc-121919-054827

Battisti, D. S., and R. L. Naylor. 2009. "Historical Warnings of Future Food Insecurity with Unprecedented Seasonal Heat." *Science,* 323(5911): 240–244. doi:10.1126/science.1164363

Byravan, S., and S. C. Rajan. 2013. *An Evaluation of India's National Action Plan on Climate Change*. SSRN 2195819. https://papers.ssrn.com/sol3/papers.cfm?abstract_id=2195819.

Cai, P., S. Zhou, and P. Huang. 2024. "Seasonally Varying SST Changes in the Joining Area of Asia and Indian-Pacific Ocean from Boreal Spring to Summer." *Climate Dynamics,* 62: 1–14. doi:10.1007/s00382-023-07037-6

Calvin, K., D. Dasgupta, G. Krinner, A. Mukherji, P. W. Thorne, C. Trisos, J. Romero, et al. 2023. In Core Writing Team, H. Lee, and J. Romero (Eds.), IPCC, 2023: Climate Change 2023: Synthesis Report. Contribution of Working Groups I, II and III to the Sixth Assessment Report of the Intergovernmental Panel on Climate Change. First. IPCC, Geneva, Switzerland: Intergovernmental Panel on Climate Change (IPCC). doi:10.59327/IPCC/AR6-9789291691647

Cassia, R., M. Nocioni, N. Correa-Aragunde, and L. Lamattina. 2018. "Climate Change and the Impact of Greenhouse Gasses: CO_2 and NO, Friends and Foes of Plant Oxidative Stress." Frontiers in Plant Science, 9. doi:10.3389/fpls.2018.00273

Cazenave, A., and W. Llovel. 2010. "Contemporary Sea Level Rise." *Annual Review of Marine Science*, 2(1): 145–173. doi:10.1146/annurev-marine-120308-081105

Ciais, Ph., M. Reichstein, N. Viovy, A. Granier, J. Ogée, V. Allard, M. Aubinet, et al. 2005. "Europe-Wide Reduction in Primary Productivity Caused by the Heat and Drought in 2003." *Nature,* 437(7058): 529–533. doi:10.1038/nature03972

Dong, Y., J. Zhang, X. Xu, Q. Dong, A. Zhang, and Z. Xiong. 2024. "Symbiotic Nitrogen Fixation Enhanced Crop Production and Mitigated Nitrous Oxide Emissions from Paddy Crops." *Field Crops Research,* 307: 109261. doi:10.1016/j.fcr.2024.109261

Dubash, N. K., D. Raghunandan, G. Sant, and A. Sreenivas. 2013. "Indian Climate Change Policy: Exploring a Co-Benefits Based Approach." *Economic and Political Weekly*, 48(22): 47–61.

https://books.google.com/books?hl=en&lr=&id=JAFEBgAAQBAJ&oi=fnd&pg=PT19&dq=IPCC+(2014). +Climate+Change+2014:+Mitigation+of+Climate+Change.+Contribution+of+Working+Group+III+to+t he+Fifth+Assessment+Report+of+the+Intergovernmental+Panel+on+Climate+Change,+eds+O.+R.+Ed enhofer,+Y.+Pichs-Madruga,+E.+Sokona,+S.+Farahani,+K.+Kadner,+A.+Seyboth,+et+al.+Cambridge:+Cambridge+Universi ty+Press.&ots=dBzHuFZaZ7&sig=xucJRj7FEvqqQlw2JFcff9Y9DHE doi:10.5194/essd-11-1783-2019

Fuhr, H. 2021. "The Rise of the Global South and the Rise in Carbon Emissions." *Third World Quarterly,* 42(11): 2724–2746. doi:10.1080/01436597.2021.1954901

Giannini, A. 2010. "Mechanisms of Climate Change in the Semiarid African Sahel: The Local View." *Journal of Climate*, 23(3): 743–756. doi:10.1175/2009JCLI3123.1

Goswami, B. N., V. Venugopal, D. Sengupta, M. S. Madhusoodanan, and P. K. Xavier. 2006. "Increasing Trend of Extreme Rain Events Over India in a Warming Environment." *Science* 314(5804): 1442–1445. doi:10.1126/science.1132027

Heavens, N. G., D. S. Ward, and M. Natalie. 2013. "Studying and Projecting Climate Change with Earth System Models." *Nature Education Knowledge*, *4*(5): 4.

IPCC. 2014. Climate Change 2014: Mitigation of Climate Change. In: O. R. Edenhofer, Y. Pichs-Madruga, E. Sokona, S. Farahani, K. Kadner, A. Seyboth, et al. (Eds.), *Contribution of Working Group III to the Fifth Assessment Report of the Intergovernmental Panel on Climate Change*. Cambridge: Cambridge University Press.

IPCC. 2022. Climate Change 2022: Impacts, Adaptation, and Vulnerability. In: [H.-O. Pörtner, D.C. Roberts, M. Tignor, E.S. Poloczanska, K. Mintenbeck, A. Alegría, M. Craig, S. Langsdorf, S. Löschke, V. Möller, A. Okem, B. Rama (eds.), *Contribution of Working Group II to the Sixth Assessment Report of the Intergovernmental Panel on Climate Change* (p. 3056). Cambridge, UK and New York, NY, USA: Cambridge University Press. doi:10.1017/9781009325844

IPCC. 2023. *Climate Change 2021 – The Physical Science Basis: Working Group I Contribution to the Sixth Assessment Report of the Intergovernmental Panel on Climate Change* (1st ed.). Cambridge University Press. https://doi.org/10.1017/9781009157896

Joshi, M., K. Shine, M. Ponater, N. Stuber, R. Sausen, and L. Li. 2003. "A Comparison of Climate Response to Different Radiative Forcings in Three General Circulation Models: Towards an Improved Metric of Climate Change." *Climate Dynamics,* 20(7–8): 843–854. doi:10.1007/s00382-003-0305-9

Jurgilevich, A., A. Räsänen, F. Groundstroem, and S. Juhola. 2017. "A Systematic Review of Dynamics in Climate Risk and Vulnerability Assessments." *Environmental Research Letters,* 12(1): 013002.

Karl, T. R., and K. E. Trenberth. 2003. "Modern Global Climate Change." *Science,* 302(5651): 1719–1723. doi:10.1126/science.1090228

Kathayat, G., H. Cheng, A. Sinha, L. Yi, X. Li, H. Zhang, H. Li, Y. Ning, and R. L. Edwards. 2017. "The Indian Monsoon Variability and Civilization Changes in the Indian Subcontinent." *Science Advances,* 3(12): e1701296. doi:10.1126/sciadv.1701296

Krishnan, R., J. Sanjay, C. Gnanaseelan, M. Mujumdar, A. Kulkarni, and S. Chakraborty. 2020. *Assessment of Climate Change Over the Indian Region: A Report of the Ministry of Earth Sciences (MOES), Government of India*. Springer Nature. https://library.oapen.org/handle/20.500.12657/39973

Kumar, P., S. E. Debele, S. Khalili, C. H. Halios, J. Sahani, N. Aghamohammadi, M. de F. Andrade, et al. 2024. "Urban Heat Mitigation by Green and Blue Infrastructure: Drivers, Effectiveness, and Future Needs." *The Innovation,* 5(2): 100588. doi:10.1016/j.xinn.2024.100588

Liu, F., H. Zhang, Y. Wang, J. Yu, Y. He, and D. Wang. 2024. "Hysteresis Analysis Reveals How Phytoplankton Assemblage Shifts with the Nutrient Dynamics during and between Precipitation Patterns." *Water Research,* 251: 121099. doi:10.1016/j.watres.2023.121099

Long, S. P., E. A. Ainsworth, A. Rogers, and D. R. Ort. 2004. "Rising Atmospheric Carbon Dioxide: Plants Face the Future." *Annual Review of Plant Biology,* 55: 591–628. doi:10.1146/annurev.arplant.55.031903.141610

Maity, S., S. Nayak, K. S. Singh, H. P. Nayak, and S. Dutta. 2021. "Impact of Soil Moisture Initialization in the Simulation of Indian Summer Monsoon Using RegCM4." *Atmosphere,* 12(9): 1148.

Manabe, S. 2019. "Role of Greenhouse Gas in Climate Change." *Tellus A: Dynamic Meteorology and Oceanography,* 71 (1): 1620078. doi:10.1080/16000870.2019.1620078

Marquardt, J., A. Fünfgeld, and J. P. Elsässer. 2023. "Institutionalizing Climate Change Mitigation in the Global South: Current Trends and Future Research." *Earth System Governance,* 15: 100163.

Mimura, N. 2013. "Sea-Level Rise Caused by Climate Change and Its Implications for Society." *Proceedings of the Japan Academy. Series B, Physical and Biological Sciences,* 89(7): 281–301. doi:10.2183/pjab.89.281

Mittler, R., and E. Blumwald. 2010. "Genetic Engineering for Modern Agriculture: Challenges and Perspectives." *Annual Review of Plant Biology,* 61(1): 443–462. doi:10.1146/annurev-arplant-042809-112116

Murshed, M. 2020. "An Empirical Analysis of the Non-Linear Impacts of ICT-Trade Openness on Renewable Energy Transition, Energy Efficiency, Clean Cooking Fuel Access and Environmental Sustainability in South Asia." *Environmental Science and Pollution Research,* 27(29): 36254–36281. doi:10.1007/s11356-020-09497-3

Murshed, M. 2022. "Pathways to Clean Cooking Fuel Transition in Low and Middle Income Sub-Saharan African Countries: The Relevance of Improving Energy Use Efficiency." *Sustainable Production and Consumption,* 30: 396–412. doi:10.1016/j.spc.2021.12.016

Nayak, S., M. Mandal, and S. Maity. 2021. "Assessing the Impact of Land-Use and Land-Cover Changes on the Climate over India Using a Regional Climate Model (RegCM4)." *Climate Research,* 85: 1–20.

Piguet, E. 2022. "Linking Climate Change, Environmental Degradation, and Migration: An Update after 10 Years." *WIREs Climate Change,* 13(1): e746. doi:10.1002/wcc.746

Prakash, S., A. K. Mitra, I. M. Momin, E. N. Rajagopal, S. Basu, M. Collins, A. G. Turner, K. A. Rao, and K. Ashok. 2015. "Seasonal Intercomparison of Observational Rainfall Datasets Over India During the Southwest Monsoon Season: Comparison of Monsoon Rainfall." *International Journal of Climatology,* 35(9): 2326–2338. doi:10.1002/joc.4129

Ratwatte, P., H. Wehling, R. Phalkey, and D. Weston. 2023. "Prioritising Climate Change Mitigation Behaviours and Exploring Public Health Co-Benefits: A Delphi Study." *International Journal of Environmental Research and Public Health,* 20(6): 5094.

Ray, R. D., and B. C. Douglas. 2011. "Experiments in Reconstructing Twentieth-Century Sea Levels." *Progress in Oceanography,* 91(4): 496–515.

Salas, R. N., and K. L. Ebi. 2021. "The Health Benefits of Urgent Upstream Action on Climate Change." *Annals of Internal Medicine,* 174 (11): 1612–1613. doi:10.7326/M21-3347

Sharif, A., S. Mishra, A. Sinha, Z. Jiao, M. Shahbaz, and S. Afshan. 2020. "The Renewable Energy Consumption-Environmental Degradation Nexus in Top-10 Polluted Countries: Fresh Insights from Quantile-on-Quantile Regression Approach." *Renewable Energy,* 150: 670–690.

Sharma, R., A. Sinha, and P. Kautish. 2020. "Examining the Impacts of Economic and Demographic Aspects on the Ecological Footprint in South and Southeast Asian Countries." *Environmental Science and Pollution Research,* 27 (29): 36970–36982. doi:10.1007/s11356-020-09659-3

Singh, D., S. Ghosh, M. K. Roxy, and S. McDermid. 2019. "Indian Summer Monsoon: Extreme Events, Historical Changes, and Role of Anthropogenic Forcings." *WIREs Climate Change,* 10(2): e571. doi:10.1002/wcc.571

Sipayung, S. B., A. Nurlatifah, B. Siswanto, and L. Slamet S. 2018. "Analysis of Climate Change Impact on Rainfall Pattern of Sambas District, West Kalimantan." *IOP Conference Series: Earth and Environmental Science,* 149: 012029. doi:10.1088/1755-1315/149/1/012029

Tol, R. 2020. *Sources of Greenhouse Gas Emissions.* Department of Economics, University of Sussex Business School. https://econpapers.repec.org/paper/sussusvid/2060.htm

US EPA, OAR. 2021. *Climate Change Indicators: Heat Waves.* Reports and Assessments. https://www.epa.gov/climate-indicators/climate-change-indicators-heat-waves

Usman, M., and D. Balsalobre-Lorente. 2022. "Environmental Concern in the Era of Industrialization: Can Financial Development, Renewable Energy and Natural Resources Alleviate Some Load?" *Energy Policy,* 162: 112780.

Von Storch, H., E. Zorita, and U. Cubasch. 1993. "Downscaling of Global Climate Change Estimates to Regional Scales: An Application to Iberian Rainfall in Wintertime." *Journal of Climate,* 6(6): 1161–1171.

Voosen, P. 2018. "The Earth Machine." *Science,* 361(6400): 344–347. doi:10.1126/science.361.6400.344

Wang, F., J. D. Harindintwali, K. Wei, Y. Shan, Z. Mi, M. J. Costello, S. Grunwald, et al. 2023. "Climate Change: Strategies for Mitigation and Adaptation." *The Innovation Geoscience,* 1(1): 100015–100095. doi:10.59717/j.xinn-geo.2023.100015

2 Food Security and Changing Climate

Tanvi Choudhary, Saurav Shekhar, Pramod N. Kamble, D. Bhagawan, Rajesh Kumar, and Jeetendra Kumar Vaishya

2.1 INTRODUCTION

2.1.1 Climate Change Patterns and Future Implications

Global climate change's significant effects on food security are expected due to projections of a rapidly expanding population. With 360 million undernourished and 300 million impoverished people, however, the largest concentration of hungry and impoverished people in the world is found in India (Ahmad et al. 2011).

Asia is seriously concerned about climate change and unpredictability because various climate models show that the global average temperature will rise by 1.5°C between 2030 and 2050 if current trends continue. Less winter rainfall would result in increased aridity throughout South and Southeast Asia. Due to climatic anomalies, the average global temperature variability worldwide will increase the sea level. The land average annual mean temperature, the global average annual mean temperature, the land average annual hottest daily maximum temperature (TXx), and the average annual coldest daily minimum temperature (TNn) time series of observed temperature anomalies have been provided in Figure 2.1. This clearly shows the continuous positive anomalies after 1970, with increasing magnitude in every extending year. Evidence suggests that global and regional warming are consistent with the anticipated increase in the frequency and intensity of hot extremes, consistent with a decrease in the occurrence and intensity of cold extremes. Extreme land temperatures typically rise faster than the global mean temperature because of the land-sea warming differential and occasionally localised feedback (Figure 2.1). It indicates that increasing industrialisation and developmental activities are responsible for the increasing temperature in all forms with varying magnitudes.

The Arctic Ocean has been predicted to experience one summer in 100 years without ice if the average world temperature rises by 1.5°C. However, ice-free summers in the Arctic might occur every ten years if warming reaches 2°C. At least 70% of coral reefs would be destroyed by warming of 1.5°C, but more than 99% would be gone at 2°C. That would wipe out communities that depend on reefs for food, livelihoods, and fish habitats.

Suppose the global temperature rise stays at 1.5°C; around 700 million people (9% of the world's population) will be subjected to extreme heat waves at least once every

 DOI: 10.1201/9781003492733-2

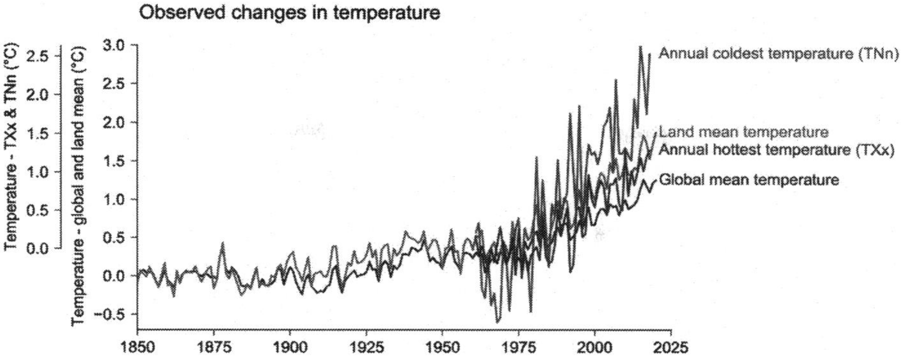

FIGURE 2.1 The world average annual mean temperature (**black**), the land average annual mean temperature (**grey**), the land average annual hottest daily maximum temperature (TXx, **dark grey**), and the land average annual coldest daily minimum temperature (TNn, **light grey**) are time series of observed temperature anomalies.

20 years. In that case, if global temperatures stay at 2°C, more than 2 billion people (28.2%) will be affected. Sea level increase in 2100, relative to 1986–2005 levels, would be 0.4 m (1.3 feet) with a 1.5°C warming. It would be 0.46 m (1.5 feet) in 2100 at 2°C (Levin 2018). A higher rise in temperature also increases the likelihood of floods. Up to 69 million people would be at risk of floods if the global average temperature rises by 1.5°C by 2100 (assuming no adaptation and present population). A 2°C global rise in temperature is projected to expose up to 79 million people, 18% of insects, 16% of plants, and 8% of vertebrates to lose more than half of their ranges. Further, the impact will decrease half of plants and vertebrates and two-thirds of insects, with a 1.5°C rise in temperature (Levin 2018).

The Sahel, southern Africa, the Mediterranean, and the Amazon are predicted to have fewer food scarcity threats under 1.5°C warming than they would under 2°C. As temperatures rise, economic losses increase, with middle-income countries expected to be most affected. For instance, by 2100, losses to the world GDP will be 0.3% if warming is kept at 1.5°C. The losses with 2°C warming would be 0.5% of the world GDP (Levin 2018).

Globally, the development of future weather data is encouraged, and the adverse effects of climate change on agriculture are quantified through the use of simulation models such as Representative Concentration Pathways (RCPs) and General Circulation Models (GCMs) (ur Rahman et al. 2018). To understand agronomic practices and climate variability's impact on plants' development and growth, researchers used DSSATv4.6 and APSIMv7.5 (Keating et al. 2003). Additionally, field data was utilised to validate these models. Baseline and projected climatic data were used to evaluate the effects of climate variability on crops like rice and wheat. Crop models can characterise internal processes by accounting for temperature rise and high CO_2 during critical crop growth stages (Challinor et al. 2018). There are currently no state-of-the-art methods or tools available to ascertain how climate fluctuations impact crop and cattle productivity, except from the modeling approach. (Asseng et al. 2015).

2.2 EFFECT OF CLIMATE CHANGE ON AGRICULTURE AND CROP YIELD

Extreme climatic events and natural calamities, such as wildfire (23%), droughts (19%), floods (37%), and insect pest infestation (9%), account for agricultural productivity losses (crops and livestock) in Asia and have a detrimental effect on farmer's revenue. However, without adaptation, predicted crop yield losses range from 7% to 23% in the worst-case climate change scenario (Rezaei et al. 2023). In the event of an infestation, warmer weather may create an environment more favourable to the growth and spread of diseases and pests, which could lead to crop losses. Farmers may need to adjust by modifying what they cultivate or how they grow it if climate change causes adjustments in the appropriate growth zones for crops. This may lead to the emergence of new crops, the displacement of established ones, and corresponding modifications in agricultural techniques. Modified precipitation patterns may also disrupt crop growth seasons. Also, due to increased greenhouse gases, climate change will have various consequences on crop yields worldwide. In particular, agricultural production in mid- and high-latitude areas experiences positive impacts, whereas those in low-latitude regions suffer negative consequences. However, beneficial effects may be observed at higher latitudes within a specific climatic range. If this range is exceeded, even high mid-latitudes will be adversely affected by agricultural climate change (Ramalho et al. 2023). Due to changes in rainfall patterns, floods and droughts are predicted more frequently, and future dry seasons are predicted to become drier. Because of outdated technologies, emerging nations are more susceptible to climate variability than developed ones, which can mitigate climate-related extremes using contemporary technologies (Lybbert and Sumner 2012). Climate change hurts the qualitative attributes of wheat crops, namely the protein content and the proportions of carbohydrates and sugar in the grains. Wheat grains' development characteristics increase with high temperatures and rising carbon dioxide levels, but their protein content diminishes (Asseng et al. 2015). In the same way, drought stress has lowered the amount of dissolved sugar and protein in the wheat crop and has a detrimental effect on several phases of reproduction (Hussein et al. 2022; Rakszegi et al. 2019). Furthermore, the amount of starch in wheat grains is declining (Noori and Taliman 2022). Kheir et al. (2019) anticipate that the North Nile Delta of Egypt could see a 17.6% decrease in wheat yield for every 4°C increase in temperature. Crop phenology in China has changed due to climate unpredictability and agriculture management strategies (Liu et al. 2018). The increasing aphid population in wheat fields due to elevated temperatures has decreased production (Tian et al. 2019), and there is a clear link between diseases and climate change. For instance, wheat crops suffer from *Fusarium* head blight caused by *Fusarium* sp. These species are more likely to be attacked under hot, humid conditions (Shah et al. 2018).

2.3 VULNERABLE REGIONS AND POPULATIONS UNDER CLIMATE CHANGE

South Asia is the most food-insecure region in the world, with 262 million underweight people (Rasul et al. 2021). South Asia consists of India, Pakistan, and Bangladesh, particularly in rural areas where these countries suffer from many problems. Some other areas and groups that are mainly at risk, like sub-Saharan Africa (countries such as South Sudan, Somalia, and Yemen), often experience acute food crises exacerbated by conflict and displacement. Southeast Asian countries (such as Cambodia, Laos, and Myanmar), Central America, and the Caribbean (such as Haiti, Guatemala, and Honduras) are experiencing an enhanced frequency of extreme weather events such as hurricanes and droughts. In addition, conflict-affected areas (such as Syria, Yemen, South Sudan, and parts of the Sahel in Africa), urban slums, marginalised communities, children, and women are also severely affected.

2.4 CHALLENGES OF CLIMATE CHANGE AND VARIABILITY ON LIVESTOCK FARMING AND FISHERIES

The farm animal industry is particularly vulnerable to rising temperatures and falling precipitation in arid to semi-arid areas (Balamurugan et al. 2017; Rojas-Downing et al. 2017). Less water will be available for crops and cattle, impacting irrigation systems and leading to a water shortage. This may significantly affect areas reliant on seasonal precipitation or snowmelt for water supplies. In addition, heat stress can lower animal fertility and productivity in climate change scenarios, lowering milk and meat output (Habeeb et al. 2018). Pasture supply and quality modifications may impact animal management techniques and grazing patterns. Domestic livestock can tolerate temperatures between 10°C and 30°C, with each degree of temperature increase resulting in a 3%–5% drop in the feed they consume. Similarly, a 59% increase in feed would be needed in the event of a temperature decline. The establishment and spread of heat stress and drought variations in the climate influence numerous animal diseases. For example, rising temperatures and higher precipitation-induced Rift Valley Fever (RVF) are causing epidemics of tick-borne diseases (TBDs) in sheep, livestock, buffalo, and camels Different livestock breeds react quite seriously to rising temperatures and a shortage of water. Thermal stress has been creating dire consequences on the animal reproductive system in India, which has led to stunted growth and higher rates of poultry mortality (Balamurugan et al. 2017; Chen et al. 2021). Severe feed scarcity would result from excessive rainfall variability combined with drought stress in arid regions of Asia (Arunrat et al. 2018). There is a drop in protein, iron, zinc, and vitamins B_1, B_2, B_5, and B_9 when there is an increasing CO_2 content in the feed (Ebi and Loladze 2019). Impending climatic situations would have a major impact on biodiversity, pastures, grasslands, and the quantity and quality of feedstuffs. Future climate situations could have an impact on livestock productivity by changing how rangeland supports and buffers ecosystems, regulates grazing, modifies animal feed, and emits greenhouse gases. Climate extremes such as cyclones, heat stress, salt, droughts, floods, unpredictable rainfall, cyclones, ocean acidification, and rising sea levels have severely harmed aquaculture throughout Asia (Ahmad et al. 2019). Though the primary danger to fish and algae is rising ocean acidity, however, drought is more damaging than floods and temperature rises because it produces insufficient rainfall, which is especially detrimental to aquaculture (Adhikari et al. 2018). Similarly, among other things, fish attaining late maturity for spawning have been influenced by temperature swings, storms, and irregular rainfall

2.5 POSSIBILITIES FOR FISHERIES AND AQUACULTURE IN INCREASING FOOD SECURITY

Aquaculture and artificial stockings exploit the irrigation and water storage systems to improve the living circumstances and food security of low-income rural populations (Ogello et al. 2013). Asia's rice produce rises when nutrients are supplied, and the rice-fish culture is altered so fish actively munch the rice stem borer (Poonam et al. 2019). Because pond fish culture incorporates the use of leftovers from many systems, it can increase food productivity when combined with crop-livestock systems (Ahmed et al. 2014; Dagar and Yadav 2017; Garlock et al. 2022; Prein 2002).

2.6 WAYS TO REDUCE THE ADVERSE EFFECTS OF CLIMATE CHANGE THROUGH INNOVATIONS IN CLIMATE-SMART FARMING PRACTICES

There are also some adaptation strategies for climate-resilient agriculture, including techniques such as integrated farming systems or agroforestry (intercropping, crop rotation, organic farming), adjustment of crop rotation with legumes, mixed livestock systems, early warning systems, soil, water, energy, carbon sequestration, biodiversity enhancement, breeding crops for resilience to stresses caused by climate change, shifting crop production to new seasons or regions, reduction of greenhouse gas emissions, and insurance against risks. Adaptation technology can potentially boost overall production and profitability through modifications to crop management techniques, including planting timing and density, as well as nitrogen and irrigation application, nutrient management technologies, and crop insurance. Aside from this, the advantages of hydroponics and vertical farming are revolutionising conventional farming methods. The technique of producing crops in layers that are piled vertically is known as vertical farming. Usually, regulated indoor settings with artificial lighting are used for this. It permits agricultural expansion in places with little arable land, such as deserts or regions affected by climate change, as well as in urban areas, lowering transportation costs and carbon emissions. It also permits year-round production.

Hydroponics uses precisely managed nutrient-rich water solutions to allow plants to thrive without soil and achieve optimal growth. This is a desirable alternative for areas with limited water resources since it permits a higher plant density per square foot while utilising much less water (Maurya et al. 2023). Aquaponics is a hybrid of hydroponics and aquaculture (fish farming), in which plant waste supplies nutrients to the plants and fish waste filters and purifies the water for the fish. In aeroponics, plants are cultivated in an atmosphere of air or mist without the use of soil. The exposed roots receive direct delivery of nutrient solutions.

2.7 CONCLUSION

The agricultural sector faces significant difficulties with heat waves, wildfires, droughts, erratic rainfall, storms, floods, and pest infestations becoming increasingly common. Moreover, the livestock industry also faces formidable challenges due to climate change-induced impacts, particularly in dry regions. Furthermore, the uneven distribution of climate change impacts, with varying effects on crop yields across different latitudes, highlights the complexity of the challenge. Even regions traditionally less affected may face unforeseen challenges if climatic conditions exceed certain thresholds. The stark reality of food insecurity, especially in regions like South Asia, sub-Saharan Africa, Southeast Asia, Central America, and conflict-affected areas, demands urgent and concerted action. The adaptation strategies and agricultural advancements for climate-resilient crops offer promising avenues to enhance productivity and profitability in the face of climate change. Therefore, the complex issues brought about by climate change require coordinated action at the local, national, and international levels, ultimately safeguarding food security and livelihoods for present and future generations. To address the multifaceted problems posed by climate change in food production and food security, cooperation between governments, non-profit organisations, academic institutions, and firms is essential for fostering innovation, exchanging knowledge, engaging assets, and global partnership and international cooperation.

REFERENCES

Adhikari, S, CA Keshav, G Barlaya, R Rathod, R Mandal, S Ikmail, G Saha, H De, I Sivaraman, and A Mahapatra. 2018. "Adaptation and Mitigation Strategies of Climate Change Impact in Freshwater Aquaculture in Some States of India." *Journal of Fisheries Sciences. Com* 12 (1): 16–21.

Ahmad, J, D Alam, and MS Haseen. 2011. "Impact of Climate Change on Agriculture and Food Security in India." *International Journal of Agriculture, Environment and Biotechnology* 4 (Citeseer): 129–137.

Ahmad, S, G Abbas, M Ahmed, Z Fatima, MA Anjum, G Rasul, MA Khan, and G Hoogenboom. 2019. "Climate Warming and Management Impact on the Change of Phenology of the Rice-Wheat Cropping System in Punjab, Pakistan." *Field Crops Research* 230: 46–61.

Ahmed, N, JD Ward, and CP Saint. 2014. "Can Integrated Aquaculture-Agriculture (IAA) Produce 'More Crop per Drop'?" *Food Security* 6: 767–779.

Arunrat, N, N Pumijumnong, and R Hatano. 2018. "Predicting Local-Scale Impact of Climate Change on Rice Yield and Soil Organic Carbon Sequestration: A Case Study in Roi Et Province, Northeast Thailand." *Agricultural Systems* 164: 58–70.

Asseng, S, F Ewert, P Martre, RP Rötter, DB Lobell, D Cammarano, BA Kimball, MJ Ottman, GW Wall, and J W White. 2015. "Rising Temperatures Reduce Global Wheat Production." *Nature Climate Change* 5 (2): 143–147.

Balamurugan, B, V Tejaswi, K Priya, R Sasikala, T Karuthadurai, M Ramamoorthy, and J Dayanidhi. 2017. "Effect of Global Warming on Livestock Production and Reproduction: An Overview." *Journal of Veterinary Science and Technology* 6 (3): 12–18.

Challinor, AJ, C Müller, S Asseng, C Deva, KJ Nicklin, D Wallach, E Vanuytrecht, S Whitfield, J Ramirez-Villegas, and A-K Koehler. 2018. "Improving the Use of Crop Models for Risk Assessment and Climate Change Adaptation." *Agricultural Systems* 159: 296–306.

Chen, S, Y Yong, and X Ju. 2021. "Effect of Heat Stress on Growth and Production Performance of Livestock and Poultry: Mechanism to Prevention." *Journal of Thermal Biology* 99: 103019.

Dagar, JC, and RS Yadav. 2017. "Climate Resilient Approaches for Enhancing Productivity of Saline Agriculture." *Journal of Soil Salinity and Water Quality* 9 (1): 9–29.

Ebi, KL, and I Loladze. 2019. "Elevated Atmospheric CO_2 Concentrations and Climate Change Will Affect Our Food's Quality and Quantity." *The Lancet Planetary Health* 3 (7): e283–e284.

Garlock, T, F Asche, J Anderson, A Ceballos-Concha, DC Love, TC Osmundsen, and RBM Pincinato. 2022. "Aquaculture: The Missing Contributor in the Food Security Agenda." *Global Food Security* 32: 100620.

Habeeb, AA, AE Gad, and MA Atta. 2018. "Temperature-Humidity Indices as Indicators to Heat Stress of Climatic Conditions with Relation to Production and Reproduction of Farm Animals." *International Journal of Biotechnology and Recent Advances* 1 (1): 35–50.

Hussein, HAA, SO Alshammari, SK Kenawy, FM Elkady, and AA Badawy. 2022. "Grain-Priming with L-Arginine Improves the Growth Performance of Wheat (Triticum Aestivum L.) Plants under Drought Stress." *Plants* 11 (9): 1219.

Keating, BA, PS Carberry, GL Hammer, ME Probert, MJ Robertson, D Holzworth, NI Huth, JN Hargreaves, H Meinke, and Z Hochman. 2003. "An Overview of APSIM, a Model Designed for Farming Systems Simulation." *European Journal of Agronomy* 18 (3–4): 267–288.

Kheir, AM, AE Baroudy, MA Aiad, MG Zoghdan, MAA El-Aziz, MG Ali, and MA Fullen. 2019. "Impacts of Rising Temperature, Carbon Dioxide Concentration and Sea Level on Wheat Production in North Nile Delta." *Science of the Total Environment* 651: 3161–3173.

Levin, K. 2018. "*Half a Degree and a World Apart: The Difference in Climate Impacts Between 1.5 C and 2 C of Warming*." *World Resources Institute* 7.

Liu, Y, Q Chen, Q Ge, J Dai, Y Qin, L Dai, X Zou, and J Chen. 2018. "Modelling the Impacts of Climate Change and Crop Management on Phenological Trends of Spring and Winter Wheat in China." *Agricultural and Forest Meteorology* 248: 518–526.

Lybbert, TJ, and DA Sumner. 2012. "Agricultural Technologies for Climate Change in Developing Countries: Policy Options for Innovation and Technology Diffusion." *Food Policy* 37 (1): 114–123.

Maurya, PK, RY Karde, AS Bayskar, N Charitha, and S Rai. 2023. "Innovative Technologics Such as Vertical Farming and Hydroponics to Grow Crops in Controlled Environments." *Advanced Farming Technology* 84.

Noori, MS, and NA Taliman. 2022. "Effect of Drought Stress and Chemical Fertilizers on Wheat Productivity and Grain and Quality." *Plant Production and Genetics* 3 (1): 43–56.

Ogello, E O, F Mlingi, B Nyonje, H Charo-Karisa, and J Munguti. 2013. "Can Integrated Livestock-Fish Culture Be a Solution to East African's Food Insecurity? A Review." *African Journal of Food, Agriculture, Nutrition and Development* 13 (4): 8058–8078.

Poonam, A, S Saha, P Nayak, D Sinhababu, P Sahu, B Satapathy, M Shahid, G Kumar, N Jambhulkar, and M Nedunchezhiyan. 2019. "Rice-Fish Integrated Farming Systems for Eastern India." *NRRI Research Bulletin* 17: 33.

Prein, M. 2002. "Integration of Aquaculture into Crop–Animal Systems in Asia." *Agricultural Systems* 71 (1–2). Elsevier: 127–146.

Rakszegi, M, É Darkó, A Lovegrove, I Molnár, L Láng, Z Bed\Ho, M Molnár-Láng, and P Shewry. 2019. "Drought Stress Affects the Protein and Dietary Fiber Content of Wholemeal Wheat Flour in Wheat/Aegilops Addition Lines." *PloS One* 14 (2): e0211892.

Ramalho, Q, M M Vale, S Manes, P Diniz, A Malecha, and JA Prevedello. 2023. "Evidence of Stronger Range Shift Response to Ongoing Climate Change by Ectotherms and High-Latitude Species." *Biological Conservation* 279: 109911.

Rasul, G, N Neupane, A Hussain, and B Pasakhala. 2021. "Beyond Hydropower: Towards an Integrated Solution for Water, Energy and Food Security in South Asia." *International Journal of Water Resources Development* 37 (3): 466–490.

Rezaei, EE, H Webber, S Asseng, K Boote, J-L Durand, F Ewert, P Martre, and DS MacCarthy. 2023. "Climate Change Impacts on Crop Yields." *Nature Reviews Earth & Environment* 4 (12): 831–846.

Rojas-Downing, MM, AP Nejadhashemi, T Harrigan, and SA Woznicki. 2017. "Climate Change and Livestock: Impacts, Adaptation, and Mitigation." *Climate Risk Management* 16: 145–163.

Shah, L, A Ali, M Yahya, Y Zhu, S Wang, H Si, H Rahman, and C Ma. 2018. "Integrated Control of Fusarium Head Blight and Deoxynivalenol Mycotoxin in Wheat." *Plant Pathology* 67 (3): 532–548.

Tian, B, Z Yu, Y Pei, Z Zhang, E Siemann, S Wan, and J Ding. 2019. "Elevated Temperature Reduces Wheat Grain Yield by Increasing Pests and Decreasing Soil Mutualists." *Pest Management Science* 75 (2): 466–475.

ur Rahman, M H, A Ahmad, X Wang, A Wajid, W Nasim, M Hussain, B Ahmad, I Ahmad, Z Ali, and W Ishaque. 2018. "Multi-Model Projections of Future Climate and Climate Change Impacts Uncertainty Assessment for Cotton Production in Pakistan." *Agricultural and Forest Meteorology* 253: 94–113.

3 Climate Change
Threat to the Marine Biota

Hassan Mhanna, Reem Al-Shikh Rasheed, and Sajeed S. Shaikh

3.1 INTRODUCTION

Climate change, currently in progress, is already affecting numerous ecological systems worldwide, impacting biodiversity both directly and indirectly. Global mean surface temperature is rising, sea levels are rising, Arctic sea ice is diminishing, mountain glaciers are retreating, and extreme weather events are increasing (IPCC 2014). Changing patterns of salinity, increasing ocean acidity, and rising sea levels are all examples of "climate change stressors" that change the environment in big ways, making it easier for marine organisms to take in pollution and become sick (Hoegh-Guldberg et al. 2007). The synergistic effects of climate change and pollutants exacerbate global pollution issues, endangering marine ecosystems and biodiversity (Doney et al. 2012). Global warming, mostly induced by anthropogenic activities such as fossil fuel combustion, results in the accumulation of greenhouse gases, which trap heat and elevate the Earth's temperature. The depletion of the stratospheric ozone layer and acid rain exacerbate global warming by means of UV radiation and the degradation of the ozone layer (Monastersky 2015). Climate change forecasts predict enormous alterations in maritime environments, significantly affecting marine fisheries, ecosystems, and biodiversity (Hoegh-Guldberg and Bruno 2010). Increasing temperatures impact ocean productivity, modifying the distribution and abundance of phytoplankton (Moore et al. 2018). Climate variability, encompassing extreme weather phenomena, affects marine biology, fisheries, and livelihoods (Weatherdon et al. 2016). Regional variations in oceanic CO_2 absorption rates influence acidification levels (Feely et al. 2004). Variations in water temperature and precipitation patterns affect ocean currents, river flow, wetland extent, and the distribution and productivity of fish stocks (Cheung et al. 2010). Extreme occurrences affect fishing operations, coastal infrastructure, and biodiversity, highlighting the severity of these environmental changes and emphasizing the urgent necessity to address the multifaceted effects of climate change on marine ecosystems and biodiversity (Sumaila et al. 2011).

Marine and nearshore habitats are among the most crucial regions, both environmentally and economically, on Earth. These are the most prolific yet delicate ecosystems. Currently, they are confronting threats stemming from the detrimental impacts of climate change, unprecedented in the last millennium (Hoegh-Guldberg and Bruno 2010).

 DOI: 10.1201/9781003492733-3

Warming escalated by 0.6°C in the last century; however, the current scenario is more critical, as forecasts suggest that warming will rise by approximately 0.5°C each decade of the 21st century, roughly equivalent to the total temperature increase observed in the 20th century (Solomon et al. 2007; Neelmani et al. 2019).

3.2 CONTRACTION

The cryosphere includes regions of snow or ice that maintain temperatures below 0°C, such as continental ice sheets, glaciers, permafrost, and frozen oceanic areas. Glaciers in renowned locations such as the Himalayas are projected to face extinction by 2100 as a result of present emission levels (Bolch et al. 2012). This delicate element of the Earth's system performs essential functions, reflecting solar radiation and providing freshwater worldwide. Nonetheless, its deterioration results in several repercussions (Abram et al. 2019).

The ramifications of cryosphere reduction are numerous. Water scarcity and inter-nation tensions intensify, while coastal wetlands diminish, adversely affecting biodiversity. Coastal cities confront inundation, prompting distress migration. Minor island nations confront imminent dangers. Alterations in meteorological patterns intensify these problems. Thawing snow increases arable land in high-latitude areas but reduces it along coastlines due to rising sea levels and saltwater inundation. Groundwater supplies are experiencing salinization, significantly impacting coastal areas. Hydroelectric power generation declines, increasing reliance on fossil fuels. The significant decline of animal populations due to habitat loss will elevate the number of species categorized as threatened or extinct.

3.3 OCEAN ACIDIFICATION

The oceans play a crucial role in regulating atmospheric carbon dioxide concentrations, absorbing roughly one-third of anthropogenic CO_2 emissions and alleviating the impacts of climate change. Ocean acidification, caused by heightened absorption of carbon compounds from the atmosphere, results in decreased ocean pH and carbonate ion concentration, making oceans less alkaline. This phenomenon has resulted in a 0.1 pH unit decrease in ocean surface waters since the Industrial Revolution, equating to a 26% rise in ocean hydrogen ion concentration. Notwithstanding the ocean's natural alkalinity, its pH presently approximates 8.0, rendering it difficult to attain a pH below 7.0, which chemically signifies acidity. Multiple factors, including acid rain and eutrophication, can locally affect CO_2 interactions with seawater and intensify the impacts of ocean acidification. Acid rain, with a pH between 1 and 6, profoundly influences surface ocean chemistry at regional scales. Eutrophication happens when too many nutrients, like nitrogen from fertilizer runoff and sewage, enter the water. This causes plankton blooms, which lower the oxygen level in the water and raise the CO_2 level, making acidification worse (Caldeira and Wickett 2003; Feely et al. 2004; Orr et al. 2005). The rising acidity of ocean waters significantly impacts marine life by hindering calcification processes essential for creatures such as corals, mollusks, crabs, and several plankton species that rely on carbonate ions for shell and skeleton development (Havenhand et al. 2008). High levels of CO_2 make it harder for carbonate ions to dissolve, which stops calcification and could have terrible effects on marine ecosystems and species that are important for trade (Hoegh-Guldberg et al. 2007). Mitigation measures encompass the reduction of CO_2 emissions through governmental legislation, the enhancement of energy conservation, and the transition to renewable energy sources such as wind and solar power. Also, ocean acidification changes the saturation horizons because deep, cold ocean waters naturally have low levels of carbonate ions, which break down the shells of species that form calcium carbonate. As ocean acidification elevates the saturation horizon, an increasing number of calcifying organisms become susceptible to shell dissolution. Compared to the 1800s, the saturation points of both calcite and aragonite have gone up. This shows that the acidification of the oceans is still having an impact on marine life and carbonate structures (Orr et al. 2005; Kumar et al. 2011). Although the natural equilibrium of CO_2 absorption and emission is sustained over extended periods, anthropogenic CO_2 emissions disturb this balance, intensifying ocean acidification and presenting considerable environmental difficulties for future generations (NOAA 2024).

3.4 INCREASE IN SEA LEVEL

The rise in sea level results from thermal expansion and glacial melting, endangering coastal regions. Observations since the early 1990s indicate a concerning increase, with substantial effects anticipated by the Intergovernmental Panel on Climate Change (IPCC). A global temperature rise of 3°C–4°C may displace millions owing to flooding, while warmer oceans exacerbate the intensity of tropical storms. In the absence of effective emission reductions to maintain temperatures well below 2°C above pre-industrial levels, as stipulated in the 2015 Paris Climate Agreement, a sea-level increase of 30–60 cm by 2100 is anticipated, with a potential rise of 1.1 m also threatening. Despite effective emission limitations, a significant rise is still anticipated (Oppenheimer et al. 2019). It is already evident that the destruction of coastal wetlands is having an impact on biodiversity. Comprehending alterations in sea levels facilitates climate research and coastal development strategies. Nevertheless, severe flooding and economic damages are anticipated to raise significantly in the absence of sufficient adaptation strategies.

3.4.1 Variations in Global Sea Level

Seasonal fluctuations in sea level typically range from 5 to 6 cm annually. Short-term sea-level fluctuations arise from a complex interplay of elements, including marine water density, atmospheric pressure, and ocean current velocities. In the past century, anthropogenic global warming has caused the thermal expansion of ocean water, leading to a sea-level rise of around 10–15 cm. Moreover, the melting of Antarctic ice sheets, which account for around 3% of the overall ice volume, has intensified this global phenomenon. The possibility of considerable global sea-level increases over 100 m is predominantly associated with scenarios of extensive ice sheet melting or significant modifications in the amount of the world's mid-oceanic ridge.

This information is supported by studies conducted by Nicholls et al. (1999) and Oppenheimer et al. (2019).

3.5 DISSOLVED OXYGEN IN AQUATIC HABITATS

Dissolved oxygen is vital for sustaining aquatic organisms. The average concentration of dissolved oxygen is significantly lower than that in air, influenced by different causes. Oxygen permeates aquatic systems via exchange at the air-water interface and through the photosynthetic processes of aquatic vegetation. Nevertheless, variables such as temperature influence oxygen solubility, as warmer water retains less oxygen. Elevated temperatures also enhance decomposer activity, hence further diminishing oxygen levels. When dissolved oxygen concentrations fall below critical limits, generally 3–5 ppm, it can lead to extensive mortality among aquatic creatures (Missaghi et al. 2017; Mahaffey et al. 2020).

3.6 OCEAN DEOXYGENATION

Ocean deoxygenation, marked by the proliferation of oxygen minimum zones (OMZs), results from anthropogenic carbon dioxide emissions. OMZs arise from physical and biological processes that diminish oxygen levels, resulting in anoxic regions. Because oxygen is less soluble at higher temperatures, deoxygenation gets worse. This is made worse by temperature-induced stratification (Keeling et al. 2010). The consequences of deoxygenating in seas are complex, with substantial impacts on marine ecosystems. Decreased oxygen levels exacerbate ocean acidification, compromising shell integrity and disrupting elemental cycles vital for organisms. A reduction in phytoplankton quantity results in a decline of food sources for fish, causing extensive death episodes. The cascading impacts highlight the essential significance of maintaining oxygen levels in marine ecosystems (Limburg et al., 2017; Oschlies 2019).

3.7 EFFECTS ON BIODIVERSITY AND MARINE ECOSYSTEMS

The International World Wildlife Fund (WWF) has expressed concerns about the vulnerability of species worldwide, from tropical areas to polar regions, due to climate change. Numerous species may find it challenging to adjust swiftly to the altering conditions caused by increasing temperatures, potentially resulting in catastrophic losses in natural habitats. Marine ecosystems are especially vulnerable, confronting threats from rising sea temperatures and elevated levels of dissolved carbon dioxide, which can alter ocean circulation and induce ocean acidification. This phenomenon significantly endangers shell-forming organisms, corals, and their associated ecosystems (Hoegh-Guldberg et al. 2007).

3.7.1 Influence of Temperature on Aquatic Organisms

Aquatic creatures possess a limited temperature tolerance range, rendering them particularly vulnerable to even slight temperature fluctuations in contrast to terrestrial organisms. Because water temperatures are generally constant, even slight fluctuations can significantly threaten aquatic life, underscoring the importance of monitoring and addressing temperature-related effects (Brierley and Kingsford 2009). The worldwide ocean has absorbed a significant amount of extra heat produced by climate change, leading to rapid ocean warming. Marine heat waves have increased in frequency and intensity, affecting marine biodiversity and ecological health. Increasing temperatures lead to oxygen depletion in surface waters, intensifying dead zones and disturbing ocean currents. Confronting these problems is essential, given that the enduring effects of ocean warming are anticipated to last for generations, even with prompt reductions in greenhouse gas emissions (Bindoff et al. 2019).

3.7.2 Mangroves

Mangroves are plants and shrubs that inhabit tropical and subtropical sheltered coastlines. They are highly adapted to the soggy and hyper saline conditions of those regions. To achieve this, they produce specialized structures such as prop roots to withstand water flow, pneumatophores for oxygen supply to the root system, and salt glands to regulate osmolality, among others. They serve as a coastal bio-shield, playing a crucial role in mitigating coastal erosion and safeguarding land from tsunamis, hurricanes, and floods. Furthermore, they maintain coastal coastlines and offer breeding habitats for fish. They are augmenting nutrient recycling and promoting floral and faunal diversity. Mangroves provide resources such as timber, medicinal flora, and job possibilities for local residents (Shaikh and Chavan 2010). Climate change significantly threatens mangroves, chiefly through sea-level rise, which can submerge ecosystems and jeopardize species incapable of withstanding saltwater intrusion. Coastal wetlands, such as mangrove forests, function as essential carbon sinks that reduce greenhouse gas emissions; nevertheless, destruction exacerbates climate change by releasing more carbon into the atmosphere. Replantation initiatives, although implemented, frequently exhibit insufficient species diversity, thereby constraining their efficacy in mitigating mangrove habitat degradation. An IPCC assessment underscores the increasing hazards associated with sea-level occurrences and coastal development patterns, forecasting significant rises in yearly flood damages by 2100 in locations such as Mumbai. Notwithstanding these hurdles, the preservation and restoration of mangrove habitats are essential for coastal resilience and climate change mitigation. A comprehensive assessment published in "Global Change Biology" assessed the sensitivity of mangrove ecosystems to climate change stressors, such as increasing sea levels, temperature fluctuations, and extreme weather events. The study emphasized the importance of understanding the complex interactions between climate change and mangrove ecosystems to develop effective conservation and adaptation strategies. Furthermore, a study by the Mangrove Action Project, in partnership with local people in Southeast Asia, investigated the resilience of mangrove forests to climate change and suggested community-centric strategies for mangrove restoration and management. These studies highlight the critical necessity for interdisciplinary research and community involvement to protect mangrove ecosystems in the face of climate change (Alongi 2015; Godoy and de Lacerda 2015; Ward et al. 2016; Bindoff et al. 2019).

3.7.3 Coral Reefs

Coral reefs, consisting of living organisms in a symbiotic association with Zooxanthellae algae, provide essential ecological functions. These algae facilitate nutrient production in corals by performing photosynthesis, supplying energy, and promoting calcification. Coral tissues are transparent, acquiring their hue from Zooxanthellae. Coral polyps offer a safeguarded habitat for Zooxanthellae and carbon dioxide for the process of photosynthesis. Corals are classified into two categories: reef-building rigid corals and soft corals, such as sea fans. Coral reefs function as protective barriers against erosion, sequester nutrients from plankton, and offer habitat for a variety of marine organisms. Nonetheless, they encounter risks from both natural and human-induced factors, including bleaching, pollution, overfishing, and climate change (Baker et al. 2008). High temperatures and irradiance mess up the enzyme systems in Zooxanthellae. This makes photosynthetic processes more difficult and may even cause the coral-algae symbiosis to end. Extended absence of Zooxanthellae may lead to coral mortality. Solar energy, namely photosynthetically active and UV radiation, contributes significantly (Brierley and Kingsford 2009; Authority 2007). Moreover, exposure to air during extreme low tides or substantial sea-level declines can induce bleaching, while sedimentation and freshwater influx from storm-induced rainfall can intensify stress on coral reefs. Climate change, including phenomena such as El Niño, has led to significant coral bleaching, resulting in the creation of efforts like the condition and conservation of coral reefs to globally monitor and mitigate reef health.

A comprehensive assessment by the National Oceanic and Atmospheric Administration (NOAA) evaluates the susceptibility of coral reef ecosystems to climate change stressors, including bleaching events, ocean acidification, and sea-level rise. This study underscores the urgent necessity for conservation initiatives and adaptive management measures to mitigate the effects of climate change on coral reefs.

3.7.4 Plankton

Plankton constitute essential components of marine ecosystems. They are a heterogeneous group of organisms that either float or swim passively in aquatic ecosystems, encompassing a broad spectrum of both microscopic and macroscopic species. Plankton encompasses both phytoplankton and zooplankton. Temperature and several environmental conditions affect photosynthetic output in plankton, exhibiting increased rates until a threshold is reached, beyond which returns decline. Salinity similarly influences primary production in conjunction with differences in light and temperature. The grazing rate of zooplankton is essential in regulating phytoplankton biomass and productivity. Marine phytoplankton display a heterogeneous global distribution, characterized by elevated concentrations at high latitudes and notable exceptions in upwelling regions along continental shelves. Tropical and subtropical climates generally have reduced concentrations, possibly due to elevated solar UV-B radiation levels. Phytoplankton productivity is restricted to the euphotic zone, shaped by wind and wave dynamics. Phytoplankton constitute the foundation of aquatic food webs, sustaining a variety of marine organisms ranging from zooplankton to whales. They play a crucial role in the carbon cycle by facilitating the passage of carbon dioxide from the atmosphere to the ocean through photosynthesis. The "biological carbon pump" yearly distributes significant carbon to deep ocean layers, influencing atmospheric carbon dioxide levels and, subsequently, global surface temperatures (Hays et al. 2005; Murphy et al. 2020).

3.7.5 Increasing Ocean Temperatures and Their Impact on Fish

The contemporary globe is seeing substantial alterations in marine and aquatic ecosystems as a result of increasing sea temperatures and climate change. These shifts affect fish populations and their ecosystems, whether in freshwater or marine environments. Research indicates that fluctuations in temperature result in substantial alterations in fish stock distribution and the deterioration of aquatic ecosystems. This alteration also impacts living organisms, as they encounter various threats, including interruption of life cycles and heightened disease prevalence. These shifts encompass the coerced relocation of fish species from their indigenous habitats to novel regions, impacting biodiversity. Moreover, elevated water temperatures and heightened salinity jeopardize ecological equilibrium, facilitating the invasion of non-native species in new regions and endangering indigenous species. Climate change and increasing temperatures substantially affect marine ecosystems and the creatures dependent on them. Consequently, it is essential to examine these alterations and their impact on biodiversity and human livelihoods. The migration of fish from various regions to the Mediterranean Sea is a significant phenomenon caused by climate change and increasing water temperatures. The rise in temperatures has prompted the migration of fish species from the Red Sea and other areas to the Mediterranean Sea in pursuit of suitable homes. The inauguration of the Suez Canal enabled fish migration to the Mediterranean, resulting in the movement of over 104 fish species (Golani et al. 2020). Increased water temperatures substantially influence fish, affecting their growth, reproduction, migration, and overall health. Below are a few possible effects:

- **Alterations in ichthyic distribution and migration patterns**: As water temperatures increase, fish may be compelled to move in pursuit of cooler regions. This may modify fish migration patterns and their impacts on local ecosystems.
- **Impact on development**: Elevated temperatures can influence fish development rates by enhancing metabolism, resulting in faster but smaller growth.
- **Reproductive impact**: Alterations in reproductive seasons and life cycles may occur owing to elevated temperatures, influencing reproductive success and species viability.
- **Reduced oxygen levels**: Elevated temperatures diminish water's ability to retain oxygen, resulting in a deficiency of oxygen accessible to fish and other marine animals, which may lead to asphyxiation.
- **Elevated disease risk**: Fish may exhibit heightened vulnerability to diseases as a result of stress induced by increased temperatures, adversely affecting their immune systems.

- **Behavioral and habitat alterations**: Increasing temperatures can influence fish behavior and activity patterns, as well as modify their preferred habitats, prompting migration to other regions.
- **Disruptions in the food chain**: Increasing water temperatures impact phytoplankton and zooplankton, which are the foundation of the oceanic food chain. This may hinder the availability of food for fish and affect their diet.

Generally, increasing water temperatures can profoundly impact fish health and welfare, potentially leading to significant alterations in marine ecosystems. To evaluate the effects of climate change on fish in the Mediterranean, *Alosa fallax* specimens were selected and gathered from the Syrian coast in the eastern Mediterranean for a duration of two years. Morphological measures (length and weight) were conducted to assess the impact of climate change on their growth, considering their significance in world fisheries (Occhipinti-Ambrogi 2007; Cheung et al. 2009; Marras et al. 2015). As per Fish Base (Froese and Pauly 2024), these fish can attain a length of 60 cm and a weight of 1.5 kg. A total of 1,200 fish were captured, comprising 487 females and 713 males, which represent 40.58% and 59.42%, respectively. Samples were collected monthly during a two-year period from September 2021 to September 2023 in the Ras al-Bassit region. Subsequent to collecting, the fish were conveyed to the laboratory for requisite measurements and analyses (Pravdin 1966).

The investigation examined the morphological and standard traits of *A. fallax* specimens collected from the Ras al-Bassit region. Both female and male sardines exhibited similar sizes, with females averaging 2.70 ± 14.05 cm in length and 29.68 ± 44.55 g in weight, whereas males averaged 13.72 ± 2.75 cm in length and 43.73 ± 29.79 g in weight. The analysis of the measures revealed a substantial influence of climatic change on *A. fallax* fish, which adversely affected the sizes of individuals reaching five years of age and reproducing at these sizes. This illustrates the evident adverse impacts of climate change on Mediterranean fish, particularly sardines.

3.7.5.1 Spawning

The reproductive patterns and techniques of animals residing in marine and freshwater ecosystems are predominantly influenced by evolutionary mechanisms. These organisms adapt to prevailing environmental circumstances, including temperature, salinity, currents, and food availability, to facilitate effective reproduction and the fulfillment of their life cycle. Spawning times and places have adapted to enhance larval survival and subsequent development into reproductive adults while also reducing possible disturbances from unpredictable environmental events (Crozier et al. 2011). Moreover, environmental variables, including temperature, significantly influence particular elements of spawning behavior. Research indicates that temperature affects the age of sexual development in species such as Atlantic salmon and Atlantic cod (Brander 2010).

3.7.5.2 Variations in Abundance

Fluctuations in the numbers and biomass of marine populations arise from alterations in recruitment and growth rates, eventually indicating the productivity of their habitat. Temperature variations can directly affect fish quantity and biomass by inducing physiological stress on individuals, leading to changes in their distribution or potential mortality. Furthermore, temperature might indirectly impact the fish population by affecting growth and recruitment processes (Beaugrand et al. 2002; Dutil and Brander 2003; Niles et al. 2018).

3.7.5.3 Invasion of Alien Species

The introduction of species outside their normal range due to human activities frequently results in the creation of invasive populations that have considerable adverse effects on native biodiversity, economies, and human health. Climate change influences alien species dynamics, modifying maritime ecosystems via temperature variations, hydrographical alterations, and acidity, in conjunction with human activities such as shipping and aquaculture. These modifications can enable the introduction of non-native species into new locations, exploit newly accessible habitats resulting from melting ice, and undermine the resistance of native groups to invasion. The effects of invasive species are anticipated to escalate as climate change continues to alter marine ecosystems, highlighting the need for coordinated mitigation strategies (Lodge et al. 2006; Parmesan 1996; Hellmann et al. 2008).

3.8 CONCLUSION

Marine ecosystems and their biodiversity are significantly threatened by climate change, risking harm and extinction. The physicochemical features of the marine environment are changing due to ocean acidification, rising sea levels, and a diminishing cryosphere. These alterations significantly impact the resident organisms. Insufficient dissolved oxygen levels and ocean deoxygenation are imposing increased stress on marine organisms. Conversely, elevated temperatures are accountable for alterations in species distribution, growth rates, and reproductive patterns. Ecologically sensitive systems such as mangroves, coral reefs, and plankton populations are significantly affected by the repercussions of climate change. A significant number of fish populations are exhibiting modified migration patterns and heightened vulnerability to illnesses. Furthermore, indigenous species are threatened by habitat destruction and the incursion of non-native species. All these factors are disrupting the natural equilibrium and intensifying the problem. In light of these issues, there is an urgent necessity for coordinated global initiatives to alleviate climate change. This can be accomplished by diminishing greenhouse gas emissions, engaging in plantation activities in coastal regions, and advocating for sustainable development methods. These efforts may not entirely prevent the advancement of climate change, but they will significantly contribute to mitigating its effects and affording the marine environment an opportunity to adapt and flourish. Through these collective efforts, we can enhance the resilience of marine ecosystems and so contribute to the preservation of our planet's marine biodiversity.

REFERENCES

Abram, N, J-P Gattuso, A Prakash, L Cheng, MP Chidichimo, S Crate, H Enomoto, et al. 2019. "Framing and Context of the Report." In *IPCC Special Report on the Ocean and Cryosphere in a Changing Climate*, edited by H-O Pörtner, DC Roberts,

V Masson-Delmotte, P Zhai, M Tignor, E Poloczanska, K Mintenbeck, et al., pp. 73–129. Cambridge University Press, Cambridge, UK and New York, NY, USA. https://doi.org/10.1017/9781009157964.003

Alongi, DM. 2015. "The Impact of Climate Change on Mangrove Forests." *Current Climate Change Reports* 1 (1): 30–39.

Authority, Great Barrier Reef Marine Park. 2007. *Great Barrier Reef Climate Change Action Plan 2007–2012*. Great Barrier Reef Marine Park Authority, Townsville, Australia.

Baker, AC, PW Glynn, and B Riegl. 2008. "Climate Change and Coral Reef Bleaching: An Ecological Assessment of Long-Term Impacts, Recovery Trends and Future Outlook." *Estuarine, Coastal and Shelf Science* 80 (4): 435–471.

Beaugrand, G, PC Reid, F Ibanez, JA Lindley, and M Edwards. 2002. "Reorganization of North Atlantic Marine Copepod Biodiversity and Climate." *Science* 296 (5573): 1692–1694.

Bindoff, NL, WWL Cheung, JG Kairo, J Arístegui, VA Guinder, R Hallberg, N Hilmi, et al. 2019. "Changing Ocean, Marine Ecosystems, and Dependent Communities." In *IPCC Special Report on the Ocean and Cryosphere in a Changing Climate*, edited by H-O Pörtner, DC Roberts, V Masson-Delmotte, P Zhai, M Tignor, E Poloczanska, K Mintenbeck, et al., pp. 447–587. Cambridge University Press, Cambridge, UK and New York, NY, USA.

Bolch, T, A Kulkarni, A Kääb, C Huggel, F Paul, JG Cogley, H Frey, et al. 2012. "The State and Fate of Himalayan Glaciers." *Science* 336 (6079): 310–314.

Brander, K. 2010. "Impacts of Climate Change on Fisheries." *Journal of Marine Systems* 79 (3–4): 389–402.

Brierley, AS, and MJ Kingsford. 2009. "Impacts of Climate Change on Marine Organisms and Ecosystems." *Current Biology* 19 (14): R602–R614.

Caldeira, K, and ME Wickett. 2003. "Anthropogenic Carbon and Ocean pH." *Nature* 425 (6956): 365–365.

Cheung, WWL, VWY Lam, JL Sarmiento, K Kearney, R Watson, and D Pauly. 2009. "Projecting Global Marine Biodiversity Impacts Under Climate Change Scenarios." *Fish and Fisheries* 10 (3): 235–251.

Cheung, WWL, VWY Lam, JL Sarmiento, K Kearney, R Watson, D Zeller, and D Pauly. 2010. "Large-Scale Redistribution of Maximum Fisheries Catch Potential in the Global Ocean under Climate Change." *Global Change Biology* 16 (1): 24–35.

Crozier, LG, MD Scheuerell, and RW Zabel. 2011. "Using Time Series Analysis to Characterize Evolutionary and Plastic Responses to Environmental Change: A Case Study of a Shift Toward Earlier Migration Date in Sockeye Salmon." *The American Naturalist* 178 (6): 755–773.

Doney, SC, M Ruckelshaus, JE Duffy, JP Barry, F Chan, CA English, HM Galindo, et al. 2012. "Climate Change Impacts on Marine Ecosystems." *Annual Review of Marine Science* 4: 11–37.

Dutil, JD, and K Brander. 2003. "Comparing Productivity of North Atlantic Cod (Gadusmorhua) Stocks and Limits to Growth Production." *Fisheries Oceanography* 12 (6): 502–512.

Feely, RA, CL Sabine, K Lee, W Berelson, J Kleypas, VJ Fabry, and FJ Millero. 2004. "Impact of Anthropogenic CO_2 on the $CaCO_3$ System in the Oceans." *Science* 305 (5682): 362–366.

Froese, R., and D. Pauly. 2024. *FishBase*. World Wide Web Electronic Publication. www.fishbase.org, version (02/2024). Stockholm, Sweden.

Godoy, MD, and LD de Lacerda. 2015. "Mangroves Response to Climate Change: A Review of Recent Findings on Mangrove Extension and Distribution." *Anais Da Academia Brasileira de Ciências* 87 (2): 651–667.

Golani, D, R Fricke, and B Appelbaum-Golani. 2020. "Zoogeographic Patterns of Red Sea Fishes–Are They Correlated to Success in Colonization of the Mediterranean via the Suez Canal?" *Marine Biology Research* 16 (9): 774–780.

Havenhand, JN, FR Buttler, MC Thorndyke, and JE Williamson. 2008. "Near-Future Levels of Ocean Acidification Reduce Fertilization Success in a Sea Urchin." *Current Biology* 18 (15): R651–R652.

Hays, GC, AJ Richardson, and C Robinson. 2005. "Climate Change and Marine Plankton." *Trends in Ecology & Evolution* 20 (6): 337–344.

Hellmann, JJ, JE Byers, BG Bierwagen, and JS Dukes. 2008. "Five Potential Consequences of Climate Change for Invasive Species." *Conservation Biology* 22 (3): 534–543.

Hoegh-Guldberg, O, and JF Bruno. 2010. "The Impact of Climate Change on the World's Marine Ecosystems." *Science* 328 (5985): 1523–1528.

Hoegh-Guldberg, O, PJ Mumby, AJ Hooten, RS Steneck, P Greenfield, E Gomez, N Knowlton, et al. 2007. "Coral Reefs under Rapid Climate Change and Ocean Acidification." *Science* 318 (5857): 1737–1742.

International Panel on Climate Change (IPCC) 2007. "Climate Change 2007: The Physical Science Basis". In *Contribution of Working Group I to the Fourth Assessment Report of the Intergovernmental Panel on Climate Change*, edited by S Solomon, D Qin, M Manning, Z Chen, M Marquis, KB Averyt, M Tignor, and HL Miller, p. 996. Cambridge University Press, Cambridge, United Kingdom and New York, NY, US.

IPCC. 2014 *Climate Change 2014: Synthesis Report. Contribution of Working Groups I, II and III to the Fifth Assessment Report of the Intergovernmental Panel on Climate Change*, edited by Core Writing Team, RK Pachauri, and LA Meyer, p. 151. IPCC, Geneva, Switzerland.

Keeling, RF, A Körtzinger, and N Gruber. 2010. "Ocean Deoxygenation in a Warming World." *Annual Review of Marine Science* 2: 199–229.

Kumar, A., .T. Schei, A. Ahenkorah, R. Caceres Rodriguez, J.-M. Devernay, M. Freitas, D. Hall, Å. Killingtveit, Z. Liu, et al. 2011. "Hydropower." In *IPCC Special Report on Renewable Energy Sources and Climate Change Mitigation*, 437–496. Cambridge University Press. https://doi.org/10.1017/cbo9781139151153.009

Limburg, KE, D Breitburg, and LA Levin. 2017. "Ocean Deoxygenation–a Climate-Related Problem." *Frontiers in Ecology and the Environment* 15 (9): 479–479.

Lodge, DM, S Williams, HJ MacIsaac, KR Hayes, B Leung, S Reichard, RN Mack, et al. 2006. "Biological Invasions: Recommendations for US Policy and Management." *Ecological Applications* 16 (6): 2035–2054.

Mahaffey, C, M Palmer, N Greenwood, and J Sharples. 2020. "Impacts of Climate Change on Dissolved Oxygen Concentration Relevant to the Coastal and Marine Environment around the UK." *MCCIP Science Review* 2002: 31–53.

Marras, S, A Cucco, F Antognarelli, E Azzurro, M Milazzo, M Bariche, M Butenschön, et al. 2015. "Predicting Future Thermal Habitat Suitability of Competing Native and Invasive Fish Species: From Metabolic Scope to Oceanographic Modelling." *Conservation Physiology* 3 (1): cov059.

Missaghi, S, M Hondzo, and W Herb. 2017. "Prediction of Lake Water Temperature, Dissolved Oxygen, and Fish Habitat under Changing Climate." *Climatic Change* 141 (4): 747–757.

Monastersky, R. 2015. "The Human Age." *Nature* 519 (7542): 144–147.

Moore, JK, W Fu, F Primeau, GL Britten, K Lindsay, MC Long, SC Doney, NM Mahowald, JT Randerson, and L Resplandy. 2018. "Sustained Climate Warming Drives Declining Marine Biological Productivity." *Science* 359 (6381): 1139–1143.

Murphy, GE, TN Romanuk, and B Worm. 2020. "Cascading Effects of Climate Change on Plankton Community Structure." *Ecology and Evolution* 10 (5): 2170–2181.

NOAA. What is coral bleaching? National Ocean Service website, https://oceanservice.noaa.gov/facts/coral_bleach.html, 12/04/24.

Neelmani, R, M Chandra Pal, V Sarman, UD Vyas, and TN Muniya. 2019. "Impacts of Climate Change on Marine Biodiversity." *Journal of Entomology and Zoology Studies* 2: 425–430.

Nicholls, RJ, FMJ Hoozemans, and M Marchand. 1999. "Increasing Flood Risk and Wetland Losses Due to Global Sea-Level Rise: Regional and Global Analyses." *Global Environmental Change* 9: S69–S87.

Niles, MT, R Ahuja, T Barker, J Esquivel, S Gutterman, MC Heller, N Mango, et al. 2018. "Climate Change Mitigation beyond Agriculture: A Review of Food System Opportunities and Implications." *Renewable Agriculture and Food Systems* 33 (3): 297–308. https://doi.org/10.1017/S1742170518000029

Occhipinti-Ambrogi, A. 2007. "Global Change and Marine Communities: Alien Species and Climate Change." *Marine Pollution Bulletin* 55 (6–12): 342–352.

Oppenheimer, M, BC Glavovic, J Hinkel, R van de Wal, AK Magnan, A Abd-Elgawad, R Cai, et al. 2019. "Sea Level Rise and Implications for Low-Lying Islands, Coasts and Communities." In *IPCC Special Report on the Ocean and Cryosphere in a Changing Climate*, edited by H-O Pörtner, DC Roberts, V Masson-Delmotte, P Zhai, M Tignor, E Poloczanska, K Mintenbeck, et al., pp. 321–445. Cambridge University Press, Cambridge, UK and New York, NY, USA. https://doi.org/10.1017/9781009157964.006

Orr, JC, VJ Fabry, O Aumont, L Bopp, SC Doney, RA Feely, A Gnanadesikan, et al. 2005. "Anthropogenic Ocean Acidification over the Twenty-First Century and Its Impact on Calcifying Organisms." *Nature* 437 (7059): 681–686.

Oschlies, A. 2019. *Ocean Deoxygenation from Climate Change*. IUCN, Gland, Switzerland.

Parmesan, C. 1996. "Climate and Species Range." *Nature* 382: 765–766.

Pravdin, G. 1966. *Methods in Ichthyology*. Moscow: High school.

Shaikh, SS, and NS Chavan. 2010. "Morphological Anomalies in the Species of a Mangrove Genus Avicennia L. Along the Coast of Maharashtra and Goa States of India." *The Ecoscan* 4: 77–80.

Solomon, S, D Qin, M Manning, Z Chen, M Marquis, KB Averyt, M Tignor, and HL Miller. 2007. Climate Change 2007: The Physical Science Basis. In *Contribution of Working Group I to the Fourth Assessment Report of the Intergovernmental Panel on Climate Change*. Cambridge University Press, Cambridge, United Kingdom and New York, NY, US.

Sumaila, UR, William WL Cheung, Vicky WY Lam, Daniel Pauly, and Sarah Herrick. 2011. "Climate Change Impacts on the Biophysics and Economics of World Fisheries." *Nature Climate Change* 1 (8): 449–456.

Ward, RD, DA Friess, RH Day, and RA Mackenzie. 2016. "Impacts of Climate Change on Mangrove Ecosystems: A Region by Region Overview." *Ecosystem Health and Sustainability* 2 (4): e01211.

Weatherdon, LV, AK Magnan, AD Rogers, UR Sumaila, and WWL Cheung. 2016. "Observed and Projected Impacts of Climate Change on Marine Fisheries, Aquaculture, Coastal Tourism, and Human Health: An Update." *Frontiers in Marine Science* 3: 48.

4 Biodiversity and Climate Change

P. R. Vasagadekar and P. D. Raut

4.1 INTRODUCTION

4.1.1 Overview of Biodiversity

For millions of years, the vast diversity of life on Earth has met man's requirements. Every civilization has relied on this diversity of living things as a support system for its expansion and advancement. Those who wisely and sustainably utilized this "bounty of nature" prevailed. Those who abused or overused it fell apart. For more than a century, science has tried to describe and categorize the variety in nature. As a result, we now understand how it is arranged into plant and animal communities. This knowledge has been essential to the process of "development" and has assisted in harnessing the biological diversity of the planet for the benefit of humankind. Better harvests, improved health care, and the utilization of these living forms as raw materials for industrial growth have all contributed to the industrialized world's increased standard of living.

But this also gave rise to the contemporary consumerist culture, which has negatively impacted the variety of biological resources that serve as its foundation. Because of the diversity of life on Earth, if we use biodiversity sustainably, we can continue to use it to create new products for years to come. This is achievable only if we preserve biodiversity as a valuable resource and prohibit species from becoming extinct.

4.1.2 Understanding Climate Change

"Climate" is the standard weather of a place in the world. In uncomplicated terms, climate is articulated as probable wind temperature and precipitation based on chronological interpretation.

"Climate change" refers to change occurring over elongated periods of time. "Climate change" is a long-term process that is influenced by factors such as the Earth's cycle, solar radiation output, energy activity, and land distribution, resulting in alternating cold and warm weather. These times occurred when glaciers covered most of the Earth's surface and oceans were higher than they are today. The current epoch, called the Holocene, is characterized by a warm and stable climate from the end of the Mesozoic until 11,700 years ago, when human civilization developed.

The main reasons for climate changes are human interest, together with burning fossil fuels, clearing forests, and cultivating cattle, which are causing Earth's weather to change. This is known as "anthropogenic climate change." The foremost

DOI: 10.1201/9781003492733-4

greenhouse gases (GHGs), such as carbon dioxide, methane, halocarbons, and nitrous oxide, are unconstrained in the atmosphere. These gases are concentrated within the environment, permitting daylight to pass through but trapping heat. This phenomenon is referred to as the "greenhouse impact" and is just like a greenhouse, in which sunlight enters but warmth is trapped for plant growth. "The global warming," an end result of the greenhouse effect, is a significant weather alternate that drives different climate changes, consisting of changes in rainfall patterns and major climate events like heat waves, storms, and droughts. At the same time as frequently used interchangeably, weather exchange encompasses other discovered climate changes and global warming. Scientists debate the devastating impacts of climate change on natural and human structures, and its existential danger to human civilization poses a great threat.

We are good at detecting temporary environmental changes, but not long-standing climate change. Climatic discipline depends on long-term climatic records and monitoring, statistics, and reconstruction of earlier climatic period, with prediction of upcoming climate by models related to the climate system. The 19th century witnessed the formation of contemporary scientific perspective on climate change as data from previous glaciations demonstrated an unstable Earth's climate. Recognizing natural climate change led to the thought that humans could modify it. Anthropogenic climate change acquired scientific traction in the 1960s and 1970s, following the 1896 proposal of chemist Svante Arrhenius that human emissions of carbon dioxide would amplify the Earth's natural greenhouse effect and raise its temperature.

The scientific community came to an agreement over global warming in the 1980s. An important first step was taken in 1988 when the Intergovernmental Panel on Climate Change (IPCC) was established by the United Nations Environment Programme (UNEP) and the Arena Meteorological Agency. The IPCC studies and assesses international scientific, technological, socioeconomic data relevant to our understanding of weather variability. It produces evaluation reviews and summaries rather than conducting its own monitoring or studies. Six of those assessment reports have been published thus far: in 1990, 1995, 2001, 2007, 2013–2014, and 2021–2022. In addition to the IPCC, the United Nations Framework Convention on Climate Change (UNFCCC) was established in 1992, offering a framework for international treaty and agreement negotiations to stabilize the levels of GHGs. The Kyoto Protocol was an official requisite agreement that industrialized countries had to abide by when it came into effect in 1997.

The Paris Agreement: A total of 196 nations ratified this historic agreement in 2015 with the target of keeping global warming considerably below 2°C (preferably, to 1.5°C) over pre-industrial levels. The framework also focuses on enhancing financial flows and adaptation capacities to direct climate accomplishment in developing countries. These are a few of the well-known programs and frameworks that genuinely assist with climate change adjustment, mitigation, and diminution of GHG emissions. Examining the aforementioned frameworks and treatments is crucial to understanding every aspect of climate change. Expertise in climate change includes spotting its anthropogenic origins, the profound and various influences it has on Earth, and the necessity of coordinated global movement to mitigate and adapt to its results. At the same time as the challenges are full-size, the capacity for innovation and collective attempt gives desire for a more sustainable future.

4.1.3 The Interconnection between Biodiversity and Climate Change

For a variety of reasons, biotic and abiotic components of Mother Earth's system are intricately linked to one another. Man, or the human, is a member of Mother Earth's biotic element and is entirely dependent on other elements to survive in the biosphere. Humans are a part of biodiversity, and there is a close association between biodiversity and climate change. The allocation of species, the timing of biological processes, and ecosystem services including the food web, food chain, and energy flow are all significantly affected due to climate change. On the other hand, biodiversity is essential for controlling climate change and strengthening ecological resilience. In the event that some elements go extinct, the ecological services that nature provides may be impacted. The changes in abiotic elements such as temperature, rainfall, humidity, and solar radiation can change the weather pattern on any part of the Earth and can cause extinction or threat to biodiversity. Consequently, knowledge of how climate change affects biodiversity is essential.

4.1.4 Impact of Climate Change on Biodiversity

4.1.4.1 Species Distribution and Habitat Loss

The habitat is the most essential part of any living organism. The changes in habitat owing to climate change can directly affect biodiversity. The uniqueness of habitat can be altered due to any changes in natural atmosphere due to GHG concentrations, climate change, and global warming. The range shifts and habitat fragmentations are caused by climate change. Range shifts refer to the movement of species' geographic distribution in response to changing environmental conditions, primarily driven by climate change. As global temperatures rise, species are compelled to migrate to more favorable habitats to survive. This phenomenon has significant ecological, economic, and conservation implications. Range shifts are driven by factors, for instance, temperature changes, changes in rainfall patterns, seasonal variability, and intense weather events.

Species typically move toward the poles or higher altitudes to maintain optimal temperature, while changes in rainfall and drought conditions affect water availability and habitats. Many species are shifting their geographic ranges toward the poles or to higher altitudes in response to changing temperatures. Range shifts can disrupt predator-prey relationships, leading to population imbalances and changes in predator-prey dynamics. Increased competition for resources can also result in the displacement of native species. Biodiversity and extinction risks are also a concern, with species that cannot migrate fast enough or find suitable habitats facing increased extinction risks. Endemism species, such as polar bears, are particularly vulnerable to range shifts and may face extinction if they cannot find suitable new habitats. Changes in species distribution can also affect agriculture, thereby affecting crop yields and food security. Changes in fish distributions can affect commercial and subsistence fisheries, potentially leading to economic losses and conflicts over fishing rights. Overall, range shifts have significant impacts on ecosystem interactions, biodiversity, and human well-being. The extreme climate events like wild fires, floods, hurricanes can directly damage the habitat of species, which can lead to habitat loss.

Marine invertebrates (Thomas and Lennon 1999), butterflies (Parmesan 1996), and alpine plants (Grabherr et al. 1994) have all been shown to exhibit variety shifts in places experiencing regional warming developments. The research by Thomas and Lennon (1999) assessed modifications mutually at the northern and southern boundaries. Of the 35 European non-migratory butterfly species sampled, 63% had tiers that moved 35–240 km north during the course of the previous century, and the best 3% moved south (Parmesan 1996). The change in range is in line with the 0.8°C warming that has occurred over Europe over the past century, which has caused the climatic isotherms to move with a mean 120 km northward (Beniston and Tol 1998). Range shifts due to climate change necessitate adaptive conservation strategies. Coordination of scientific research, conservation planning, and policy implementation are crucial for understanding and addressing these shifts. Protecting biodiversity is essential for preserving ecosystems and maintaining ecological services supporting human well-being.

The process of dividing vast, continuous environments into smaller, isolated sections—often as a result of human activity—is known as habitat fragmentation. This phenomenon has significant ecological implications, affecting biodiversity, ecosystem functions, and species survival. Climate change exacerbates habitat fragmentation, making it harder for species to migrate and find suitable habitats. Forests, wetlands, and coral reefs are particularly vulnerable. Habitat fragmentation is a significant issue in ecosystems, resulting from various factors such as urbanization, agricultural development, infrastructure development, resource extraction, and climate change. Urbanization converts natural habitats into residential, commercial, and industrial areas, while agricultural development creates isolated patches of natural habitats. Infrastructure development creates physical barriers, while resource extraction and climate change lead to fragmentation of ecosystems, leaving a scarcity of natural habitats.

Because of smaller populations and scarcer resources, habitat fragmentation causes biodiversity loss, species extinction, and genetic diversity loss. Isolated populations are more susceptible to inbreeding, reducing their adaptability. Fragmentation disrupts ecological processes, affecting migration and dispersal, particularly for migratory species. It also affects pollination and seed dispersal, affecting plant reproduction. Edge effects, such as microclimate changes and invasions by non-native species, can also occur. Ecosystem stability is also impacted by fragmentation, which modifies food webs, predator-prey interactions, and community structure. Therefore, with the goal to safeguard the survival and sustainability of species, it is imperative that these challenges be addressed.

Habitat fragmentations are reported in European forests. This has also affected North American prairies, where extensive agricultural conversion has created isolated patches of native grasslands. Pollinator species like bees and butterflies have been negatively affected, while European forests have been fragmented by road networks and urban expansion. Large mammals like the European lynx and brown bear face difficulties in finding mates and expanding their territories due to fragmented habitats. Habitat fragmentations pose a noteworthy threat to global biodiversity and ecosystem well-being. Addressing this issue requires a multifaceted approach involving habitat conservation, restoration, and sustainable land-use practices. By implementing effective conservation strategies and fostering ecological connectivity, we can mitigate the adverse effects of fragmentation and promote resilient ecosystems capable of supporting diverse species and ecological processes.

4.1.4.2 Alteration of Ecosystem Functions

Ecosystem functions encompass biological, geochemical, and physical processes essential for ecosystem health and human services like clean water, fertile soil, and pollination. Phenological changes and ecosystem productivity are the two main characteristics of alteration of ecosystem functions. The study of periodic biological events in plant and animal life cycles and how they are impacted by seasonal and interannual climate variations is known as phenology. Alterations in the life cycle's timing activities including blossoming, migration, breeding, and hibernating are referred to as phonological changes. These shifts, which have important ecological, agricultural, and economic impact, are mostly caused by climate change. There is a shift in the timing of biological activities like migration, flowering, and breeding. For example, because of the warmer weather, many plants are flowering earlier in the spring.

Climate change impacts plant phenology, including flowering time, leaf-out, fruit ripening, migration, breeding, and hibernation. Warmer temperatures can lead to earlier flowering dates, earlier leaf-out, and accelerated fruit ripening, affecting harvest timing and food availability. Animal phenology, such as bird migration, breeding, and hibernation, also varies due to temperature and food availability. Phenological changes are primarily driven by rising global temperatures, which can significantly impact biological events. Changes in precipitation patterns, particularly for species dependent on moisture conditions, can also influence phenology. Photoperiod, which interacts with temperature, can also influence phenology, determining the time of life cycle events for some species. Phenological changes can have significant ecological impacts, including mismatches in plant-pollinator interactions, altered food webs, range shifts, and population dynamics. These changes can affect plant-pollinator interactions, predator-prey dynamics, food availability, and trophic interactions. Species may also shift their ranges to match phenological needs, potentially leading to changes in community composition and habitat use. Phenological changes can also affect agricultural productivity, pest and disease management, and economic costs. Case studies include cherry blossoms in Japan, European birds, and North American amphibians. These changes can affect tourism, bird populations, and ecosystem services. The economic costs of these changes can be significant, affecting agricultural yields, tourism, and ecosystem services.

Habitat fragmentation and climate change significantly impact ecosystem functions, including nutrient cycling, primary productivity, pollination, decomposition, soil formation, water regulation, and climate regulation. Fragmented habitats disrupt nutrient cycling, leading to increased temperatures and altered precipitation patterns.

Pollination is also affected, with smaller patches less accessible to animal dispersers. Decomposition and soil formation are also affected, with warmer temperatures and altered moisture levels affecting nutrient availability. Climate change effects include changes in rainfall patterns, increased evaporation rates, and intense weather events. The loss of ecosystems, such as the Amazon Rainforest, coral reefs, grasslands, and savannas, further accelerates global warming. Climate regulation is also affected by fragmentation, with fragmented forests and ecosystems reducing carbon storage capacity, contributing to higher atmospheric CO_2 levels. The loss of large, continuous forests also impacts local climates by altering albedo and evapotranspiration rates. Climate change exacerbates the loss of ecosystems that regulate the climate, creating feedback loops that further accelerate global warming.

The Amazon Rainforest, a vital ecosystem, is significantly affected by deforestation and fragmentation. This disrupts nutrient cycling, primary productivity, and climate regulation, contributing to higher atmospheric CO_2 levels and climate change. Coral reefs, which rely on a symbiotic relationship between corals and algae, also suffer from fragmentation due to bleaching and physical damage. These reefs help regulate coastal water quality by filtering nutrients and providing habitat for species contributing to nutrient cycling. Fragmentation also reduces the carbon storage capacity of coral reefs, impacting global carbon cycles. Additionally, fragmented, savannas and grasslands experience decreased plant variety and changed community composition. The general health of the ecosystem may be impacted by these variations in decomposition rates and soil nutrient dynamics.

Productivity of Ecosystems: the pace at how fast energy is transformed into biomass in an environment is referred to as ecosystem productivity. It is a crucial component of ecosystem functioning and influences the availability of resources for all organisms within the ecosystem. There are different types of productivity, including primary productivity (both gross and net), secondary productivity, and overall ecosystem productivity. Ecosystem productivity includes gross primary productivity (GPP), net primary productivity (NPP), and secondary productivity. GPP represents energy captured by autotrophs through photosynthesis, NPP represents energy remaining after respiration, and secondary productivity refers to biomass generation by heterotrophic organisms. Changes in temperature and precipitation affect primary productivity and nutrient cycling, which can alter food web dynamics and ecosystem services.

Ecosystem productivity is influenced by various factors such as climate, water availability, light availability, nutrient availability, CO_2 concentration, soil quality, and disturbances. Warmer temperatures increase metabolic rates, while adequate water and light availability affect plant growth and photosynthesis. Higher CO_2 levels can enhance photosynthesis and NPP, while soil quality and microbial activity affect nutrient cycling and water retention. Tropical rainforests have the highest NPP due to warm temperatures, high precipitation, and sunlight. Savannas and grasslands have varying productivity based on seasonal rainfall. Temperate forests have moderate to high NPP. Deserts have low productivity due to limited water availability. Aquatic ecosystems vary in productivity based on nutrient availability and water flow.

Climate change impacts ecosystem productivity through temperature changes, altered precipitation patterns, increased CO_2 levels, nutrient cycling disruptions, and extreme weather events. Climate change can extend growing seasons, stress plants, and disrupt nutrient availability. Conservation and management practices like sustainable land use, restoration ecology, and protected areas can enhance productivity and maintain ecosystem health. Mitigation strategies include reducing GHG emissions and implementing adaptive management techniques to lessen the adverse impact of climate change on ecosystem productivity. Ecosystem productivity is crucial for biodiversity, carbon cycling, and resource availability. Understanding factors and global patterns is essential for conservation and management, especially in climate change-affected environments.

4.1.4.3 Increased Extinction Risk

Effective conservation techniques are required as a result of human actions that are increasing the risk of species extinction worldwide. These activities include habitat degradation, overexploitation, environmental damage, climate change, and exotic species. Pollution, habitat degradation, climate change overexploitation, hunting, poaching, logging and timber extraction, and invasive species are all contributing factors to increased extinction risk. The species vulnerability and ocean acidification are the prominent reasons that keep biodiversity in extinction risk.

Climate changes, such as rising sea levels and temperature, can lead to species that cannot migrate or adapt quickly enough. Changes in precipitation can disrupt habitats and food sources, affecting species sensitive to moisture levels. Hurricanes, wildfires, and heat waves serve as illustrations of severe weather phenomena that have the potential to devastate habitats, killing a lot of people. Habitat fragmentation, such as deforestation and urbanization, can also lead to species loss. Agricultural expansion reduces available habitats for wildlife, leading to monocultures lacking biodiversity. Chemical pollution, plastic pollution, and nutrient pollution can poison wildlife and disrupt reproductive success. Overfishing, illegal hunting, and timber extraction can also lead to species decline. Invasive species can compete for resources, cause predation, and introduce new diseases that native species lack immunity to. These factors contribute to the extinction of many species worldwide.

Climate change poses significant threats to biodiversity, affecting species through environmental changes, with species vulnerability influenced by sensitivity, exposure, and adaptive capacity. Species vulnerability is influenced by various factors, including physiological sensitivity, life history traits, geographic range, microhabitat availability, adaptive capacity, genetic diversity, and behavioral flexibility. Species with narrow thermal tolerance ranges, specialized habitat requirements, low reproductive rates, or long generation times are more sensitive to climate change. Species with limited geographic ranges or isolated ecosystems are more vulnerable. Adaptive capacity, such as dispersal ability, genetic diversity, and behavioral flexibility, can also impact species vulnerability. An increased danger of extinction exists for species with restricted ranges, specific habitat needs, or limited migration abilities. The International Union for Conservation of Nature (IUCN) Red List emphasizes how many endangered species are being threatened by climate change. Ocean Acidification—Increased CO_2 levels lead to ocean acidification, harming marine organisms, corals, and shellfish that depend on calcium carbonate for their skeletons and shells.

4.2 ROLE OF BIODIVERSITY IN CLIMATE REGULATION

4.2.1 Carbon Sequestration

Biodiversity is acting as a crucial part in climate regulation, primarily through the process of carbon sequestration. The technique of capturing and storing atmospheric carbon dioxide (CO_2) through chemical, biological, or physical methods is known as carbon sequestration. As significant carbon sinks, ecosystems like wetlands, forests, grasslands, and oceans contribute to lessen the consequences of climate change. Wetlands and Forests: Both wetlands and forests, particularly tropical rainforests, are important carbon sinks. They absorb CO_2 from the atmosphere, helping to mitigate climate change. Tropical forests, such as the Amazon Rainforest, are highly productive and biodiverse, sequestering large amounts of carbon through photosynthesis and storing it in biomass and soil. For instance, the biomass and soil of the Amazon Rainforest contain enormous amounts of carbon storage. The temperate forests are located in high latitudes, store significant amounts of carbon in their soils and permafrost. Grasslands and savannas sequester carbon through extensive root systems and soil organic matter, with proper management enhancing their carbon sequestration capabilities. Buotte et al. (2020) report that by 2099, high-productivity, low-vulnerability forests absorb up to 5,450 Tg CO_2 equivalent (1,485 Tg C). This is equivalent to approximately six years' worth of current regional fossil gasoline emission or 20% of worldwide alleviation potential earlier recognized for all boreal and temperate forests. The wetlands including peatlands, marshes, and mangroves store carbon in waterlogged soils and help slow down organic matter decomposition. Oceans, the largest carbon sinks, absorb about 25% of CO_2 emitted by human activities annually. Marine ecosystems, such as mangroves, seagrasses, and salt marshes, sequester carbon through plant growth and sediment accumulation.

Marine ecosystems play an essential part in climate mitigation via sequestering carbon dioxide (CO_2) with regulating the Earth's climate. Biodiversity within these ecosystems enhances their ability to capture and store carbon, making them decisive in the battle next to climate change. The marine habitats involved in carbon sequestration include mangroves, seagrasses, salt marshes, and coral reefs, collectively known as "blue carbon" ecosystems and sequester substantial amounts of carbon, known as "blue carbon." Because blue carbon is accumulating in biomass and sediments, it is necessary for reducing the impact of climate change. Blue carbon is captured by oceanic and coastal ecosystems such as mangroves, seagrass beds, and salt marshes.

4.2.2 Water Cycle Regulation

Water cycle regulation involves various mechanisms, including transpiration and evapotranspiration, soil water retention and infiltration, and water filtration and quality improvement. Biodiversity enhances evapotranspiration rates, which help cool the air and modulate weather conditions. Diverse root systems improve soil structure, while organic matter contributes to soil moisture retention and plant growth. Wetlands filter pollutants and sediments, while riparian buffers trap sediments and nutrients.

Vegetative cover stabilizes soil and reduces erosion, while water absorption helps mitigate floods. Biodiversity enhances ecosystem resilience, climate mitigation, and adaptive capacity, allowing them to adapt to changing conditions. Sustainable water management practices involve incorporating biodiversity conservation, ensuring sustainable water use and availability. By transpiring, plants contribute to the water cycle, which in turn influences precipitation patterns and cloud formation. Forests, for instance, influence local rainfall and hydrology. Diverse ecosystems such as forests, wetlands, grasslands, and riparian zones contribute to key hydrological processes, including transpiration, water retention, filtration, and flood control. Protecting and restoring these biodiverse ecosystems is essential for maintaining their climate regulation functions and ensuring sustainable water management.

4.2.3 Soil Health and Erosion Control

Biodiverse ecosystems, such as grasslands and forests, maintain soil structure and fertility through complex root systems and organic matter decomposition. This reduces soil erosion and enhances water retention. Biodiversity plays a crucial role in erosion control. Ground cover plants and vegetation protect soil from rainfall, preventing soil particle displacement. Forest canopy trees and understory plants intercept rainfall, reducing surface runoff and soil erosion. Plant roots bind soil particles, enhancing stability on slopes and erosion-prone areas. Deep-rooted trees and shrubs act as erosion barriers, stabilizing soil on slopes and along riverbanks. Biodiverse plant communities enhance soil structure, allowing better water infiltration and reduced surface runoff. Vegetative buffer strips along waterways trap sediments, protecting water quality and preventing soil erosion.

4.2.4 Feedback Mechanisms

Feedback mechanisms are systems that control how one component behaves by using its conditions to control how another or the same component behaves. These systems, such as ecological and climatic systems, depend on these mechanisms to be stable, or in a state of homeostasis. Both positive and negative feedback loops have different effects on the dynamics of systems.

4.2.4.1 Loops of Positive Feedback

Positive feedback loops magnify modifications or departures from the starting point, which increases the effect even more. These loops have the potential to quicken procedures and push systems closer to a tipping point, where even a slight alteration can have profound and occasionally permanent consequences. If there are no negative feedback systems to counteract positive feedback, it might be disruptive. An example is ice-albedo feedback in climate systems. In consequence of the melting of snow and ice, the Earth's albedo, or reflectivity, diminishes as global temperatures rise. More solar energy is absorbed by darker surfaces, such as water and land, which raises temperatures and accelerates ice melt. The impact of this feedback loop adds to global climate change and quickens the warming of the Arctic.

4.2.4.2 Loops of Negative Feedback

By opposing changes, negative feedback loops help a system maintain balance and stability. By lessening the impact of deviations and returning the system to its initial condition or within a predetermined range, these loops aid in the maintenance of homeostasis. An example is cloud cover and albedo in climate systems. Higher global temperatures cause more water to evaporate, which results in the formation of more clouds. The Earth's surface can cool when incoming solar energy is reflected by clouds (high albedo). The impact of this is that by reflecting more solar radiation back into space, this negative feedback helps to stabilize global temperatures.

4.3 IMPORTANCE OF GREEN ENERGY IN REDUCING CARBON FOOTPRINT

Green energy is power produced in conjunction with solar power and other natural resources. While clean energy no longer releases pollutants into the environment, such as carbon dioxide, renewable power comes from non-exhaustible source, for example, solar, hydroelectric, wind energy (Androniceanu and Popescu 2017). Green electricity resources include solar, wind, geothermal, hydroelectricity, tidal and oceanic electricity, and biomass. Depending on the generation and environmental impact, these sources of power are clean and renewable (Bilgen et al. 2004). The carbon footprint, or total amount of GHGs emitted by human actions, is commonly quantified as carbon dioxide equivalents (CO_2^e). Green energy plays a critical part in reducing this amount. Photovoltaic (PV) cells are used in the popular renewable energy source known as solar energy to convert solar light into electric current. It is environmentally friendly and clean, as it does not produce GHGs or air pollution. Wind energy uses the Earth's motion to generate electricity in highland or offshore locations by turning turbines. Offshore wind power has increased gradually, offering a solution to noise and intrusiveness concerns.

Hydroelectric power produces energy by rotating a generator's turbine blades, similar to wind power. Larger dams divert natural water sources, negatively impacting animals. Less harm is done to the surrounding ecosystem by smaller hydroelectric power facilities, which generate 17% of all electricity worldwide. Hydropower alone supplies this electricity and has an efficiency of about 90%. Geothermal energy has the ability to create cleaner, more environmentally friendly systems by utilizing thermal energy that is deposited immediately underneath the surface of the Earth. Conversely, there are worries about possible ecological damage, contamination of the air and water, and changes to the habitats of wildlife. Biomass energy is generated from plant and animal-based organic materials, including leftover wood, trees, and agricultural products. By burning this biomass, heat is produced, which drives a steam generator to produce energy. Another well-liked technique for turning biomass into energy to produce heat and power is gasification. While properly gathered biomass can be sustainable, in many other situations it is not regarded as clean or green energy. All things considered, renewable energy sources that include biomass, geothermal, hydropower, solar, and wind offer viable customs to assemble our energy desires without having a negative environmental impact. To reduce carbon emissions and stop global warming, green energy utilization is essential. By utilizing renewable energy source including biomass, geothermal, hydro, wind, and solar power, we can mitigate GHG emissions, improve air quality, and promote sustainable development.

Green energy is becoming more feasible and affordable despite obstacles such as infrastructure requirements, initial expenses, and intermittency. These obstacles can be addressed with the help of supportive legislation and technical breakthroughs. The transition to cleaner, more sustainable energy is being fueled by international efforts and national commitments, which is crucial to protecting the environment for coming generations.

4.4 BIODIVERSITY: DEFINITION AND SIGNIFICANCE

Biodiversity, also known as biological diversity, is best understood as "the variety of life," encompassing variation at all levels of biological organization. Numerous recognized definitions of biological diversity, also known as biodiversity, have been offered, expanding upon this basic definition. Undoubtedly the most important and all-encompassing definition of these is the one included in Article 2 of the Convention on Biological Diversity (CBD).

As per the Convention, "Biological diversity" encompasses the variation among living organisms originating from various sources, such as terrestrial, marine, and other aquatic ecosystems, as well as the ecological complexes they form. It also includes diversity within species, between species, and among ecosystems. The diversity of life in all of its form is known as biodiversity. It functions as a broad unifying concept as it embraces all manifestations, intensities, and combinations of natural variety. The term "biological diversity," also known as "biodiversity," refers to that aspect of nature that consists of variations in the genes of individual members of a species; richness and variety of all plant and animal species at various spatial scales, including locally, regionally, nationally, and globally; and different kinds of terrestrial and aquatic ecosystems within a specific area.

The term "Biodiversity" was coined in the 1980s by biologist Edward O. Wilson. The result of billions of years of evolution shaped through both natural processes and increasing human action is the biodiversity that exists today. It creates the web of life that we are completely dependent upon and of which we are a part. At present, there are almost 2.1 million known species, most of them are small animals like insects. While the UNEP believes that while the number of species on Earth ranges from 9.0 to 52 million, experts estimate that there are actually roughly 13 million species (Mora et al. 2011).

4.4.1 Types of Biodiversity

Evolution, along with the genesis of biodiversity, climate change, favorable conditions, and changes in the Earth's structure, causes the emergence of a variety in living things. The changes in elements cause the formation of a variety of biodiversity. Figure 4.1 shows three major categories: genetic, species, and ecosystem diversity are typical of the distribution of biodiversity.

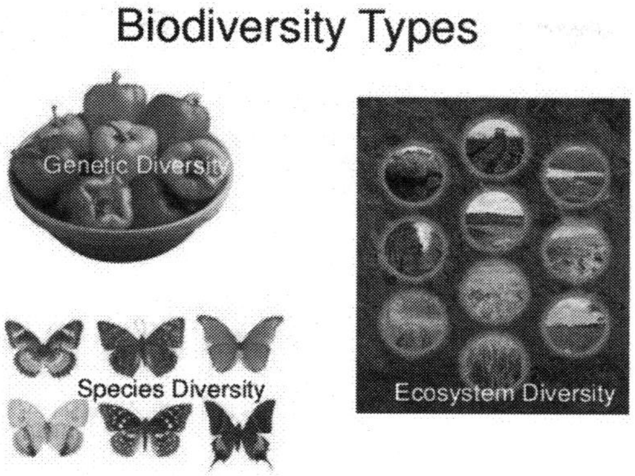

FIGURE 4.1 Types of biodiversity.

Source: Google images.

The genetic composition of all animal and plant species is referred to as genetic variety, and it is necessary for a healthy reproductive population. It can be defined as the degree of genetic diversity found in a population or as the basic composition of all living things on Earth. Less inbreeding and genetic dissimilarity arise from fewer breeding individuals, which may lead to species extinction. Domestic animals and crops have evolved from the "gene pool" created by wild species, and modern biotechnology alters genes to produce commercial products and potent pharmaceuticals. Species diversity is mainly the well-known element of biodiversity, often the first topic covered in biodiversity studies. It is one of the three foremost layers of biological diversity, along with species, ecosystem, and genetic diversity within species. Because different locations give the same species different names, measuring species richness at the global or large regional scale sometimes encounters synonymy issues.

The range of biotic community and ecological and habitat processes that exist in the biosphere is referred to as ecosystem diversity. It can be defined for a specific geographical region or political entity, such as a country or state. Natural ecosystems are considered degraded when productivity declines due to exploitation or excessive utilization. The establishment of ecosystems is greatly influenced via factors such as climate, geographic location, and availability of natural resources; tropical regions are globally recognized as hotspots.

4.4.2 Ecosystem Services Provided by Biodiversity

Biodiversity is a crucial resource for preserving ecological processes, nutrient fixation, soil formation, air and water circulation, global life support, water balance, watershed protection, erosion control, and flood reduction. It is critical for human development and long-term survival. The consumptive use value of biodiversity includes the direct utilization of natural resources for daily life, such as timber, food, fuel wood, and fodder. The productive use value defines the market value of biodiversity, with modern human making every aspect of living things marketable goods. Traditional societies have preserved biodiversity as a life-supporting resource due to their small population and resource requirements. Biodiversity conservation is an ethical issue, involving protecting all forms of life on Earth. It is important for awareness and appreciation of biodiversity, as people have worshipped symbols from wild animals for thousands of years.

4.4.3 Threats to Biodiversity

Diversity of life on Earth, known as biodiversity, is crucial for ecosystem health, human well-being, and the planet's overall functioning. On the other hand, a number of threats confront it, including pollution, overindulgence, invasive species, illness, altered land use, unsustainable forestry and agriculture practices, climate change, and acidification of the ocean. Destroying and fragmenting habitats limits the amount of space that species may have, obstructs migration paths, and isolates populations, which lowers genetic diversity and increases species' susceptibility to extinction.

The main factor causing climate change is an increase in GHG emissions, which raises global temperatures, modifies rainfall pattern, and increases the occurrence of extreme weather proceedings. Because of this, species experience changes in distribution, altered patterns of reproduction and migration, and a higher risk of extinction as a result of their inability to quickly adjust to changing environmental factors. Ecosystems that are especially vulnerable include Arctic ecosystems and coral reefs.

Pollution of water, air, and soil from industrial, agricultural, and urban sources poses a serious danger to biodiversity. Overexploitation involves the unsustainable harvesting of resources, leading to population declines and potential extinction. Invasive species can alter ecosystem dynamics, reduce native biodiversity, and cause significant economic and ecological damage. Land-use change, driven by urbanization, agriculture, and industrial development, leads to habitat loss, fragmentation, and degradation, significantly impacting biodiversity and ecosystem services. Unsustainable agriculture and forestry practices can lead to soil degradation, deforestation, and loss of biodiversity. To address these threats, coordinated global action, sustainable practices, robust policies, and increased public awareness are needed to protect and conserve the planet's biodiversity for future generations.

4.4.4 Conservation Strategies

Biodiversity conservation involves the protection, preservation, and management of ecosystems, species, and genetic diversity to ensure natural processes continue and human beings benefit from nature's resources. Strategies for biodiversity conservation include protected areas, sustainable use practices, ecosystem restoration, ex situ conservation, legislation and policy, community involvement, climate change adaptation and mitigation, research, monitoring, education and awareness, and financial mechanisms.

The preservation of habitats and species depends on protected places like national parks, animal reserves, and marine sanctuaries. Sustainable practices in agriculture, forestry, and fisheries help conserve biodiversity while meeting human needs.

Ecosystem restoration involves restoring degraded ecosystems through methods like reforestation, afforestation, wetland restoration, and coral reef restoration.

Ex situ conservation involves protecting species outside their natural habitats through institutions like zoos and aquariums, botanical gardens, and seed banks. Strong legal frameworks and policies, such as the Endangered Species Act (ESA) and the CBD, are essential for protecting biodiversity. Community involvement in conservation efforts ensures sustainable management of natural resources and supports livelihoods.

Addressing climate change is crucial for biodiversity conservation through carbon sequestration projects, climate-smart agriculture, and habitat corridors. Scientific research and monitoring are essential for evaluating conservation strategies, and education and awareness campaigns can foster support and action. Financial mechanisms for biodiversity conservation include conservation trust funds, payment for ecosystem services, and green bonds and investments. Effective conservation efforts depend on global cooperation, robust policies, adequate funding, and active participation from all sectors of society.

4.5 CLIMATE CHANGE: CAUSES AND EFFECTS

Climate change is among the most imperative issues of the 21st century, which has reflective effects on society, the economy, and the environment. The main cause of it is human activity, which raises GHG emissions. Lasting changes in temperature, precipitation, wind patterns, and other components of the Earth's climate system are referred to as climate change. When specific gases in the Earth's atmosphere trap heat and keep the temperature high enough to support life, it is known as the "greenhouse effect" in nature. By raising the concentration of GHGs, human action has exacerbated this effect, causing more heat to be stored and a warmer world that is more sensitive to the phenomenon known as global warming. The lasting increase in Earth's average surface temperature brought on by human activity is known as global warming. This rise is mostly caused by the release of GHGs, which include carbon dioxide (CO_2), methane (CH_4), and nitrous oxide (N_2O). Fossil fuel combustion (coal, oil, natural gas), deforestation, industrial activities, and agricultural practices are the main causes of these emissions. Global temperature records, glacier retreat, ice melting, increasing sea levels, altered weather patterns, and ocean acidification are examples of climate change evidence. Marine life, coastal communities, and ecosystems are all under risk from these influences.

4.5.1 Anthropogenic Activities and GHGs

Anthropogenic activities are the human acts responsible for the alteration of the natural environment. Climate change is a result of these activities significantly augmenting the atmosphere's concentration of GHGs. Main sources of GHGs include industrial operations, deforestation, and the combustion of fossil fuels. The main GHG, carbon dioxide (CO_2), makes up about 75% of human emissions and can linger in the atmosphere for decades. A GHG called methane (CH_4) has a longer half-life in the atmosphere than carbon dioxide (CO_2), but it is more potent over a 100-year period. With a lengthy atmospheric lifetime of around 114 years, nitrogen oxide is a GHG that traps heat in the atmosphere 300 times more effectively than CO_2. Fluorinated gases, hydrofluorocarbons (HFCs), perfluorocarbons (PFCs), and sulfur hexafluoride have high global warming potentials and can remain in the atmosphere for long periods. Major anthropogenic activities contributing to GHG emissions include energy production, transportation, aviation, shipping, industry, agriculture, land use, and waste management.

4.5.2 Climate Change's Effects on Human and Natural Systems

Climate change has far-reaching impacts on natural ecosystems and human societies, threatening biodiversity, water resources, food security, health, infrastructure, and economies. Climate change has significant impacts on natural systems, including ecosystem disruption, species interactions, habitat alteration, water scarcity, and changes in the hydrological cycle. It also threatens coastal and marine environments, with rising sea levels and ocean acidification posing threats to communities and infrastructure. Rising temperatures can also aggravate extreme weather events, like heat waves, intense precipitation, and tropical cyclones, which can result in heat-related diseases and fatalities. These impacts are exacerbated by climate change's effects on water availability, flood risks, and groundwater recharge.

Climate change impacts human systems in various ways, including food security, water resources, health, infrastructure, and economic impacts. Climate change affects crop yields, livelihoods, water availability, and quality, leading to food shortages and price volatility. It also increases the risk of heat-related illnesses and vector-borne diseases. Infrastructure damage and urban flooding are also significant. Economic impacts include industry disruptions, increased insurance costs, and displacement and migration. Social disruptions include displacement and migration, and vulnerable populations, such as small-income communities and marginalized group, are more vulnerable. In order to address these effects, immediate and coordinated action is needed to reduce GHG emissions, improve resilience at the international, national, and local levels, and adapt to changing circumstances. By taking proactive measures to mitigate climate change and build adaptive capacity and by working together, society can lessen the negative effects and build a supplementary resilient and sustainable future for each person.

4.5.3 Mitigation and Adaptation Strategies

The two complementary approaches of adaptation and mitigation are utilized to address the issues brought on by climate change. While adaptation refers to making adaptations for the consequences that are already being felt or are predicted in the future, mitigation is the process of reducing GHG emissions in order to lessen the effect of climate change.

The conversion to green energy, or renewable energy, carbon capture and storage, low-carbon transportation, sustainable land-use and forestry practice are some of the mitigation techniques that lower GHG emissions. Non-exhausted sources like geothermal, hydro, wind, and solar energy are promoted, while energy efficiency is improved in buildings, transportation, and industries. Sustainable land-use and forestry practices, such as agro-forestry, conservation tillage, and organic farming, reduce emissions and increase carbon sequestration. Before being released into the atmosphere, CO_2 emissions from factories and power plants are captured, stored using carbon capture and storage technology. Low-carbon transportation is promoted through the adoption of electric vehicles (EVs) and public transportation systems. In order to reduce waste creation and increase resource efficiency, enterprises are using greener manufacturing methods and adopting circular economy ideas as part of their energy shift.

Climate-resilient infrastructure, water management, agriculture, ecosystem restoration, and disaster preparedness are key strategies for tackling climate change. Ecosystem restoration is one of the main actions humanity has taken to combat climate change. It should focus on restoring degraded ecosystems and establishing protected areas to conserve critical habitats and species. Health and disaster preparedness should be strengthened through healthcare infrastructure and disaster preparedness plans. Infrastructure design should focus on withstanding extreme weather events, while natural infrastructure, such as wetlands and green roofs, can enhance resilience. Water conservation and flood management should be implemented to manage water resources efficiently. Agriculture should diversify crop varieties and invest in irrigation systems and water storage infrastructure. These tactics seek to guarantee a sustainable future while reducing the effects of climate change.

A multilateral organization called the UNFCCC was created through the goal of stabilizing GHG concentrations in order to stop humans interfering with the climate system. As a result, accords to combat global climate change—the Paris Agreement and the Kyoto Protocol—have been adopted. With efforts to keep the increase to 1.5°C, the Paris Agreement seeks to keep global warming to 2°C over pre-industrial levels. Meanwhile, the Kyoto Protocol, which was approved in 1997, provided wealthier nations legally binding targets for cutting their emissions.

To preserve the ozone layer and lessen its impacts, the UNEP adopted the Montreal Protocol in 1987. Its goal is to gradually phase out the production and consumption of chemicals depleting the ozone layer. It establishes schedules for the phase-out of specific ozone-depleting substances (ODSs), CFCs, halons, carbon tetrachloride, and methyl chloroform, through control measures, promotion of alternative substances and technologies. The Montreal Protocol is a significant achievement in international environmental governance, demonstrating the effectiveness of global cooperation in addressing environmental challenges. It has protected the ozone layer and safeguarded Earth's environment for present and future generations. However, its success will depend on continued commitment to its objectives, addressing remaining challenges, and exploring synergies with climate change mitigation.

With the objective to achieve a climate-neutral world by 2050, non-party stakeholders (sub-national governments, businesses, organizations, and individuals) are urged to take voluntary action by participating in the UNFCCC's (2015) Climate Neutral Now Initiative. It encourages organizations to sign the Climate Neutral Now Pledge, follow three steps (Measure, Reduce, Contribute), and report on their actions per annum. Governments resolved to mobilize more robust climate action to meet the objectives of the Paris Agreement at the United Nations climate change conference, known as COP21. Governments, towns, regions, companies, and investors must take initiative. The accord recognizes the pressing necessity of an international response and pledges from all parties involved. Other than this, Clean Development Mechanism (CDM), Paris Agreement, Green Climate Fund (GCF) are some of the initiatives taken by the UNFCCC.

4.6 GREEN ENERGY: AN OVERVIEW

Because of advancements in welfare, health, and human, social, and economic spheres, there is an increasing need for power and related services globally (Gielen et al. 2019). According to Kaygusuz (2012) and Papadis and Tsatsaronis (2020), sustainability will ensure the delivery of electricity and lessen the influence of the energy industry on climate change. In discussions on the financial, environmental, and social aspects of sustainable development, green power is a key topic (Dincer 1999). According to studies like those by Edenhofer (2015), inexperienced strength resources have the potential to significantly lower GHG emissions caused by the burning of fossil fuels in electrical plants, which would ultimately lead to a reduction in climate change.

According to Masini and Menichetti (2013), the global power sector is heavily dependent on market diversity and the ongoing advancement of green energy technology in the current global context. More than 90% of nations worldwide are pursuing the expansion of renewable energy sources (Owusu and Asumadu-Sarkodie 2016). Present-day governments across the board are being compelled to embrace a bright future with regard to renewable energy resources. Reducing greenhouse fuel emissions and improving supply safety are two major goals of renewable energy (Mohsin et al. 2021; Qayyum et al. 2021). Controlling the cost of energy raw materials like crude oil and natural gas is one of the primary objectives of power regulations. It is also one main factor influencing the adoption of new assets and the significant shift in the composition of energy resources consumed, along with the importance of protecting the environment (Zakeri et al. 2022).

Inexhaustible energy, such as solar and wind, is a clean, green, and renewable natural resource that is non-polluting, environmentally benign, and self-renewing. Pratiwi and Juerges (2020), Nengroo et al. (2022), Gupta and Grubb (2000), and Shen et al. (2022) have all contributed to the debate on the distinction between renewable energy and other forms of energy. They contend that the optimal clean energy mix is produced when green energy and renewable sources such as solar and wind are integrated, highlighting the significance of taking into account the minute distinctions between renewable energy and conventional energy sources. The energy industry emits the largest carbon footprint due to resource use, according to a number of studies conducted by the experts. Additionally, they came to the determination that the production of energy derived from green source has a less carbon footprint than traditional methods. Making the conversion to green energy will therefore assist humanity in reducing climate change's impacts and the irrevocable harm done to Mother Earth, biodiversity, and humankind. Keeping this in mind, the UNFCCC, IPCC, and all countries want to switch from traditional to green energy sources. The main objective of worldwide zero emission efforts was to employ renewable energy sources to lower carbon footprints. The new approach to combating climate change is thus to adapt to these green technologies and resources.

4.6.1 Types of Green Energy

Green power is any energy kind that is generated from natural resources, which include sunlight, wind, or water. They are mainly of five types as follows.

4.6.1.1 Solar Energy

Solar energy, which is produced by photovoltaic PV cells, is a popular renewable energy source. Theses devices are made to convert sunlight into electric current. Potentially, the sun can provide a significant portion of our energy needs. In addition to being clean energy sources, solar energy structures are also environmentally benign because they don't emit any GHGs or air pollutants (Rabaia et al. 2021).

4.6.1.2 Wind Energy

Wind energy is used to power Earth's air circulation, turning turbines to generate electricity. The best locations for this type of energy consumption are offshore or in highlands. On land, wind turbines must be positioned on plains, open spaces, and hilltops—areas known for strong winds. Offshore wind power has steadily grown over time. Most concerns about wind farms being noisy or obtrusive on land are dispelled by the fact that they offer an excellent way to generate electricity. Indeed, operating offshore has drawbacks of its own due to the hostile environment in which the turbines must function. The decrease in GHG emissions is a noteworthy advantage that is receiving a lot of financial and technological support. It also represents an outstanding solution due to the fact wind energy structures do no longer emit any gaseous pollution, including CO_2, oxides of sulfur, oxides of nitrogen, and particulate matter (PM), which include dust or any other air pollution at the same time as they're in operation. A wind energy project's daily carbon footprint is recovered in less than six months, after which the plant generates electricity for several years without emitting any emissions (Mehrjerdi 2020).

4.6.1.3 Hydropower

Hydroelectric power produces energy by rotating a generator's turbine blades, just like wind power does. The fact that many of the larger dams divert natural water sources has a detrimental effect on animals since it makes it harder for them to receive the water they require. Smaller hydroelectric power plants (less than 40 MW) do not have the same detrimental effects on the surrounding ecosystem because they only divert a minute water flow percentage, provided they are controlled properly. Hydropower alone accounts for 17% of the worldwide electricity delivered, and among all different technologies, it gives a performance of about 90% (Javed et al. 2020).

4.6.1.4 Biomass Energy

The organic material comes from plants and animals, such as leftover wood, lumber, and harvests, and makes up biomass energy. By burning this biomass, heat is produced, which fuels a steam generator to produce electricity. Gasification is another well-liked method for turning biomass into energy to provide heat and electricity. Such methods are required by the non-renewable energy resource alternative. Although biomass can be renewable if it is gathered ethically, there are many different situations when biomass is not seen as a simple or environmentally friendly material.

4.6.1.5 Geothermal Energy

The thermal energy deposited just beneath the Earth's surface is used by this environmentally friendly power source. Despite the fact that drilling is necessary to obtain this resource, raising questions about its possible environmental effects, once it is utilized, the resource has enormous potential. Actually, geothermal energy has been used for many decades to produce steam, which powers turbines and produces electricity, and to heat hot springs for swimming. Even though each technology has negative environmental impact consequences that have been extensively researched by numerous researchers, geothermal energy structures are nevertheless significantly cleaner and more environmentally friendly than traditional power structures (Hosseini and Wahid 2020; Hoang and Nguyen 2021). There are certain worries about air and water pollution, ecosystem degradation, and the potential for changing plant and animal habitats when it comes to geothermal power systems (Shortall et al. 2015).

4.6.2 Benefits of Green Energy

Because it is clean, renewable, and produces little GHG emissions, renewable green energy is a sustainable replacement for fossil fuels. It is affordable, easily obtainable, and less susceptible to global crises, price increases, or disruptions in the supply chain. Green energy furthermore creates jobs and helps neighborhoods where employees work. Eleven million new employment were generated or converted in renewable energy sector in 2018; as more states strive toward zero emissions objectives, this figure is anticipated to increase. The localized production of energy from resources such as wind and solar power increases the adaptability and resilience of energy network infrastructure to weather-related climate change. Furthermore, there exist certain constraints associated with renewable energy sources. Androniceanu and Sabie (2022) have provided data that illustrates the primary benefits and drawbacks of these sources, which are listed in Table 4.1.

4.6.3 Challenges in Implementing Green Energy

Renewable energy is currently facing numerous challenges, including intermittency and variability due to weather conditions or time of day, high initial capital investments, and a competitive levelized cost of electricity (LCOE) with fossil fuels. In order to connect remote resources to population centers and upgrade current grids to allow bidirectional power flow, the switch to renewable energy will require significant infrastructure development. To overcome these obstacles, policy backing and public approval are essential. The expansion of renewable energy can be hampered by the NIMBY (Not In My Backyard) mentality, which is concerned about noise, aesthetic effects, and environmental effects. To overcome

TABLE 4.1
Primary Benefits and Drawbacks of Renewable Green Energy (RGEs) Sources

Benefits	Drawbacks
Inexpensive RGEs are inexpensive. An economically viable or even free energy generation source is made available through the utilization of RGEs.	**Variation and unpredictability in the weather** The weather has an impact on RGEs. They are not always the same. Since many RGE sources depend on the environment and the weather, their dependability may occasionally be compromised.
Energy independence For developing nations, increased access to energy from green, renewable sources provides an additional advantage.	**RGE systems demand a large amount of room to set up and operate** RGEs could require a considerable area to operate and set up. Large energy storage systems are required to ensure there is always enough electricity when production drops because renewable resources have variable output capacities.
Lower electricity cost For both the financial sector and the general public. Using green, renewable energy sources can also result in lower monthly energy bills.	**RGE technologies advance slowly** Since technologies are just getting started when it comes to storing RGEs, an additional adaptable logistics of supply is required to make up for declines in output from a given source. They cost a lot of money as well.
Clean energy Certain RGEs are regarded as clean energy sources as well. The fact that a significant amount of renewable energy sources are also acceptable as clean and green energy is one of their biggest benefits.	**Community protest** Community demonstrations may result from the exploitation of RGEs. Some residents may object and protest if they want to live distant from renewable green energy sources like wind and solar farms.
Environmentally friendly One of the many benefits to the environment and economy that comes with using green and clean energy is lower air pollution.	**High implementation costs** The expense of implementing RGEs is substantial. The cost of mounting, installing, and maintaining technology and equipment (infrastructure) is high.
Reducing the quantity of major pollutants used in energy generation RGEs reduce the long-term expenses related to power provision. Additionally, a significant supply of renewable energy lowers the need for imported fuels and the resulting costs to the environment and economy.	**Political influence** There is political influence over RGEs.
Inexhaustible resources Unlike oil or coal, which needs to be obtained and transported, the resources for green and clean energy naturally replenish themselves. By doing this, a large expense that would have been incurred is avoided.	**Storage-related issues** Specialized equipment (accumulator-type batteries) with a restricted capacity and additional expenditures are needed for the storage of green energy.
Cutting down on waste RGEs aid in waste reduction.	**High expenses associated with RGE development and research** The RGE sector incurs substantial and rising costs for research and development.
Safety in operation RGEs can be used with a high level of safety.	

these challenges, stable policy frameworks and public acceptance are essential. Despite these challenges, renewable energy remains a promising alternative to traditional energy sources.

The changes in nation's policies, changing politics, and lack of funding could be major challenges in developing countries. The site selection, sentiments of local people, modernization, and lack of technology are some key challenges for the establishment of any green energy projects. The economy also plays an important role in the maintenance and establishment of projects. The required investment, skilled employees, subject expert are other major challenges to underdeveloped countries. While the developing countries are still figuring out the safety measures regarding green energy, the magnitude of disasters caused by geothermal, hydropower, and biomass energy is still under investigation. Therefore, safety measures of such projects are yet to be defined and should be standardized carefully. The expansion of such projects is also concerning as it involves the natural resources and economy.

4.7 GREEN ENERGY AND BIODIVERSITY CONSERVATION

The natural world as we know it and human existence are both in grave danger due to climate change. To prevent the irreversible impact of global warming and keep it below 1.5°C, decisive action is required this decade. Loss of biodiversity puts mankind and the life-support systems that provide us with food, water, health care, and so much more at risk of an existential crisis. It is essential that these two problems—which are inherently connected—be addressed jointly. A healthy Earth is essential for maintaining human well-being, stable climate, and provision of life-sustaining services to the natural world. Natural regions with high natural carbon uptake and storage can be preserved. These kinds of solutions can help lower atmospheric carbon dioxide (CO_2) levels and protect and restore biodiversity in support of the UN CBD expansive new aims, which were recently decided at the CBD COP15 in Montreal.

Universal hard work to move toward a net-zero energy system will not be successful if nature and the products and services it offers are not taken into consideration. If this isn't done, energy paths may be created that compromise health, environmental, poverty alleviation, and climate change objectives, resulting in a globe in perpetual decline. Thankfully, there are energy-related solutions that can work together to address both the biodiversity and climate challenges. We need to keep in mind that not every source of low-carbon energy is made equal as we move toward a net-zero future. Promoting renewable technologies that have the least negative effects on nature can be inexpensive and provide significant advantages. Research studies that consider the entire spectrum of effects make it abundantly evident that switching to renewable energy with a concentration on solar and wind power can drastically lessen environmental effects. When compared to other forms of renewable energy, yet other renewable might be the best option based on the local conditions and sustainability standards used.

As per Baruch-Mordo et al. (2019), the present global mapping of solar and wind energy sites suggests that there exists sufficient energy on low-conflict, environmentally damaged areas to fulfill the International Energy Agency's (IEA) projections for a power system that will allow for a maximum increase of 1.5°C in the average global temperature. New technologies that have the potential to significantly alter our energy system include floating solar and floating wind, albeit more study is required to fully comprehend their effects.

4.7.1 Impact of Green Energy Projects on Biodiversity

Climate, environment, and energy are all intimately tied to one another (Figure 4.2). The preservation of biodiversity and the integrity of ecosystems cannot come at the price of effectively combating climate change. It's crucial to prevent the destruction or loss of natural habitat. If you don't, it might greatly undercut efforts to mitigate climate change. As nations work to fulfill their climate objectives under the Paris Agreement, more than 10 million hectares of natural lands—an area the size of Iceland—may be removed for wind and solar development due to poor siting (Baruch-Mordo et al. 2019). Additionally, the transmission lines and related roads may fragment and alter natural habitats. The objectives of SDG 15, which aim to stop and halt the loss and degradation of habitat, clash with these effects.

From supply chains to project life, transitioning to renewable energy must consider the effects on the environment and society. The global electrification effort requires clean energy technologies requiring metals and minerals, which already have significant environmental and social impacts. Strategic planning and rigorous mechanisms are needed to ensure ethically sourced materials and circular resource use.

Energy paths entail trade-offs between land, seawater, and material resources, as well as environmental and social costs. Comparing energy types requires considering biodiversity impacts, pollution, and species. Life cycle assessments (LCAs) account for prospective effects during the course of a project, extraction of raw materials to waste management, and recycling. According to LCA results, the technology selection has a significant impact on how much energy policies harm the environment. When all environmental effects are taken into account, using renewable energy sources to generate and store energy is much less harmful to the environment than using fossil fuels (Luderer et al. 2019). Apart from its advantages for the environment

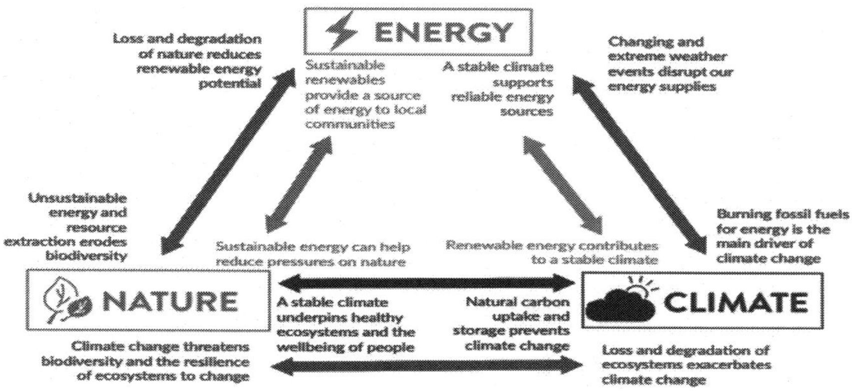

FIGURE 4.2 The connections between energy, climate, and environment.

and human health, switching to renewable energy can lead to considerably less of an influence on the environment when using wind and solar (Gibon et al. 2017). These include fewer negative effects on species as well as noticeably less pollution, eco toxicity, and overall negative effects on freshwater. However, the consequences of solar and wind energy are primarily limited to specific habitats or individuals; by contrast, the threat of fossil fuel-induced climate change to our globe threatens civilization and will irreversibly alter a multitude of habitats and species.

Since solar and wind energy are abundant and widely distributed geographically, they offer more sitting flexibility than fossil fuels. Due to this flexibility, previously converted or damaged land as well as offshore locations away from extremely sensitive areas like bird migration routes and marine mammal habitats can be used. Areas where natural, wind, and solar developments are managed can aid in improving the local biodiversity in the area surrounding the solar panels or turbines. For instance, solar farms on previously farmed area can present chances to reestablish natural flower meadows surrounding solar arrays.

4.7.2 Strategies for Minimizing Negative Impacts

Balancing green energy development with biodiversity conservation involves implementing best practices that curtail environmental impacts while maximizing ecological and social benefits. The best practices suggested by the expert to generate green energy without being a threat to biodiversity are environmental impact assessment (EIA), adaptive management, stakeholder engagement, policy integration, technology innovations, strategic site selection, habitat restoration and enhancement, sustainable operation and maintenance (O&M), capacity building and research, and legal and institution framework (Figure 4.3).

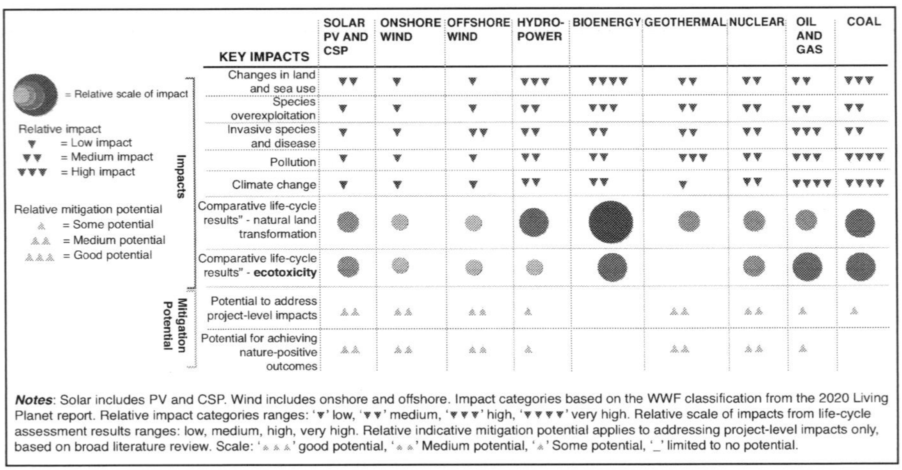

FIGURE 4.3 An essential study of the effects of major renewable energy sources and their potential mitigating factors (Gibon et al. 2017; Luderer et al. 2019; Hertwich et al. 2015).

4.7.2.1 Environmental Impact Assessments

To evaluate the potential effects of planned developments on the environment, EIAs are essential. The process includes scoping, baseline environmental studies, impact prediction, mitigation measures, impact evaluation, public participation, reporting, review and decision-making, monitoring and compliance, and post-project analysis. Scoping involves identifying key environmental issues and involving stakeholders, while baseline environmental studies examine the current state of the environment. Potential environmental repercussions, including direct, indirect, and cumulative effects, are predicted through impact prediction. Mitigation measures include avoiding, reducing, or compensating for negative impacts. Impact evaluation assesses the magnitude, extent, duration, frequency, and reversibility of predicted impacts. Public participation involves holding public meetings, workshops, and hearings to discuss the project and its potential impacts. Reporting documents the findings in a comprehensive report, including an executive summary, project description, environmental baseline, impact assessment, mitigation plan, stakeholder engagement, and conclusion.

Review and decision-making ensure the EIA report meets regulatory requirements and supports informed decision-making. Monitoring and compliance ensure compliance with environmental standards and the effectiveness of mitigation measures. Post-project analysis evaluates the long-term impacts of project and success of mitigating actions.

4.7.2.2 Adaptive Management

A methodical approach to enhancing resource management through knowledge from management outcomes is called adaptive management. It involves a cycle of planning, implementation, monitoring, and evaluation, allowing for continual learning and adjustment. Key principles include an iterative process, flexibility, learning-oriented approach, and stakeholder involvement. Problem assessment, design, development, implementation, monitoring, evaluation, and adjustment are all steps in the process. Examples of adaptive management in practice include wildlife conservation, water resource management, and forest management. Benefits of adaptive management include enhanced resilience, improved outcomes, cost-effectiveness, and stakeholder buy-in.

Challenges of adaptive management include complexity, uncertainty, institutional resistance, and monitoring costs. Strategies to overcome these challenges include capacity building, collaborative networks, incremental implementation, and secure funding. Investing in training and capacity building for managers and stakeholders can help overcome these challenges. Collaborative networks can share knowledge and resources, and incremental implementation can demonstrate success. Ultimately, adaptive management can lead to better environmental and resource management outcomes.

4.7.2.3 Stakeholder Engagement

An essential component of environmental management is stakeholder participation and decision-making processes, particularly in green energy projects and biodiversity conservation. It involves identifying and including all relevant stakeholders, maintaining transparency, being responsive, and treating them with respect. Key principles of stakeholder engagement include inclusiveness, transparency, responsiveness, and respect. Stakeholder identification involves identifying stakeholders, mapping their interests, planning the engagement process, and defining objectives. Communication is essential, with appropriate channels and methods used. Involving stakeholders in decision-making processes, joint problem-solving, and providing feedback are also essential. Monitoring and evaluating the engagement process is crucial, with regular assessments and adjustments made based on feedback. Benefits of stakeholder engagement include enhanced project acceptance, improved decision-making, risk management, and a social license to operate.

Challenges include diverse interests, resource intensiveness, communication barriers, and managing expectations. Strategies to overcome these include customized approaches, capacity building, ongoing engagement, and clear communication. Tools for stakeholder engagement include public meetings, focus groups, surveys, panels, and online platforms. Case studies of successful stakeholder engagement include wind farm projects in Denmark and conservation initiatives in Kenya. By following best practices and addressing challenges proactively, strong, cooperative relationships with stakeholders can lead to more sustainable and accepted outcomes.

4.7.2.4 Policy Integration

Policy integration is a crucial process for integrating green energy development and biodiversity conservation. It involves considering environmental, social, and economic factors in policy-making across different sectors to create coherent and synergistic policies. Key principles include a holistic approach, cross-sectoral coordination, a sustainability focus, and consistency across different levels of government and sectors. Policy integration involves policy review, analysis, stakeholder involvement, policy formulation, and implementation. It involves clear objectives, incentives, and regulations, inter-agency cooperation, cross-sectoral committees, action plans, capacity building, and regular reviews. Benefits of policy integration include synergies, resource efficiency, conflict reduction, and sustainable outcomes.

Challenges include institutional silos, complexity, conflicting interests, and limited capacity. Strategies to overcome these challenges include strong leadership, capacity building, institutional frameworks, and pilot projects. By addressing these challenges, integrated policies can lead to more harmonious development, resource efficiency, conflict reduction, and sustainable outcomes.

4.7.2.5 Technology Innovations

Technology innovations are crucial in balancing green energy development and biodiversity conservation by minimizing environmental impacts and increasing the effectiveness of renewable energy systems. These innovations include wildlife-friendly renewable energy technologies, habitat-compatible installations, advanced monitoring and management systems, energy storage and grid integration, biotechnological innovations, and marine renewable energy technologies. Wildlife-friendly renewable energy technologies include bird and bat-friendly wind turbines, anti-reflective coatings on solar panels, and habitat-compatible installations like floating solar panels. These technologies reduce the risk of collisions for birds and bats, detect approaching wildlife, and provide real-time data for adaptive management. Habitat-compatible installations include floating solar panels on water bodies, dual-purpose reservoirs, and agrovoltaics, which combine solar power generation with agriculture.

Advanced monitoring and management systems, such as remote sensing and Geographic Information System (GIS), can help identify optimal sites for renewable energy projects that minimize biodiversity impacts. Energy storage and grid integration technologies, such as battery storage systems and smart grids, help stabilize the grid and reduce habitat impact. Biotechnological innovations, bioenergy with carbon capture and storage (BECCS), produce bioenergy from sustainable biomass sources and sequester carbon, contributing to climate change improvement and dropping pressures on natural ecosystems. Artificial intelligence (AI) and machine learning can also play a role in predicting wildlife impacts, optimizing operations, and making data-driven decisions. Successful technology innovations in renewable energy sectors include bird-friendly wind turbine designs in Norway, floating solar farms in Japan, and agrovoltaics in Germany.

4.7.2.6 Strategic Site Selection

Strategic site selection is a crucial process in balancing green energy development and biodiversity conservation. It involves choosing locations for renewable energy projects that minimize environmental impacts while maximizing energy production. Key principles of strategic site selection include minimizing environmental impact, maximizing energy efficiency, involving stakeholders, and ensuring regulatory compliance. Steps in strategic site selection include preliminary screening using GIS, exclusion criteria, detailed site assessment, EIA, impact assessment, stakeholder consultation, site selection criteria, decision-making using Multi-Criteria Analysis (MCA), scenario planning, monitoring and adaptive management, and using tools like GIS, habitat mapping, remote sensing, biodiversity databases, and decision support systems.

Case studies of strategic site selection include wind farm development in the UK, solar energy projects in California, and hydropower projects in Norway. Benefits of strategic site selection include environmental protection, project viability, conflict reduction, and regulatory compliance. Challenges in strategic site selection include data limitations, complexity, dynamic environments, and cumulative impacts. To overcome these challenges, strategies include investing in data collection, capacity building, collaborative planning, and adaptive management. By following best practices, utilizing advanced tools and techniques, and engaging stakeholders, it is possible to identify sites that minimize impacts on biodiversity while maximizing renewable energy potential. This approach contributes to a balanced and sustainable development pathway, addressing both energy needs and conservation goals.

4.7.2.7 Habitat Restoration and Enhancement

Habitat restoration and enhancement are essential practices in mitigating the impacts of green energy projects on biodiversity. These practices aim to restore degraded ecosystems and enhance habitats to support greater biodiversity. Principles of habitat restoration include ecological integrity, biodiversity conservation, sustainability, and community involvement. Steps in habitat restoration and enhancement include baseline assessment, goal setting, site selection, design and implementation, species selection, phased approach, regular monitoring, adaptive management, long-term maintenance, and microhabitat creation. Techniques include reforestation, afforestation, wetland restoration, grassland and prairie restoration, controlled burns, coastal and marine restoration, invasive species management, and microhabitat diversity.

Case studies of successful habitat restoration include the Everglades Restoration Project in Florida, the Great Green Wall in Africa, and the Coral Triangle Initiative in Southeast Asia. Successful projects have resulted in significant improvements in water quality, increased populations of native species, and the return of key ecological functions. Benefits of habitat restoration include biodiversity conservation, ecosystem services, resilience to climate change, and community benefits such as improved livelihoods, recreation opportunities, and education.

Challenges in habitat restoration include funding and resources, technical expertise, invasive species, and monitoring and maintenance. Strategies to overcome challenges include public-private partnerships (PPPs), capacity building, integrated management, and community involvement. By following best practices, utilizing advanced techniques, and engaging stakeholders, it is possible to restore and enhance habitats in a way that supports biodiversity conservation and sustainable energy development, contributing to a healthier, more resilient environment for both nature and human communities.

4.7.2.8 Lasting Maintenance and Operation (O&M)

Sustainable O&M of green energy projects is crucial for minimizing environmental impacts, enhancing efficiency, and ensuring the longevity of energy infrastructure and ecosystems. Key principles include minimizing environmental impact, maximizing efficiency, ensuring longevity, involving local communities and stakeholders in O&M activities, and continuously improving O&M practices based on monitoring, feedback, and new technologies. Sustainable O&M practices for different renewable energy types include wind energy, solar energy, hydropower, biomass energy, and energy efficiency. Wind energy involves wildlife monitoring and mitigation through bird and bat deterrents, turbine efficiency through regular inspections, and vegetation management. Solar energy involves panel maintenance through cleaning, inspections, and dry cleaning techniques. Hydroelectric energy involves flow management, fish passage, and turbine upgrades to improve efficiency. Biomass energy sources include certified sustainable forestry or agricultural practices, waste biomass, emissions control, and advanced filters and scrubbers.

Energy efficiency involves monitoring systems, predictive maintenance, waste management, and water management. Worker safety and training are essential, along with environmental monitoring through biodiversity surveys, water and soil quality monitoring, performance metrics, and reporting.

Challenges in sustainable O&M include resource availability, technical complexity, regulatory compliance, and climate variability. Strategies to overcome these challenges include PPPs, capacity building, innovation and technology, and adaptive management. Case studies of successful sustainable O&M include wind farms in Denmark, solar parks in India, and hydropower plants in Norway. By adopting best practices, leveraging advanced technologies, and engaging stakeholders, it is possible to achieve a balance between renewable energy development and biodiversity conservation, supporting sustainable energy goals while also contributing to broader environmental and community benefits.

4.7.2.9 Capacity Building and Research

Building capacity and conducting research are essential to ensuring the execution and sustainability of sustainable energy projects, which in turn ensures biodiversity preservation. Through these initiatives, stakeholders are given the information, tools, and resources they need to manage renewable energy projects in an ethical and efficient manner. Key principles of capacity building include education and training programs, community engagement, institutional strengthening, and technology transfer. Stakeholder analysis is essential in identifying key stakeholders and addressing gaps in knowledge, skills, and resources. Training programs can cover technical aspects of renewable energy systems, environmental training, project management, financial planning, and regulatory compliance. Workshops, seminars, expert lectures, on-the-job training, mentorship programs, and resource development are also essential. Research involves developing innovative solutions that address energy needs and biodiversity conservation, data-driven decision-making, interdisciplinary approaches, and collaboration between research institutions, industry, government, and local communities. Research priorities include stakeholder input, literature review, research methodology, field studies, laboratory experiments, data analysis and interpretation, and dissemination of findings.

Examples of capacity building and research initiatives include renewable energy training centers, research on wildlife-friendly wind turbines, and community-based solar energy projects. Benefits of capacity building include enhanced skills and knowledge, informed decision-making, innovation, and community empowerment. Challenges in capacity building and research include limited funding, access to technology, coordination and collaboration, and long-term commitment. Strategies to overcome these challenges include developing funding mechanisms, technology transfer programs, collaborative networks, and sustained support from governments, international organizations, and the private sector.

4.7.2.10 Legal and Institutional Frameworks

Legal and institutional frameworks are essential for integrating green energy projects with biodiversity conservation. These frameworks provide the necessary guidelines, standards, and oversight mechanisms to ensure sustainable and environmentally responsible renewable energy development. Key components of these frameworks include legislation, permitting processes, standards, and guidelines for renewable energy projects. Institutional frameworks include government agencies, interagency coordination, independent bodies, and national policies. Key elements of effective legal and institutional frameworks include EIAs, biodiversity protection laws, renewable energy targets, grid access, land-use planning, compliance monitoring, and enforcement mechanisms. To develop and implement effective frameworks, stakeholders should engage in consultation processes, foster collaborative decision-making, develop capacity-building programs, and strengthen institutions responsible for enforcing legal and regulatory frameworks. Integrated policies should align renewable energy goals with biodiversity conservation objectives, and adaptive management approaches should be incorporated. Benefits of robust legal and institutional frameworks include environmental protection, sustainable development, investor confidence, and public trust. Challenges include coordination and integration, resource constraints, policy coherence, and enforcement and compliance. Strategies to overcome these challenges include establishing interagency committees, seeking international funding and technical assistance, ensuring policy coherence, and strengthening monitoring and enforcement mechanisms.

Therefore, developing and implementing effective legal and institutional frameworks is crucial for balancing renewable energy development with biodiversity conservation. By engaging stakeholders, building capacity, and implementing clear guidelines, it is possible to create a harmonious relationship between renewable energy development and biodiversity protection.

4.8 CASE STUDIES OF SUCCESSFUL GREEN ENERGY PROJECTS

Germany has been a pioneer in renewable energy, particularly solar thermal projects. Launched in 1999, the "Solar thermite 2000 plus" program intended to enhance the amount of solar thermal energy used in industrial, commercial, and residential buildings for heating and cooling. Key components included research and development, demonstration projects, market development, technological focus on solar collectors, thermal storage systems, and system integration. The initiative contributed to increased deployment of solar thermal systems across Germany, spurred technological innovation, and laid the foundation for subsequent renewable energy initiatives. The success of the initiative laid the foundation for the Renewable Energy Sources Act (EEG) and the Energy Transition (Energiewende), demonstrating the viability of solar thermal technology as a clean and sustainable energy solution.

Spain has been a major player in wind energy development, with the Atlántida Wind Farm in Tenerife, the Canary Islands, being one of the largest projects. With a 50 MW installed capacity overall, the project was created by Iberdrola and makes use of wind turbines from top producers like Siemens Gamesa and Vestas. The project has created jobs, stimulated economic growth, and contributed to the local economy of Tenerife. It has also helped Spain meet its renewable energy targets and minimize its dependence on fossil fuels. The project contributes to clean energy generation, diversification of energy sources, regional development, and renewable energy leadership. However, challenges include grid integration, community engagement, technological innovation, and policy support. Despite these challenges, the Atlántida Wind Farm exemplifies Spain's commitment to renewable energy development and its potential to harness abundant wind resources in regions like the Canary Islands.

The Kurnool Ultra Mega Solar Park, located in Andhra Pradesh, is a solar energy project created by India that has a 1,000 MW installed capacity. The park, located in Kurnool district, uses thousands of PV solar panels to convert sunlight into electricity using semiconductor materials. As part of India's energy transformation, the park's electricity production is linked to the local power system. The park's development required extensive land acquisition, facilitated by the state government of

Andhra Pradesh. EIAs were conducted to minimize disturbance and protect sensitive habitats. The project has stimulated economic growth by creating jobs, attracting investment, and supporting local businesses. It supports India's attempts to lower GHG emissions and meet its goals for renewable energy. Challenges include grid integration, land acquisition, technological innovation, and policy support. However, the Kurnool Ultra Mega Solar Park exemplifies India's commitment to renewable energy expansion and its potential to harness solar power to meet the growing energy demand sustainably.

Austria is actively pursuing renewable energy projects, including solar thermal initiatives, to shrink GHG emissions and transition toward a more sustainable energy future. Solar thermal technology harnesses sunlight to generate heat for diverse applications, including space heating, water heating, and industrial processes. Austria is at the forefront of solar thermal technology adoption, with a focus on both residential and commercial applications. Notable solar thermal projects include "Solar City" Linz in Linz, large-scale solar thermal installations in cities like Graz and Vienna, and research and innovation in Austrian universities, research institutes, and industry organizations. Austria's renewable energy policies and incentives support the deployment of solar thermal systems across residential, commercial, and industrial sectors. Benefits of solar thermal projects include reduced GHG emissions, energy cost savings, job creation, and improved energy security. Challenges include upfront capital costs, variability in solar radiation, integration challenges, and competing priorities and policy focus on other renewable energy technologies.

The Bhira Hydroelectric Project, also known as the Bhira Dam, is a major hydroelectric power plant in Maharashtra, India, with an installed capacity of 300 MW. Located on the Kundalika River, it was developed by the Maharashtra State Electricity Board (MSEB), now known as Maharashtra State Electricity Distribution Company Limited (MSEDCL). The project utilizes the natural flow of water from the Kundalika River to generate electricity. The project comprises a concrete gravity dam that diverts water from the Kundalika River into a penstock, which channels water to turbines in the powerhouse. The electricity generated is connected to the Maharashtra State Electricity Grid, contributing to the overall power supply in the state. The project generates clean and renewable electricity, reduces reliance on fossil fuels, provides reliable power supply, manages water resources, supports the local economy, and highlights the hydropower potential of Maharashtra's rivers. However, the project faces environmental, social, and technical challenges, necessitating careful planning, management, and mitigation measures.

Denmark leads the world in developing renewable energy, with a particular emphasis on projects pertaining to energy efficiency, biomass, and wind power. The green energy projects include offshore wind farms like Horns Rev and Anholt, onshore wind farms like Middelgrunden, and community-owned wind farms like Horsens Wind Turbine Cooperative. Denmark also has biomass-fired power stations like Skaerbaek and Viborg Biomass CHP Plants, which reduce GHG emissions and promote energy efficiency. Copenhagen's district heating system lowers the carbon footprint of the city by utilizing waste heat from industrial activities along with renewable energy sources. Denmark has implemented energy renovation programs to improve building energy efficiency. Green transportation initiatives include

Copenhagen's bicycle infrastructure and EV adoption through tax breaks, subsidies, and charging infrastructure development. Denmark's commitment to sustainable development, renewable energy deployment, and climate action demonstrates its commitment to a low-carbon future.

4.9 POLICY AND REGULATORY FRAMEWORK

The global push toward green energy is backed by international policies, regulatory frameworks, and agreements aimed at promoting renewable energy adoption, decreasing GHG emissions, and increasing energy efficiency. Key frameworks include the Paris Agreement (2015), United Nations Sustainable Development Goals (SDGs), the International Renewable Energy Agency (IRENA), Renewable Energy Policies for the 21st Century (REN21), regional policies like the European Union's RED II, the Asia-Pacific Economic Cooperation (APEC), and national policies of countries such as the US, China, India, and the GCF. Financial mechanisms include the GCF, the Global Environment Facility (GEF), and the Clean Energy Ministerial (CEM), which promote clean energy technology and innovation. Trade and market mechanisms include carbon pricing, carbon taxes, and green bonds, which provide financing for renewable energy projects.

To conclude, the global policy and regulatory framework for green energy is characterized by a combination of international agreements, regional policies, national targets, and financial mechanisms. These frameworks collectively endeavor to advance sustainable energy adoption, enhance energy efficiency, and lower GHG emissions, helping to battle climate change worldwide and shift to sustainable energy sources in the future.

4.10 TECHNOLOGICAL INNOVATIONS IN GREEN ENERGY

Green energy innovations are crucial for enhancing efficiency, reducing costs, and overcoming the limitations of traditional energy sources. Key innovations include solar PV technology, wind energy technology, energy storage solutions, hydrogen energy, bioenergy technology, geothermal energy, and smart grid technology.

4.10.1 Solar Energy Innovations

Through the use of semiconductors, solar PV technology produces electricity by converting solar energy into DC electricity. It is composed of solar panels, which are composed of many silicon-based solar cells. The basic components of a PV system include solar cells, inverters, mounting systems, and the Balance of System (BOS). Technological advancements include perovskite solar cells, bifacial solar panels, tandem solar cells, and Building-Integrated Photovoltaics (BIPV). Utility-scale solar farms, off-grid solar systems, portable solar gadgets, and household and business solar electricity are just a few of the uses for PV technology.

Reduced GHG emissions, resource conservation, employment creation, and energy independence are some of the environmental advantages of these systems. Intermittency, upfront costs, government subsidies, incentives, and funding choices are some of the difficulties. A key component of the switch to renewable energy, solar PV technology will become more and more important in attaining global energy sustainability and reducing climate change. Solar PV will become more and more important in attaining global energy sustainability and reducing climate change as technology develops and its use grows. Comparing perovskite solar cells to conventional silicon-based solar cells, the former provide reduced production costs and improved efficiency. Smart turbines and floating wind turbines both maximize performance by instantly adapting to changing wind conditions. Hywind Scotland is the first floating wind farm in history, along with Tesla Powerwall and Powerpack.

4.10.2 Innovation in Wind Energy

Innovations in wind energy technology have significantly improved effectiveness, and viability of wind power is a major renewable energy source. Key innovations include floating wind turbines, advanced blade designs, vertical axis wind turbines (VAWTs), smart wind turbines, direct-drive turbines, energy storage integration, hybrid renewable energy systems, and advanced wind resource assessment. Affixed to the seabed, floating platforms hold the floating wind turbines, allowing them to be installed in deeper waters with higher wind speeds. Advanced blade designs focus on materials, aerodynamics, and structures to improve efficiency and durability. Because of their blades' rotation around a vertical axis, VAWTs are appropriate for windy, urban locations. Smart wind turbines are equipped with advanced sensors, control systems, and communication technologies to optimize performance and maintenance.

Direct-drive turbines eliminate the need for a gearbox, reducing mechanical complexity and reducing maintenance requirements. Energy storage integration is another key innovation, integrating energy storage systems with wind turbines to manage intermittency and ensure a stable energy supply. Hybrid renewable energy systems combine wind energy with other renewable energy sources, maximizing energy production and reliability. Advanced wind resource assessment utilizes technologies like LiDAR, remote sensing, and meteorological modeling to optimize turbine placement and energy capture. Innovations in wind energy technology are driving the efficiency, reliability, and scalability of wind power, making it a more viable and integral part of the global renewable energy landscape.

4.10.3 Innovation in Energy Storage Technology

Energy storage solutions are essential to managing the intermittent nature of renewable energy sources and offering a consistent and reliable energy supply. The field of energy storage technology is developing quickly, opening up new avenues for cost-effective cost reduction, efficiency gains, and grid integration of renewable energy. Solid state batteries, flow batteries, lithium-sulfur batteries, hydrogen energy storage, super capacitors, pumped hydro storage, and thermal energy storage are a few noteworthy inventions. Compared to conventional lithium-ion batteries, solid-state batteries have a solid electrolyte and a higher energy density, as well as increased safety and a longer lifespan. Flow batteries provide the benefits of scalability, longevity, and flexibility in that they store energy in liquid electrolytes housed in external tanks. Lithium-sulfur batteries have a higher theoretical energy density and are more economically viable since they use lithium as the anode and sulfur as the cathode material. Hydrogen can be stored and subsequently transformed back into electricity through the use of fuel cells or combustion. Hydrogen energy storage is the process of converting excess electricity into hydrogen through electrolysis.

Super capacitors have a high power density, a long cycle life, and complementary applications. They store energy by electrostatically separating charges. Pumped hydro storage functions by moving water via a lower reservoir to an upper reservoir when power demand is low, and turbines are utilized to create electricity when power demand is high. Thermal energy storage is the process of storing energy as heat, which can then be utilized for heating or repurposed as electricity. Sensible heat storage and latent heat storage are two examples of technologies that offer benefits including high efficiency, affordability, and grid stability. Innovations in energy storage technologies, which address issues of intermittency, scalability, and cost reduction, are essential for the broad adoption and integration of renewable energy sources.

4.10.4 Innovation in Hydrogen Energy

Hydrogen energy innovations include green hydrogen production, fuel cells, bioenergy technology, enhanced geothermal systems, and binary cycle power plants. Bioenergy technologies produce advanced biofuels from non-food biomass sources and biogas upgrading techniques, 90% reduction in GHG emissions is possible. An essential component of the transition to clean, sustainable energy is hydrogen energy, involving the use of hydrogen as a fuel for generating electricity, powering vehicles, and as a feedstock in industrial processes. Numerous techniques, including electrolysis, blue hydrogen production, biomass gasification, steam methane reforming, and photocatalytic water splitting, can be used to manufacture it. Hydrogen storage is critical for its use in various applications, including compressed hydrogen, liquid hydrogen, metal hydrides, chemical storage, energy storage, and grid balancing. Applications of hydrogen include power generation, transportation, industrial use, energy storage, and grid balancing.

Green hydrogen is currently more expensive than fossil fuels, requires significant investment in infrastructure, and has efficiency losses. Safety concerns arise due to its flammability. Future prospects include cost reduction, policy support, technological innovations, and global collaboration. Thanks to scientific breakthroughs in electrolysis and economies of scale, green hydrogen should become more affordable. The adoption of hydrogen technologies is being fueled by government regulations and incentives, and further development and research in the fields of hydrogen production, storage, and consumption will increase the technology's practicality, efficiency. International partnerships and collaborations are accelerating the deployment of hydrogen infrastructure and projects. Overcoming current challenges through technological advancements and infrastructure development will be essential to realizing the entire capacity of hydrogen energy.

4.10.5 Innovation in Bioenergy

Bioenergy innovations aim to enhance the efficiency, sustainability, and economic viability of bioenergy systems while reducing environmental impacts. Solution areas of innovation include feedstock diversification, waste-to-energy conversion, conversion technologies, biochemical conversion, bioelectrochemical systems, bioenergy integration and co-production, biorefineries, advanced biofuel production, sustainable feedstock production, precision agriculture, agroforestry, LCA, market and policy support, and international collaboration. Advancements in feedstock diversification, conversion technologies, integration approaches, and sustainability practices are crucial for advancing the evolution in the direction of a sustainable, low-carbon energy future. Advanced bioenergy sources include lignocellulosic biomass, algae, organic waste, waste-to-energy conversion, thermochemical conversion, biochemical conversion, bioelectrochemical systems, combined heat and power (CHP), biorefineries, next-generation biofuels, synthetic biology and genetic engineering, and metabolic engineering. Sustainable feedstock production, precision agriculture, agroforestry, and LCA are essential for optimizing crop management and reducing environmental impacts. Market and policy support includes incentives like feed-in tariffs, renewable fuel standards, tax credits, subsidies, and international collaboration. The transition to a low-carbon, sustainable energy future that supports economic growth, environmental stewardship, and energy security must be accelerated by bioenergy innovations. Reaching the benefits of bioenergy on a worldwide scale and reaching its full potential would require sustained research, investment, and legislative support.

4.10.6 Innovation in Geothermal Energy

Geothermal energy innovations include enhanced geothermal systems and binary cycle power plants, allowing for the utilization of lower temperature geothermal resources. Another type of renewable energy source, commonly known as geothermal energy, produces electricity and heat and cools various applications by harnessing heat from beneath the surface of the Earth. Innovations in geothermal energy focus on improving resource identification and characterization, enhancing exploration and drilling techniques, optimizing energy extraction, and expanding the range of geothermal applications. Resource identification and characterization include advanced exploration techniques like geophysical surveys, remote sensing, fluid sampling, and isotope geochemistry. Drilling and reservoir engineering include enhanced drilling technologies, reservoir stimulation techniques, and reservoir engineering models.

Examples of geothermal power generation include upgraded geothermal systems, dual cycle power plants, and real applications like district heating and cooling. Agricultural and industrial applications include greenhouse heating and aquaculture. Heat extraction and energy conversion innovations include advanced heat exchangers, organic Rankine Cycle systems, and geothermal heat pump systems. Environmental and sustainability innovations include geothermal heat pump systems and geothermal energy and carbon capture (Geothermal-CCS). Regulatory frameworks, such as incentives and streamlined permitting processes, support geothermal development and investment. International collaboration and research consortia are essential for unlocking the full potential of geothermal energy and maximizing its benefits worldwide.

4.10.7 Smart Grid Technology

Smart grid technology includes advanced metering infrastructure (AMI) and dispersed energy resources management, integrating undersized-scale renewable energy sources into the grid. The generation, distribution, and consumption of electricity are all being revolutionized by smart grid technologies. It makes use of cutting-edge information and communication technologies to improve the electrical grid's sustainability, dependability, and efficiency. The innovations include AMI, Grid Automation and Control, Integration of Distributed Energy Resources (DERs), Energy Storage Systems, Automated Demand Response (ADR), Consumer Engagement, Cyber Security and Data Privacy, Advanced Analytics and Machine Learning, Grid Modernization and Infrastructure, Renewable Energy Integration, Micro grid Controllers, Resilience Planning, and EV Integration. AMI technologies provide real-time data on electricity usage, enabling two-way communication, remote reading, and dynamic pricing. Meter Data Management Systems collect, store, and analyze data from smart meters, supporting advanced billing, outage management, and energy theft detection.

Grid automation and control include automated switches, reclosers, and capacitor banks, while Advanced Distribution Management Systems integrate data from various grid components for real-time monitoring and control. DERs, energy storage systems, and demand response systems are integrated to reduce peak demand, enhance grid stability, and lower electricity costs. Cybersecurity and data privacy are protected through advanced encryption and authentication, intrusion detection systems, and machine learning algorithms. Smart grid technology is transforming the electrical grid into a more efficient, reliable, and sustainable system. Advancements in cybersecurity, predictive analytics, and microgrid technology are enhancing grid resilience and security.

In conclusion, technological innovations in green energy are dynamic, the evolution to a sustainable and low-carbon potential, improving efficiency and cost-effectiveness of renewable energy sources, and addressing challenges of intermittency and scalability.

4.11 THE ROLE OF COMMUNITIES AND BUSINESSES

The role of energy companies is crucial; by removing financial obstacles, offering access to renewable energy sources, and offering their knowledge in energy management and distribution, energy corporations play a critical role in assisting energy communities. They provide incentives for energy saving and financial choices for the installation of renewable energy. Furthermore, they guarantee the dependability and effectiveness of energy distribution by utilizing smart grid technology and energy storage solutions.

4.11.1 Community-Led Renewable Energy Projects

First, it was discovered that communities and locals benefited financially and economically from communal energy. Renewable energy plants provide power or heat for communities, present lucrative chances for investments, and/or employment. Certain renewable energy projects utilize biomass that is generated as a byproduct of agriculture, providing advantages to local farmers. They aid in the revival of communities. Renewable energy projects allow for greater local community participation in decision-making than projects like large-scale, centralized power plants located outside the region. Communities may become less reliant on the electricity produced by these kinds of large-scale facilities if they use community energy. Projects carried out by and for the community also improve community cohesion and regeneration, which are closely related to community spirit.

Energy communities have a straightforward concept: individuals and businesses can produce their own energy and share it with the community by banding together. Installing solar panels, wind turbines, or other sustainable energy sources can do this. Community members can lower their dependency on traditional energy sources along with saving money on energy costs by producing their personal energy. Although they can take various shapes, energy communities are fundamentally made up of organizations of individuals who collaborate to produce and distribute energy. These communities can consist of local businesses, residences, and associations. Their aim is to become more independent of conventional energy sources and gain control over how much energy they use.

In order for energy communities to function, energy must normally be produced using renewable resources and distributed via a shared infrastructure. This infrastructure, which is frequently referred to as a microgrid, enables neighbors to share the energy they produce with one another. Energy communities occasionally have the option of being linked to the main power grid, which enables them to resell extra energy to the system. They frequently concentrate on energy efficiency and conservation in addition to producing and distributing energy. This can include programs like energy audits, which assist locals in locating opportunities to lower their energy usage. Energy communities might also provide incentives for their members to replace outdated equipment and appliances with more energy-efficient models.

Being part of an energy community has several advantages. Energy communities can provide a dependable and reasonably priced energy source for homes and businesses. Community members can lower their dependency on traditional energy source and save money on energy costs by producing their own energy. Additionally, they may lessen their carbon footprint, furthermore assisting in creating a future that is sustainable. By generating jobs and enhancing the local economy, villages can also be advantageous to the larger community. Energy communities can generate new employment opportunities in the renewable energy zone by investing in renewable energy sources. They may also draw new companies and sectors of the economy that are searching for environmentally friendly energy sources.

4.11.2 Corporate Sustainability and Green Energy

Modern business plans are not complete without corporate sustainability and green energy, which reduce environmental impact and improve social equity while fostering environmental stewardship, social responsibility, and economic competitiveness. Corporate sustainability strategies include environmental management systems (EMS), sustainability reporting, circular economy principles, green energy adoption, renewable energy commitments, investment in renewable projects, sustainable supply chain management, corporate social responsibility (CSR), innovative green energy technologies, regulatory and policy support, and compliance and reporting.

Companies can implement EMS, sustainability reporting frameworks, and circular economy principles to improve resource efficiency, reduce waste, and enhance regulatory compliance. They can also adopt renewable energy procurement, energy efficiency measures, and corporate renewable energy initiatives.

Sustainable supply chain management includes supplier engagement, logistics optimization, and green transport. The concept of CSR dates back to the Industrial Revolution or possibly the earlier part of the 20th century. CSR initiatives support local communities through renewable energy projects and education, while employee engagement encourages sustainability efforts.

Innovative green energy technologies include solar PV, wind energy, and energy storage solutions. Government incentives, such as subsidies and tax credits, can support renewable energy projects and energy efficiency improvements. Compliance and reporting are essential for adhering to national and international environmental standards. Leading corporations like Google, IKEA, Apple, and Tesla have implemented sustainable practices to achieve 100% renewable energy.

Corporate sustainability and green energy are crucial in modern business strategies, promoting environmental, social, and economic benefits. Companies are adopting renewable energy, enhancing energy efficiency, and integrating sustainability into operations. Innovations in renewable technologies, energy storage, and smart grid systems can help achieve sustainability goals, reduce carbon footprint, and contribute to a sustainable future.

4.11.3 Public and Private Partnerships

PPPs are partnerships linking government and private sectors that accelerate the deployment of green energy solutions, address climate change, and promote sustainable development by leveraging their strengths. PPPs are cooperative arrangements between public and private sectors for the provision of public assets and services, combining resources, expertise, and efficiencies to deliver large-scale infrastructure projects, including green energy. They can be structured into contractual agreements, such as Build-Operate-Transfer (BOT), Build-Own-Operate (BOO), Build-Operate-Own-Transfer (BOOT), and joint ventures.

The public sector provides regulatory support, land access, and initial funding or subsidies, while the private sector invests capital for project development and construction. Benefits of PPPs for green energy include financial leverage, accelerated deployment, innovation, and sustainable development. Challenges include regulatory and policy challenges, financial risks, and technical and operational risks. Key components of successful PPPs include feasibility studies, stakeholder engagement, and transparent procurement processes.

Examples of successful PPPs include Morocco's Noor Ouarzazate Solar Complex, India's Rewa Solar Park, Kenya's Lake Turkana Wind Power Project, and regional initiatives like the North Sea Wind Power Hub in Europe and New York's Reforming the Energy Vision in the United States.

Future trends and opportunities in PPPs include digitalization and smart technologies, green hydrogen, and energy storage. Smart grids integrate digital technologies to enhance grid management and efficiency, while Internet of Things (IoT) and AI use predictive maintenance and optimized energy use. Green hydrogen projects through PPPs decarbonize sectors like industry and transportation, and electrolysis powered by renewable energy produces hydrogen as a clean energy carrier. PPPs are vital for advancing green energy globally. They combine the strengths of public and private sectors, overcoming financial, technical, and operational challenges, and delivering large-scale renewable energy projects that benefit the environment, economy, and society.

Green energy entrepreneurship is a crucial driver of economic growth, focusing on modern business opportunities in renewable energy. It involves developing sustainable solutions to diminish dependence on fossil fuels and minimize environmental impact, addressing pressing global issues like climate change and resource depletion. Green energy entrepreneurship is crucial for economic growth, job creation, innovation, clean energy access, and investment attraction, especially in the face of climate change and resource scarcity.

4.12 CONCLUSION, FUTURE PROSPECTS, AND CHALLENGES

Green energy has bright future prospects thanks to regulatory support, increased public awareness of environmental issues, and technology improvements. Nonetheless, considerable obstacles still need to be overcome in order to properly utilize renewable energy. Green energy presents numerous technological advancements, including next-generation solar panels, offshore wind farms, advanced turbines, and energy storage technologies. Policy and authoritarian support for renewable energy includes government targets, financial incentives, and carbon pricing. Green jobs can be created through renewable energy sectors, creating new jobs in maintenance, manufacturing, and installation. Investment in renewable energy projects can stimulate local economies, especially in rural and underserved areas.

Climate change mitigation is crucial for renewable energy, with emissions reductions and decarbonization essential for meeting global climate targets. Renewable energy provides a diversified and resilient energy supply, reducing dependence on fossil fuels and supporting climate adaptation efforts. However, challenges include intermittency and grid integration, cost and financing, supply chain and resource availability, regulatory and policy barriers, and public acceptance and social issues. Solar and wind energy are dependent on weather conditions, leading to intermittent energy supply.

Solutions include developing energy storage systems and smart grid technologies to manage variability and maintain grid stability. Cost and financing include capital intensity, economic competitiveness, and supply chain and resource availability. Regulatory and policy barriers include policy uncertainty, inconsistent policies, and streamlined permitting and approval processes. Public acceptance and social issues include NIMBYism, local opposition, equity and access, and addressing public acceptance and social issues.

Green energy has a promising future full with prospects and major achievements. But in order to reach its full potential, issues with costs, supply chains, public acceptance, policy frameworks, and intermittency must be resolved. Governments, corporations, and communities can overcome these challenges and clear the path for a sustainable energy future by keeping up with innovation, investing, and working together.

4.12.1 Projected Trends in Green Energy

Future directions for green energy point to a dramatic change in the way that power is produced, distributed, and used worldwide. Technological developments, legislative actions, changes in the economy, and rising environmental consciousness are the main forces behind these movements. The intensification in renewable energy capacity is expected to continue due to declining costs and efficiency improvements. Advances in PV technologies, such as bifacial panels and perovskite solar cells, will drive adoption of renewable energy. Onshore and offshore wind capacity will increase, with offshore wind seeing significant investment due to advancements in turbine technology and floating platforms. Larger turbines will enhance energy output and lower the cost per kilowatt-hour (kWh).

The energy storage market will grow as battery costs continue to fall and performance improves. Smart grid technologies will be implemented to enhance grid efficiency, resilience, and flexibility. Digitalization will also be increased, with the use of digital tools, AI, and the IoT for real-time monitoring and management of energy systems. New renewable technologies, such as green hydrogen, will emerge as a key energy carrier for sectors that are hard to electrify, like heavy industry and transportation. Infrastructure development and deployment of floating solar farms will maximize space utilization and reduce land-use conflicts.

Policy and regulatory trends include enhanced climate goals, carbon pricing, renewable energy mandates, decentralization of energy systems, cost competitiveness, job creation, and integration of EVs. Renewable energy sources are gaining cost competitiveness with traditional fossil fuels, leading to investment shifts and job creation. Corporate sustainability initiatives include renewable energy commitments, corporate power purchase agreements, transparency in sustainability reporting, and integrated strategies for integrating renewable energy into broader sustainability and climate strategies.

Driven by growing sustainability commitment, regulatory support, economic incentives, and technical improvements, green energy trends are shifting toward a sustainable future. Nonetheless, issues including regulatory complexity, financial obstacles, and grid connectivity need to be resolved. These developments will safeguard global energy security, encourage economic growth, and lessen the effects of climate change.

4.12.2 Role of Education and Awareness

A sustainable future and the promotion of green energy depend heavily on awareness and education. They promote policy support, innovation, and changes in behavior in a variety of industries. Green energy education involves various forms of education, including formal education, research and development, vocational training, continuing education, public awareness campaigns, corporate and industry initiatives, policy and advocacy, and government programs. Formal education includes incorporating renewable energy topics into school and university curricula, offering specialized degree programs, conducting cutting-edge research, providing vocational training for technicians and engineers, and establishing certification programs.

Continuing education includes workshops, seminars, webinars, online courses, and public events like energy fairs and competitions. Corporate and industry initiatives include employee training on sustainability practices and green certifications, as well as public outreach and partnerships with educational institutions and NGOs. Government programs provide grants and funding for renewable energy education and training programs, while advocacy groups like NGOs and nonprofits advocate for renewable energy through education, policy recommendations, and public campaigns. Benefits of education and awareness in green energy include informed decision-making, policy support, behavioral change, innovation and entrepreneurship, and increased spending on scientific and technological research.

Accessibility to renewable energy education and training is a challenge, but solutions include expanding online education platforms, engaging the public through interactive content, and overcoming misinformation. Solutions include expanding online education platforms, offering scholarships and funding for renewable energy education, using interactive content, and providing clear, accurate, and evidence-based information through trusted sources and expert voices.

Promoting green energy requires raising people's knowledge and awareness in order to provide them the tools they need to comprehend, support, and use renewable energy sources. Governments, academic institutions, businesses, and advocacy organizations are all essential to creating a society that is knowledgeable about environmental sustainability and climate change.

As we wrap up this chapter, we have to realize that while green energy offers significant environmental benefits, it also poses challenges to biodiversity conservation that require careful management and strategic planning. By integrating innovative green energy solutions with biodiversity protection strategies, we can achieve a balanced approach to sustainable development. Continuing to collaborate between governments, businesses, and communities is essential to overcoming the challenges associated with green energy deployment, ensuring a resilient and sustainable future for all. This chapter has highlighted the symbiotic relationship between biodiversity and climate regulation, underscoring the importance of adopting green energy technologies to reduce carbon footprints. In order to facilitate this crucial shift, we must advance knowledge, propel technical advancements, and fortify legislative frameworks.

REFERENCES

Androniceanu, A., and C.R. Popescu. 2017. "An Inclusive Model for an Effective Development of the Renewable Energies Public Sector." *Administratie Si Management Public* 28: 81–96.

Androniceanu, A., and O.M. Sabie. 2022. "Overview of Green Energy as a Real Strategic Option for Sustainable Development." *Energies* 15: 8573. doi:10.3390/en15228573

Baruch-Mordo, S., J.M. Kiesecker, C.M. Kennedy, J.R. Oakleaf, and J.J. Opperman. 2019. "From Paris to Practise: Sustainable Implementation of Renewable Energy Goals." *Environmental Research Letters* 14: 024013.

Beniston, M., and R.S.J. Tol. 1998. "Europe." In *The Regional Impacts of Climate Change: An Assessment of Vulnerability*, edited by R.T. Watson and M.C. Zinyowera, 149–185. Cambridge, UK: Cambridge University Press.

Bilgen, S., K. Kaygusuz, and A. Sari. 2004. "Renewable Energy for a Clean and Sustainable Future." *Energy Sources* 26: 1119–1129.

Buotte, P.C., B.E. Law, W.J. Ripple, and L.T. Berner. 2020. "Carbon Sequestration and Biodiversity Co-Benefits of Preserving Forests in the Western United States." *Ecological Applications* 30 (2): e02039. doi:10.1002/eap.2039.

Dincer, I. 1999. "Environmental Impacts of Energy." *Energy Policy* 27: 845–854.

Edenhofer, O. 2015. *Climate Change 2014: Mitigation of Climate Change* (Vol. 3). Cambridge University Press. https://books.google.com/books?hl=en&lr=&id=JAFEBgAAQBAJ&oi=fnd&pg=PT19&dq=IPCC+(2014). +Climate+Change+2014:+Mitigation+of+Clima te+Change.+Contribution+of+Working+Group+III+to+t he+Fifth+Assessment+Repor t+of+the+Intergovernmental+Panel+on+Climate+Change,+eds+O.+R.+Ed enhofer,+Y. +Pichs-Madruga,+E.+Sokona,+S.+Farahani,+K.+Kadner,+A.+Seyboth,+et+al.+Cambr idge:+Cambridge+Universi ty+Press.&ots=dBzHuFZaZ7&sig=xucJRj7FEvqqQlw2JF cff9Y9DHE

Gibon, T., E.G. Hertwich, A. Arvesen, B. Singh, and F. Verones. 2017. "Health Benefits, Ecological Threats of Low-Carbon Electricity." *Environmental Research Letters* 12: 034023.

Gielen, D., F. Boshell, D. Saygin, M.D. Bazilian, N. Wagner, and R. Gorini. 2019. "The Role of Renewable Energy in the Global Energy Transformation." *Energy Strategy Reviews* 24: 38–50.

Grabherr, G., M. Gottfried, and H. Pauli. 1994. "Climate Effects on Mountain Plants." *Nature* 369: 448.

Gupta, J., and M.J. Grubb. 2000. *Climate Change and European Leadership: A Sustainable Role for Europe?* (Vol. 27). Springer Science & Business Media.

Hertwich, E.G., T. Gibon, E.A. Bouman, A. Arvesen, S. Suh, G.A. Heath, J.D. Bergesen, A. Ramirez, M.I. Vega, and L. Shi. 2015. "Integrated Life-Cycle Assessment of Electricity-Supply Scenarios Confirms Global Environmental Benefit of Low-Carbon Technologies." *Proceedings of the National Academy of Sciences* 112: 6277–6282.

Hoang, A.T., and X.P. Nguyen. 2021. "Integrating Renewable Sources into Energy System for Smart City as a Sagacious Strategy towards Clean and Sustainable Process." *Journal of Cleaner Production* 305: 127161.

Hosseini, S.E., and M.A. Wahid. 2020. "Hydrogen from Solar Energy, a Clean Energy Carrier from a Sustainable Source of Energy." *International Journal of Energy Research* 44: 4110–4131.

Javed, M.S., T. Ma, J. Jurasz, and M.Y. Amin. 2020. "Solar and Wind Power Generation Systems with Pumped Hydro Storage: Review and Future Perspectives." *Renewable Energy* 148: 176–192.

Kaygusuz, K. 2012. "Energy for Sustainable Development: A Case of Developing Countries." *Renewable and Sustainable Energy Reviews* 16: 1116–1126.

Luderer, G., M. Pehl, A. Arvesen, T. Gibon, B.L. Bodirsky, H.S. de Boer, O. Fricko, M. Hejazi, F. Humpenöder, and G. Iyer. 2019. "Environmental Co-Benefits and Adverse Side Effects of Alternative Power Sector Decarbonization Strategies." *Nature Communications* 10: 5229.

Masini, A., and E. Menichetti. 2013. "Investment Decisions in the Renewable Energy Sector: An Analysis of Non-Financial Drivers." *Technological Forecasting and Social Change* 80: 510–524.

Mehrjerdi, H. 2020. "Modeling and Optimization of an Island Water-Energy Nexus Powered by a Hybrid Solar-Wind Renewable System." *Energy* 197: 117217.

Mohsin, M., H.W. Kamran, M.A. Nawaz, M.S. Hussain, and A.S. Dahri. 2021. "Assessing the Impact of Transition from Nonrenewable to Renewable Energy Consumption on Economic Growth-Environmental Nexus from Developing Asian Economies." *Journal of Environmental Management* 284: 111999.

Mora, C., D.P. Tittensor, S. Adl, A.G. Simpson, and B. Worm. 2011. "How Many Species Are There on Earth and in the Ocean?" *PLOS Biology* 9 (8): e1001127. doi:10.1371/journal.pbio.1001127

Nengroo, S.H., H. Jin, and S. Lee. 2022. "Management of Distributed Renewable Energy Resources with the Help of a Wireless Sensor Network." *Applied Sciences* 12: 6908.

Owusu, P. A., and S. Asumadu-Sarkodie. 2016. "A Review of Renewable Energy Sources, Sustainability Issues and Climate Change Mitigation." *Cogent Engineering* 3 (1): 1167990. doi:10.1080/23311916.2016.1167990

Papadis, E., and G. Tsatsaronis. 2020. "Challenges in the Decarbonization of the Energy Sector." *Energy* 205: 118025.

Parmesan, C. 1996. "Climate and Species Range." *Nature* 382: 765–766.

Pratiwi, S., and N. Juerges. 2020. "Review of the Impact of Renewable Energy Development on the Environment and Nature Conservation in Southeast Asia." *Energy, Ecology and Environment* 5: 221–239.

Qayyum, M., M. Ali, M.M. Nizamani, S. Li, Y. Yu, and A. Jahanger. 2021. "Nexus between Financial Development, Renewable Energy Consumption, Technological Innovations and CO_2 Emissions: The Case of India." *Energies* 14: 4505.

Rabaia, M.K.H., M.A. Abdelkareem, E.T. Sayed, K. Elsaid, K.J. Chae, T. Wilberforce, and A.G. Olabi. 2021. "Environmental Impacts of Solar Energy Systems: A Review." *Science of the Total Environment* 754: 141989.

Shen, M., L. Tong, S. Yin, C. Liu, L. Wang, W. Feng, and Y. Ding. 2022. "Cryogenic Technology Progress for CO_2 Capture under Carbon Neutrality Goals: A Review." *Separation and Purification Technology* 299: 121734. doi:10.1016/j.seppur.2022.121734

Shortall, R., B. Davidsdottir, and G. Axelsson. 2015. "Geothermal Energy for Sustainable Development: A Review of Sustainability Impacts and Assessment Frameworks." *Renewable and Sustainable Energy Reviews* 44: 391–406.

Thomas, C.D., and J.J. Lennon. 1999. "Birds Extend Their Ranges Northwards." *Nature* 399: 213.

Zakeri, B., K. Paulavets, L. Barreto-Gomez, L.G. Echeverri, S. Pachauri, B. Boza-Kiss, C. Zimm, et al. 2022. "Pandemic, War, and Global Energy Transitions." *Energies* 15: 6114.

5 Green Energy for Lesser Carbon Footprint

Amelia Chebyala, Iddi Mapande, Oscar Pungut, and Jovine Bachwenkizi

5.1 INTRODUCTION

Changes in climate are among of the most significant concerns facing the twenty-first century, largely caused by human activity, notably our continued reliance on non-renewable sources of energy including natural gas, oil, and coal. Combustion of these natural fuels emits significant amounts of greenhouse gas (GHG), primarily methane and carbon dioxide causing global warming by trapping heat in the atmosphere (Letcher 2022). This sustained use of non-renewable energy (RE) substantially accelerates climate change and global air pollution. Transformation of current energy systems is critical to mitigating its severe repercussions since changes in the climate could be mitigated and GHG emission drastically reduced with the usage of RE source (Edenhofer et al. 2011).

Energy is central to GHG emissions, accounting for nearly three-quarters of the total (IPCC 2022); however, the share varies greatly by country. CO_2 from fuel combustion accounts for the majority of energy, around 92%, with variations depending on the country's economic structure. This condition exacerbates global warming, which has a negative impact on the ecosystem, including rising sea levels, increased frequency of extreme weather conditions, and habitat degradation (IRENA, IEA, and REN21 2018).

Globally, the leading sources of GHG emissions are electricity and heat (31%), agriculture (11%), transportation (15%), forestry (6%), and manufacturing accounts for about 12% (Lang et al. 2016).

In response to these concerns, the Paris Agreement seeks to keep global rise in temperature "well below" 2°C, preferably at 1.5°C above levels prior to the Industrial Revolution (IRENA 2019). Achieving Paris Agreement's climate targets will necessitate a fundamental restructuring of the global energy environment. This change necessitates the rapid adoption of low-carbon technology to replace traditional fossil fuel use. To comply with the Paris Agreement, energy-related CO_2 emissions should fall by 3.5% per year until 2050, with further reductions thereafter (Dubie 2019). By 2050, electrified transportation and renewable power generation could reduce CO_2 emissions by around 60%. Additional reductions from direct RE consumption boost this share to 75%, and when combined with energy efficiency measures, it exceeds 90% of the necessary emission reductions (Demarco 2017).

DOI: 10.1201/9781003492733-5

Investing in green energy infrastructure diversifies energy portfolios while concurrently encouraging employment creation, economic growth, and technical innovation.

This chapter will present an overview of green energy, various forms of green energy, technology used, RE and sustainable development, policy, funding, and implementation, mitigation potential and costs, the integration of green energy into the current and future energy systems, challenges and barriers as well as future outlook.

5.2 TYPES OF GREEN ENERGY SOURCES

RE, or green energy, comes from naturally occurring sources that are never-ending, like biomass, sunshine, water, wind, and geothermal energy. The following are detailed explanations of each type of green energy.

5.2.1 Solar Energy

Sunrays are the source of solar energy. It may be harnessed through two basic technologies including the systems that use concentrated solar power (CSP) and those using photovoltaic (PV) panels (Čotar 2012). PV panels convert sunlight directly into energy using semiconductor materials; nevertheless, CSP technologies make use of lenses or mirrors to concentrate sunlight on an area of limited size. And produce heat powering turbines to generate electricity. As a RE source, solar energy is unlimited, widely available, and emits no GHGs during operation, making it a vital participant in efforts to prevent changes in climate change and lessen dependency on fossil fuel (Arvizu et al. 2011).

Solar energy is an abundant resource, capable of surpassing the world's annual usage of energy and consumption in just one hour (Arvizu et al. 2011). Solar technologies offer a wide range of applications, including heating, energy generation, and fuel production while some household's pool heating and hot water are already cost-competitive. Furthermore, in areas with government support, large-scale solar installations, both PV and CSP, have been successfully implemented (Arvizu et al. 2011). Solar energy's variability necessitates integration with other energy sources, ideally renewable, to ensure consistent availability.

5.2.1.1 Technologies Used in Solar Energy

5.2.1.1.1 Concentrated Solar Power

This technology for RE is really promising. Power blocks, thermal storage systems, and solar collectors are the three main parts of CSP systems (Viebahn et al. 2011). Sunlight is directed by the mirrors or lenses, known as collectors, into a receiver that holds a heat transfer fluid, like molten salt, which is heated to a temperature of 600°F–1,000°F and stored in thermal tanks. Even when the Sun is not shining, this stored heat can be used to generate power through a steam turbine (Pitz-Paal 2020).

5.2.1.2 Types of CSP Systems

5.2.1.2.1 Parabolic Trough Systems

Mirrors are used in parabolic trough systems to reflect light so that heat transfer fluid is contained inside a receiver tube (Pitz-Paal 2020). After the fluid is heated to a high temperature, a steam turbine is used for the generation of power.

5.2.1.2.2 Power Tower Systems

A high-temperature heat transfer fluid within the receiver is used to power a steam turbine, producing energy. Although power tower systems can be more costly to construct and run than parabolic trough systems, they are more efficient (Pitz-Paal 2020).

5.2.1.2.3 Engine/Dish Systems

A mirror designed to resemble a parabolic dish can be used to focus sunlight onto a receiver at the focal point of the dish (Pitz-Paal 2020). A high-temperature heat transfer fluid inside the receiver is utilized to power a Stirling engine, which generates electricity. Dish/engine systems are more expensive to build and run than parabolic trough and power tower systems, despite being more efficient.

5.2.1.3 Advantages and Disadvantages of CSP

CSP's key advantage is its thermal storage technology, which allows it to generate power in the absence of sunlight, making it more reliable than solar PV. CSP has no GHG emissions or air pollution, producing less of an environmental effect than fossil fuel-powered power plants. Furthermore, CSP is capable of being combined with other RE sources, such as hydropower or wind to help in stabilizing the electricity grid. CSP also has drawbacks, such as the high cost of construction and maintenance, as well as the need for extensive acreage for mirrors or lenses. In places with limited water resources, the CSP system's enormous water requirement for steam generation poses a problem. Furthermore, because part of the sunlight is lost during the concentration process, CSP may be less efficient than solar PV installations.

5.2.1.4 Green Energy Project (CSP) in Spain

Spain has emerged as a global leader in CSP, leveraging its abundant sunlight resources to drive sustainable energy initiatives. The CSP plants in Spain have been installed over several years, starting in the mid-2000s. One of the earliest and most notable CSP plants in Spain is the PS10 solar power tower, which began operation in 2007 in Seville. Following PS10, other CSP installations, including PS20 (operational since 2009) and various parabolic trough plants like Andasol (commissioned in phases between 2008 and 2009) and Solnova (operational since 2010), have been developed across different regions of Spain (Pitz-Paal 2020).

The capacity of CSP plants in Spain varies by installation. The PS10 Solar Power Tower in Seville, which has been active since 2007, has a capacity of around 11 MW and was one of the first in the world commercial towers of solar energy. PS20 solar power tower, also in Seville, began operation in 2009 with a higher capacity of 20 MW, showcasing advancements in solar tower technology. The Andasol Solar Power Station in Granada, consisting of three phases (Andasol 1, 2, and 3), has a combined capacity of over 150 MW, using parabolic troughs and thermal energy storage for dispatchable power. Another notable plant, the Solnova Solar Power Station, also in Granada, has a total capacity of approximately 150 MW and started commercial operations in 2010, enhancing Spain's RE portfolio (Behar et al. 2013).

5.2.1.4.1 A PV system

A PV system is a RE technology that generates power from sunlight. When exposed to sunlight, it generates energy using PV cells, which are often constructed of semiconductor materials like silicon. PV systems are available in a range of sizes, from modest residential rooftop installations to massive utility-scale solar farms (Čotar and F 2012).

PV modules (solar panels): These are composed of multiple PV cells interconnected to form a panel. Electric current flows through the PV cells when sunlight concentrates them since it excites electrons in the semiconductor material.

Inverter: Transform DC (direct current) into AC (alternating current). Inverters in the PV sector are divided into two main categories.

- **Stand-alone inverters**: These inverters need batteries to run properly and are not dependent on the electrical distribution network. The DC input of the inverter receives a constant voltage supply from the batteries.
- **Grid-tied Inverters**: Since these inverters are connected to the electrical distribution system, they need to be able to provide nearly flawless sinusoidal voltages and currents.

Batteries: Most often, batteries are used in stand-alone applications to store energy for use in low-light conditions (such as at night or on a rainy day). In addition, batteries are used in many different applications, such as utilities' interactive systems and standby power. PV batteries need to be able to tolerate both irregular charging cycles and deep discharges. The batteries may need to be kept in a random state of charge for a considerable amount of time for certain applications. Lead-acid batteries are the technology that PV systems use the most frequently (Moore et al. 2021).

Charge controllers and the balance of system components (BOS): Generally, between 30% and 50% of system expenses go towards BOS components. These are all the extra components needed for a proper PV system installation. The 2005 NEC specifies the minimum specifications. Wiles (2008) provides a comprehensive overview of BOS requirements including conductors, boxes and conduits, charge controllers, maximum power point trackers, disconnects, metering devices, mount gear (support structure) ground fault protection, and overcurrent protection (fuses and breakers).

5.2.1.5 Solar Energy Project (PV System) in Arizona, USA

One notable example of the transformative RE's potential is the Agua Caliente Solar Project in Arizona, USA. This utility-scale solar PV power plant completed its construction in 2014 and has a total capacity of 290 MW, making it among the world's largest solar installations at the time. The project spans around 2,400 acres of arid area and includes millions of solar panels that use sunshine to generate clean electricity. The Agua Caliente Solar Project contributes to the diversification of the region's energy portfolio by replacing fossil fuel-based power, hence lowering carbon emissions. Additionally, the project has provided economic benefits to the local community through job creation and tax revenues, demonstrating the multifaceted advantages of RE development. This is according to the US Department of Energy.

5.2.1.6 Global Technical and Resource Potential of Solar Energy

Sun energy is abundant; nevertheless, successful harvesting requires exact information of the Sun irradiation. Outside Earth's atmosphere, it measures 1,367 W/m², decreasing to 1,000 W/m² at the surface. Wavelengths range from 0.25 to 3 μm (40% visible light, 10% UV, 50% infrared). Clouds and atmospheric conditions can reduce irradiance by up to 90%, averaging 198 W/m². Accurate measurement is crucial for optimizing solar collectors (Arvizu et al. 2011).

Solar energy potential depends on local conditions like land availability, weather patterns, and energy demands, varying globally due to solar irradiance and land use. Research for the German Environment Agency indicates that both CSP and PV rely on solar irradiation, land availability, and technology. CSP requires high direct-normal irradiance areas, with few exclusions, highlighting the need to understand regional nuances and use diverse assessment methods (Viebahn et al. 2011).

5.2.2 Wind Energy

Wind turbines harvest wind energy and transform its kinetic energy into mechanical power, which is subsequently transformed into electrical power. A turbine for wind power usually consists of rotor blades that are wired to a generator. Wind energy is clean, abundant, and widely distributed, with the capacity to provide a significant portion of global electricity demand. Compared to power facilities that rely on fossil fuels, it has a lower carbon footprint and produces no GHGs when in operation (Jacobson and Delucchi 2011).

Wind energy (in various uses) is an established source of RE effectively used in a number of countries. It has the potential to expand significantly both economically and technically, and increasing its exploitation could be crucial for efforts to reduce GHG emissions worldwide (Wiser et al. 2011). Several countries throughout the world are actively participating in wind energy generation, with China leading the charge as the world's largest producer. The United States follows closely behind, with considerable onshore wind projects in areas such as Texas and Iowa. Germany, India, Spain, and the United Kingdom are also major contributors in wind energy (IRENA 2022).

5.2.2.1 Basic Design Principles

The process for generating wind power involves converting the kinetic energy of moving air into mechanical energy, which is subsequently translated into electrical energy (Wiser et al. 2011). The engineering problem facing the wind energy sector is the development of power plants and wind turbines at a reasonable cost for this conversion. A turbine can only capture a small portion of the wind's potential kinetic energy, which increases with wind speed cube. The cut-in speed, or wind speed at which most modern large wind turbines start to capture energy, is usually between 3 and 4 m/s (IRENA 2023).

5.2.2.2 Technology and Application

Wind energy technology is largely concerned with the utilization of windmills and turbines to convert the wind's kinetic energy into usable electricity. These turbines are typically made up of blades installed on a rotor and coupled to a generator. As the wind blows, the blades revolve, producing mechanical energy that is ultimately transformed into electricity (Jacobson and Delucchi 2011). Wind turbines provide clean and renewable electricity, which may be fed into the power grid to provide sustainable energy to homes, companies, and industries. Furthermore, wind energy is appropriate for isolated or off-grid places with limited access to traditional power sources, delivering electricity to remote villages, telecommunications infrastructure, and off-grid facilities. Wind-powered water pumps are utilized for irrigation, livestock watering, and water supply in rural areas (IEA 2023).

5.2.2.3 Various Technologies Used to Harness Wind Energy

5.2.2.3.1 Onshore Wind Energy Technology

Onshore wind energy technology involves designing and installing wind turbines on land to harness wind's kinetic energy for electricity generation. Turbines, featuring blades on a horizontal axis rotor, convert wind energy into electricity via a generator. They are placed strategically in windy areas often elevated on towers. Supporting infrastructure like access roads and transmission lines is essential for grid connectivity. Many onshore wind turbine configurations, including those having horizontal and vertical axes, were studied in the 1970s and 1980s. Over time, the horizontal axis design became the most common; however, there were variations in terms of the number of blades and their orientation relative to the tower (Hevia-Koch and Klinge Jacobsen 2019).

Wind turbines are made out of rotor blades that capture wind energy and are often composed of fibreglass-reinforced polyester or epoxy. The gearbox, generator, and control electronics are housed in the nacelle. It is located at the top of the towers and is connected to rotor blades. Tower, the structure that supports the nacelle and rotor blades, is usually composed of steel or concrete. Towers can range in height from 80 to 140 m. The foundation, which supports the tower, is often made of reinforced concrete.

5.2.2.3.2 Offshore Wind Energy Technology

Offshore wind energy utilizes larger turbines over oceans, which can be fixed-bottom for shallow waters or floating for deeper waters, to generate electricity. It offers stronger, more consistent winds, higher energy yields, and reduced visual and noise impacts on communities, but faces higher costs and technical challenges due to the marine environment. Innovations such as larger turbines, floating wind farms, and hybrid projects with wave energy or hydrogen production are advancing the technology, while environmental impacts and grid integration require careful consideration (Mathew and Philip 2012).

Despite challenges, offshore wind is poised for significant growth, driven by declining costs and ambitious RE goals. As technology advances, offshore wind is becoming a key player in the global RE landscape (Hevia-Koch and Klinge Jacobsen 2019).

5.2.2.4 Classification of Wind Turbines

Wind turbines are classified based on their axis. The main types are horizontal (HAWT) and vertical axis wind turbines (VAWT), which can further be divided.

5.2.2.4.1 Horizontal Axis Wind Turbines

Wind turbines with a horizontal axis are the most commonly utilized type with the rotor shaft facing the direction of the wind. They can be categorized as single, double, or triple bladed wind turbine. The blades rotate in accordance with the force's aerodynamic lift since they are aerodynamically designed. The top and lower faces of the turbine blades create the pressure differential.

Turbine blades are connected to rotor of electrical generator. In general, HAWT captures stronger winds and is more efficient (Kamran 2023).

5.2.2.4.2 A Vertical Axis Wind Turbine

The rotation of a VAWT is perpendicular to the direction of the wind because the generator and gearbox are situated near the turbine's base; maintenance is easy. The design does not require exceptional aerodynamics since the rotor is perpendicular to the wind speed. The VAWT is smaller than the HAWT, providing various advantages (Mathew and Philip 2012).

VAWT comes in two varieties: Darrieus and Savonius. The helix-shaped blades on Savonius wind turbines mimic DNA and wrap to encircle the vertical shaft. With longer and broader wings connecting to the upper and lower ends of the rotor shaft to produce a maximum swept area, Darrieus turbines resemble eggbeater shapes (Sawant et al. 2021).

According to the International Renewable Energy Agency (IRENA), wind power has experienced rapid growth since 2000, fuelled by research and development, supportive policies, and decreasing costs. Global installed wind generation capacity, including both onshore and offshore, surged from 7.5 GW in 1997 to approximately 733 GW by 2018, marking a 98-fold increase. Onshore wind capacity expanded from 178 GW in 2010 to 699 GW in 2020, while offshore wind saw a more substantial proportional growth, albeit from a lower starting point, rising from 3.1 GW in 2010 to 34.4 GW in 2020 (IRENA, 2022). Despite this progress, offshore and onshore wind still have unrealized potential to develop and improve globally (Table 5.1).

5.2.2.5 The Role of Wind Energy in Global Transformation

The worldwide electrical sector will undergo a change led by wind and solar energy. By 2050, wind power would account for over one-third of all electricity demand. This corresponds with various institutions' energy transformation projections, highlighting the importance of increasing wind power output for greening the energy system over the next 30 years (IRENA 2019). When compared to 2016 levels, wind power's proportion of the entire producing mixture is expected to increase approximately ninefold by 2050. However, in the context of total installed capacity by 2050, much larger capacity development will be needed for solar PV (8519 GW) than for wind (6044 GW), considering the normally lower capacity factors achieved by solar PV installations (Demarco 2017).

TABLE 5.1
The Global Electricity Capacity Trends, Adopted from IRENA Energy Transition Technology Retrieved in July 2024 and Modified

Year	Onshore (MW)	Offshore (MW)	Total Electricity Capacity Per Year (MW)
	Global Wind Energy Capacity Trends		
2012	216,419	5,345	221,764
2013	292,747	7,182	2,999,929
2014	340,956	8,503	349,459
2015	404,700	11,735	416,435
2016	452,883	14,258	467,141
2017	496,084	18,846	514,930
2018	540,079	23,601	563,680
2019	594,426	28,305	622,731
2020	699,091	34,381	733,472
2021	770,043	54,278	824,321
2022	840,279	62,604	902,883
2023	944,205	73,185	1,017,390

5.2.2.6 Resource Potential and Global Technical Potential of Wind Energy

Wind energy resource potential varies according to variables including wind constancy, speed, and accessibility. High wind speed regions, such as plains and coastal areas, provide more resource potential for wind energy production. In addition, the range of feasible areas for wind power development has increased due to technological developments such as higher turbines and increased efficiency (Wiser et al. 2011).

Wind energy's technical potential worldwide is estimated based on factors such as land availability, environmental considerations, and infrastructure constraints. It considers the theoretical maximum capacity for wind energy generation using current technology and practices. While the actual deployment of wind energy may be restricted by factors including land use conflicts, regulatory barriers, and economic feasibility, the global technical potential provides insight into the scale of possible wind energy contributions to the global energy mix.

5.2.2.7 Wind Energy Project in India

India has an installed wind energy capacity of 42,633 MW, contributing to 30.08% of the country's total installed RE capacity of 125,160 MW. This makes India the fourth-largest wind power market globally, with wind energy primarily concentrated on onshore wind farms in the southern, western, and north-western regions (Raghuvanshi et al. 2008). Wind energy constitutes 34.06% of India's renewable capacity, with a 2022–2023 capacity utilization factor of 18%. Most generation occurs from May to September during the Southwest monsoon. The government aims for 140 GW of wind energy by 2030, with Gujarat and Tamil Nadu as key contributors to its 4.43% share of total electricity (Kumar 2022).

5.2.3 Hydroelectric Energy

Hydroelectric energy source generates power by driving turbines with the water's gravitational pull. It comes from a variety of sources, including rivers, streams, and ocean tides (Kumar et al. 2011). Hydroelectric power facilities, ranging from big dams to minor run-of-river installations, are an established and dependable RE source with low GHG emissions. They account for 6.3% of the country's electricity generation and have great potential for increase via pumped storage, new site development, water transportation systems, and non-powered dams.

The Water Power Programme supports hydropower by developing essential components and identifying key areas for enhancing generation, with pumped storage hydropower playing a crucial role in grid storage. Major hydropower-generating countries include China, India, New Zealand, Latin America, North America, and African countries (Edenhofer et al. 2011).

5.2.3.1 Resource Potential and Global Technical Potential

5.2.3.1.1 Resource Potential

Hydropower is among the most well-known and ancient forms of RE. The resource potential for hydropower is vast, primarily concentrated in regions with abundant water resources and significant elevation changes. Countries with large river systems and mountainous terrain, such as China, Brazil, Canada, and the United States, have substantial hydropower potential (Kumar 2011).

5.2.3.1.2 Global Technical Potential

The International Energy Agency (IEA) estimates that the world's technical potential for hydropower is roughly 16,000 TWh annually. This estimate encompasses both current hydroelectric plants and possible future locations (IEA 2021); the world has utilized only about one-third of its economically viable hydropower potential. Therefore, there remains considerable room for expansion in this sector.

5.2.3.2 Technology and Application

5.2.3.2.1 Conventional Hydropower

Conventional hydropower systems can be divided into three main categories:

1. **Run-of-river (RoR HPP)**: This type harnesses the flow of water from a river without significantly altering its natural flow. It is less impactful on the environment but is dependent on river flow, which can vary seasonally. The main source of power produced by a RoR HPP is the river's available flow. A hydroelectric plant of this type may feature hourly or daily short-term storage. As a result, generation is dependent on runoff and precipitation, which can fluctuate dramatically on a daily, monthly, or annual basis (Kaunda et al. 2012).
2. **Reservoir (or Dam)/hydropower plant with reservoir**: These systems store large amounts of water in reservoirs created by damming rivers. They offer more consistent power generation and can also provide flood control, irrigation, and water supply benefits. Since they store water for later use,

hydropower plants with reservoirs are often referred to as storage hydropower (Kaunda et al. 2012). The reservoir reduces reliance on inflow fluctuations. The generating stations are situated downstream or at the toe of the dam and are connected to the reservoir by pipelines or tunnels (Kumar et al. 2011).

3. **Pumped storage plants**: This technique entails utilizing two water storage areas positioned at different heights. During periods of low electricity demand, water is pumped into the higher reservoir. Conversely, when demand surges, water is released into the lower reservoir to generate electricity, essentially functioning as an energy storage mechanism (Edenhofer et al. 2011). Pumped storage plants function not as energy generators themselves, but rather as storage solutions. In this configuration, water is pumped from a lower reservoir to an upper one, usually during times of low demand, and then the flow is reversed to produce power during peak load hours or whenever energy is necessary. Despite the energy losses incurred during the pumping phase, these plants offer substantial benefits as large-scale energy storage systems (Kaunda et al. 2012).

5.2.3.2.2 Emerging Technologies

Emerging hydropower technologies aim to reduce environmental impacts and increase efficiency. These include:

1. **Hydrokinetic turbines**: These turbines are placed directly in flowing water (rivers, tidal flows) without the need for dams.
2. **Small and micro hydropower**: These systems are designed for smaller applications and can be installed in remote areas with minimal environmental disruption.

5.2.3.3 Environmental Impacts and Benefits of Hydropower Energy

a. **Environmental impacts**: While hydropower is a clean energy source, it does have environmental impacts, particularly large dam projects; damming rivers can disrupt local ecosystems, affect fish migration patterns, and alter water quality and flow regimes. Large reservoirs can flood extensive areas, displacing communities and wildlife habitats. Although hydropower emits far fewer GHGs compared to fossil fuels, reservoirs in tropical regions can release methane from decomposing organic matter (*Investing in* Energy 2012).

b. **Benefits**: Despite these effects, hydropower provides significant environmental and economic benefits, including low carbon footprint. Hydropower emits far fewer GHGs than fossil fuel-based energy sources. It is sustainable since it is a RE source that depends on the water cycle. Pumped storage hydropower plants provide considerable energy storage, which helps balance supply and demand (Gleick 2009).

5.2.3.4 Hydropower Energy Project in Tanzania

Hydropower is a significant and well-developed RE source in Tanzania. The country has abundant water resources, with numerous rivers and lakes suitable for hydroelectric power generation (Mdee et al. 2018). Major rivers like the Rufiji, Pangani, and Kagera have been harnessed to produce electricity, contributing a substantial portion of the national grid's power supply. For example, the Kidatu and Kihansi hydropower plants on the Rufiji River are some of the biggest in the nation, offering dependable and sustainable electricity. About 30% of people have access to electricity. Approximately 15% of the nation's entire commercially feasible hydropower potential, or 4,700 MW, has been developed to date, with 717 MW of installed hydropower capacity. Tanzania is already building Stiegler's Gorge Dam, a hydroelectric power facility that can produce 2,115 MW of electricity (Pamba 2019). The benefits of hydropower include its steady and clean electricity supply, which lessens dependency on fossil fuels and helps cut down on emissions of GHGs.

5.2.4 Geothermal Energy

Energy from geothermal sources uses hot water or steam to transfer heat from below the surface of the Earth into electrical energy. Because of the continuous natural heat replenishment and fluid recharge, it is renewable. Its efficacy is mostly unaffected by climate change, and fossil fuel plant replacement can aid in lowering GHG emissions. Geothermal energy has significant technical promise worldwide. The IRENA projects that by 2050, the potential for geothermal power worldwide will exceed 200 GW. The United Republic of Tanzania, Indonesia, the Philippines, Turkey, New Zealand, Iceland, and several East African states are among the nations with significant geothermal potential.

Mature technologies used in geothermal energy include dry steam and flash plants, with binary cycle technology enabling medium-temperature field usage. The use of geothermal energy has increased slightly increased from 10,059 MW in 2011 to 14,846 MW in 2023 (Table 5.2).

5.2.4.1 Technology and Applications

Several technologies are employed to harness geothermal energy, each suited to different geological conditions and resource types.

5.2.4.1.1 Direct Use of Technology

Direct usage serves buildings through district heating and supports applications such as fish ponds, greenhouses, bathing, wellness, and swimming pools, as well as industrial and process heat for drying, mining, and agricultural products.

The oldest form of geothermal power plant is dry steam, which uses geothermal steam directly to drive turbines and generate electricity (Lund and Toth 2021).

5.2.4.1.2 Enhanced Geothermal System

Enhanced geothermal systems (EGS) use hydraulic stimulation to enhance or generate geothermal resources, considerably increasing geothermal energy supply by tapping into hot, dry rocks that standard methods cannot reach. Heat is retrieved by cycling water through the reservoir in a closed loop, which can be used for power generation with binary cycle plants, as well as industrial or residential heating (Pruess et al. 2010).

TABLE 5.2
Global Geothermal Electricity Generation Trend

Year	Geothermal Energy (MW)
Geothermal Electricity Capacity Trends	
2012	10,518
2013	10,786
2014	11,249
2015	11,847
2016	12,173
2017	12,754
2018	13,157
2019	13,824
2020	14,157
2021	14,432
2022	14,653
2023	15,026

Source: IRENA (2024).

5.2.4.1.3 Power Plants

The two primary types of geothermal power plants are dry steam and flash steam plants. These plants include cooling systems, separators, vaporizers, de-misters, heat exchangers, turbine generators, and grid transmission step-up transformers. High-pressure hot water drawn from the earth is heated in flash steam plants to create steam, which drives a turbine. Geothermal water that has been moderately heated is used in binary cycle plants to heat a secondary fluid that has a lower boiling point (Edenhofer et al. 2011).

5.2.4.1.4 Geothermal Heat Pumps

Geothermal heat pumps are commonly utilized in both residential and commercial settings to provide buildings with hot water and heating by utilizing the relatively consistent temperature of the ground close to the surface (Daikin 2000).

5.2.4.2 Environmental Impacts and Benefits of Geothermal Energy

Geothermal energy has various advantages and effects on the environment. The primary benefits include low GHG emissions, a small land footprint, and the ability to provide a continuous and stable power supply, making it an excellent option for meeting baseload electricity demand. Additionally, geothermal plants produce very little air pollution, contributing to better air quality. However, geothermal energy does have some environmental impacts, including significant water usage for cooling and reinjection processes, potential land subsidence if geothermal fluids are not managed properly, and the risk of inducing minor seismic events, particularly with EGS (Mock et al. 1997).

5.2.4.2.1 Geothermal Energy Project in Iceland

Iceland is a prime example of the successful utilization of geothermal energy. Iceland, located on the Mid-Atlantic Ridge, is rich in geothermal resources. Geothermal energy is utilized for the production of electricity as well as for direct use (Pavolová et al. 2015). Approximately 25% of Iceland's electricity comes from geothermal power, with the Hellisheidi Power Station, one of the greatest geothermal establishments in the world, with 303 MW of installed capacity. Geothermal energy meets over 90% of Iceland's heating demand through district heating systems, providing hot water and space heating for homes and businesses. It is also used in greenhouses, fish farming, and for snow melting on roads and sidewalks. According to UNEP, the usage of geothermal energy has greatly reduced Iceland's carbon footprint, assisting the country in its transition away from fossil fuels, while also providing economic benefits by cutting heating costs for households and encouraging sectors such as greenhouse agriculture and tourism.

5.2.5 Biomass Energy/Bioenergy

Biomass energy is also referred to as bioenergy. It's made from organic materials including wood, agricultural wastes, and organic waste. Several methods, including combustion, gasification, and anaerobic digestion (AD), can turn it into power, heat, or biofuels. Biomass energy is considered renewable as long as it is sourced sustainably and does not contribute to deforestation or land degradation. While biomass energy can reduce carbon emissions by displacing fossil fuels, its environmental sustainability depends on responsible sourcing and management practices (IPCC et al. 2012).

Globally, biomass has been a vital source of energy since the beginning of civilization. For almost 50% of the world's population, biomass serves as their principal energy source. In the developing world, wood biomass is an important RE source that supplies a sizeable portion of the energy needed in rural areas (Ladanai and Vinterbäck 2009). Concerns about sustainable energy sources and the Kyoto Protocol (i.e., the additional carbon cost imposed through carbon trading increases the cost of fossil fuels and thus makes "carbon-lean" biomass more competitive, increasing prices for fossil fuels and availability of stocks of wood raw material) have had a significant impact on the promotion of wood energy policies (Hoogwijk et al. 2005).

5.2.5.1 Classification of Biomass Sources Used for Energy Production

Biomass energy has been divided into various categories: woody biomass, which includes wood residues, forest thinning, and logging residues, and dedicated energy plantations (e.g., fast-growing trees). Moreover, agricultural sources such as crop residues (e.g., husks and straw), crop energy (e.g., *Miscanthus* and switchgrass) as well as animal manure can be used to produce biomass energy. Also waste biomass such as municipal solid waste includes food waste and garden trimmings, industrial waste (e.g., paper mill sludge, black liquor, as well as sewage sludge). Lastly, aquatic biomass, such as algae and aquatic plants, is used for biomass energy production.

Numerous uses exist for biomass, including power generation, heat, liquid biofuels, and the production of bio-based products (such as traditional non-commercial biomass). Furthermore, biomass can provide a wide range of energy, which means it can provide all the energy services required by a contemporary society. In addition, biomass is among the most widely available resources globally in comparison to other RE sources.

5.2.5.2 Bioenergy Production Technologies

Biomass conversion into bioenergy can be done primarily using two basic processes:

i. Biochemical conversion technique that generates liquid fuels like biodiesel, biogas, and bioethanol.
ii. Thermochemical conversion method generates solid, liquid, and gaseous products like charcoal and syngas.

5.2.5.3 Biochemical Conversion Technology

AD and fermentation are both included in this process. While fermentation, with sufficient feedstock and bacteria, yields bioethanol, butanol, and other compounds, AD creates biogas. Microbial populations, namely bacteria and archaea, are utilized by AD to transform organic waste into gaseous fuel, or biogas (a blend of carbon dioxide and methane), which is inexpensive and widely utilized globally. During the fermentation process, yeasts are employed to transform the sugar present in feedstock into ethanol. This led to the production of diluted alcohol that required distillation, which increased plant costs and negatively impacted overall process performance (Meena et al. 2021).

5.2.5.4 Thermochemical Conversion Technologies

Gaseous or liquid intermediate products are produced by thermochemical processes, and these products can be refined into higher-value energy sources like electricity and gasoline for transportation. Thermochemical conversion technology converts biomass into fuel through a number of processes such as pyrolysis, torrefaction, gasification, liquefaction, and combustion (Meena et al. 2021). All thermochemical processes begin with pyrolysis, which is a series of chemical events that produce solid, liquid, and gaseous products when oxygen is not present. Syngas is produced through gasification, which is the semi-oxidation (degradation) of biomass with regulated oxygen addition (Sadhwani et al. 2013). Through chemical reactions with solvent mediums such as water, ethanol, and acetone, biomass is processed to produce liquid products during liquefaction (Demirkaya et al. 2019) and during burning, organic materials undergo thermal conversion with the aid of any oxidant – typically oxygen-producing CO_2 and H_2O as well as the full oxidation of biomass.

Additionally, there is a process for physical conversion called pelletization. It involves squeezing biomass into pellet form. The uniform size of this dandified biofuel increases thermal energy output above regular household wood. The process of pelletization includes the following steps: storage of feedstock, removal of undesired impurities, size reduction, transportation via screw conveyors, drying, mixing, and conditioning of biomass, manufacturing of pellets, and lastly cooling and moisture extraction from pellets (Garcia-Maraver 2015).

5.2.5.5 Bioenergy Consumption Pattern in the World

Bioenergy is the greatest RE source, making up more than two-thirds of the RE supply. It generates approximately 50 EJ of energy globally, accounting for 10% of total annual primary energy use. RE accounts for roughly 20% of worldwide energy production, with bioenergy accounting for 77% of the total. Wood biomass accounts for 87% of this bioenergy, followed by agricultural crops and by-products at 9% and municipal and industrial wastes at 4% (Lang et al. 2016). Biomass is mostly utilized for heating and cooking in most nations. More than 80% of the biomass energy produced worldwide is produced in Brazil and the US combined. India, Indonesia, China, Argentina, Australia, Canada, Colombia, Mexico, Senegal, South Africa, and Zambia are among the countries that have passed legislation promoting biofuels.

5.2.5.6 Biomass Energy Project in India

About 32% of India's primary energy usage comes from biomass; 18.63% comes from biomass power, 5.31% from bagasse cogeneration, and 2.88% from waste-to-energy. The Ministry of New and RE encourages the development of power plants utilizing biomass gasifiers that utilize locally accessible biomass sources, such as crop wastes (Prakash and Lal 2023). Among these leftovers are sawdust, bagasse, jute waste, groundnut shells, straw, cotton stalks, coconut shells, soy husk, de-oiled cakes, and coffee waste. Between 120 and 150 million metric tonnes of surplus biomass from forestry and agriculture are expected to be available in India each year, with the potential to generate about 18,000 MW of electricity in addition to an extra 7,000 MW from bagasse cogeneration (Prakash and Lal 2023). Maharashtra is the state with the greatest potential for bioenergy, behind Punjab. The top states for biomass power projects are also Chhattisgarh, Uttar Pradesh, Andhra Pradesh, and Tamil Nadu. The states that lead the nation in bagasse cogeneration projects are Andhra Pradesh, Karnataka, Maharashtra, Tamil Nadu, and Uttar Pradesh.

5.3 INTEGRATING GREEN ENERGY INTO THE CURRENT AND FUTURE ENERGY SYSTEMS

Energy and power services are required in all communities to support both productive activities and fundamental human needs such as cooking, heating, lighting, and transportation (Sims et al. 2022). Energy services must be provided in a safe manner with the least amount of adverse environmental effects possible, since continued social and economic growth depends on having affordable and dependable access to energy resource needed to offer basic and environmentally friendly energy services (Sims et al. 2011).

Whether for large or small communities in Organisation for Economic Cooperation and Development (OECD) or non-OECD countries, there are enough RE resources available in many nations for system integration to meet a sizeable portion of energy demands, either directly to the end-user industries or in an indirect manner through current and future power supply system and carriers of energy (IPCC 2022). Simultaneously, many green energy resources are intrinsically unstable and unpredictable on time scales spanning from seconds to decades, distinguishing them from nuclear systems and fossil fuels. Increasing the share of RE sources in heat, power, or gaseous and liquid fuels may hinder integration and raise costs for the system (Gowrisankaran et al. 2016).

The projected integration of RE system technologies across a wide range of current global energy and power supply networks may confront economic challenges, although there are minimal technological constraints. Current infrastructure for energy supply, markets, and other institutional frameworks may need to be modified. A portfolio of RE resources and technologies can be more thoroughly integrated, increasing RE shares and overall system efficiency. Flexible power generation of fuels, electricity, heating, and cooling, as well as the utilization of storage and demand response options across various supply networks, can all help to improve this (Okumuş et al. 2021).

Communities in developed and developing nations can receive a variety of energy services through the integration of RE sources. Significant investments in technology and infrastructure are needed to make the transition to low-carbon energy systems with high percentages of RE. These investments include flexible power grids, more extensive district heating and cooling programmes, changes to the current distribution systems to accommodate liquid fuels and gases derived from RE, energy storage systems, innovative distributed energy systems in buildings, and new means of transportation (Kanoğlu et al. 2012).

To increase RE in energy supply by 2030 and beyond, challenges in transportation, building, industry, and agriculture must be overcome, along with understanding social issues and strategic elements. Transition paths should simplify integration, varying by region and industry. Therefore, to achieve projected RE levels in the future, rapid deployment of RE technologies in electricity generation, heating/cooling, gas/fuel distribution, and autonomous energy supply systems is necessary (Sims et al. 2011).

Currently, the building industry, excluding conventional biomass, employs RE to supply about 0% of its total consumer energy usage. In the future, energy-efficient or green building designs that allow for resource and/or time-flexible energy use in conjunction with RE will make it easier to integrate RE into urban contexts (Eslamian et al. 2017). Furthermore, the industrial sector can employ several ways to include RE in both light processing and large, energy-intensive industries. While material recycling, efficiency in energy use, and, eventually, carbon dioxide capture and storage can all be integrated with RE, sectors like sugar, pulp, paper, and rice production can become net energy providers to nearby grids; biomass wastes and electro-thermal processes can also directly replace fossil fuels on-site (Kanoğlu et al. 2012).

Sector coupling, which integrates RE with other sectors like transportation (through electric vehicles), heating (via heat pumps), and industry (through electrification of processes), can enhance overall system efficiency and sustainability (Ndaki et al. 2022). Furthermore, the promotion of micro grids, which are small, localized grids that may operate independently of the main grid, can provide energy security and resilience, especially in remote or disaster-prone regions. As a result, integrating renewable sources of energy into current and future systems for energy generation is critical to worldwide shift to more sustainable and robust energy infrastructures. The implementation of RE technologies becomes not only an environmental imperative but also an economic and social need as the world struggles to address climate change.

5.4 RENEWABLE ENERGY AND SUSTAINABLE DEVELOPMENT

Growing utilization of energy and emissions of GHGs has historically been closely associated with economic development; however, sources of RE can help break this link in order to promote sustainable development. Furthermore, RE presents a chance to enhance the impoverished people's access to contemporary energy services, an essential component for achieving any one of the eight Millennium Development Goals (Ruhul et al. 2021).

When compared to natural fuels, sources of RE emit few or no GHGs. As a result, reducing carbon footprint helps to mitigate climate change by reducing global warming and its related effects, which directly contributes to SDG 13 (Climate Action). Additionally, by reducing harms and supporting the health of natural ecosystems, REs help the preservation of ecosystems and biodiversity, which is in line with SDG 15 (Life on Land). But also REs support SDG 12 (Responsible Consumption and Production) as they harness renewable natural resources like wind and direct sunlight to encourage the sustainable use of the planet's natural wealth (Bundschuh et al. 2021).

While RE projects can enhance local economies and produce steady, well-paying jobs, the RE sector is also an important source of employment, providing possibilities in manufacturing, installation, maintenance, research and development, and contributing to SDG 8 (Decent Work and Economic Growth) (Okumuş et al. 2021). By reducing reliance on imported fuels and diversifying the energy supply, green energy enhances energy security by stabilizing energy costs and shielding economies from erratic fossil fuel markets, ultimately promoting sustainable economic growth and energy independence. It also provides access to affordable and clean energy, which supports SDG 7 (Affordable and Clean Energy), thereby improving quality of life and economic opportunities. On the other hand, RE improves air quality leading to major health advantages, lowering respiratory and cardiovascular ailments, which support SDG 3 (Good Health and Well-being). As a result, RE is fundamental to sustainable development, providing a path to environmental sustainability, economic prosperity, and social equality.

5.5 MITIGATION POTENTIAL AND COSTS

Green energy has the potential to considerably cut GHG emissions, contributing to climate change mitigation efforts. RE sources, such as wind, solar, hydro, and geothermal energy, can displace fossil fuels and help reduce GHG emissions, which are a primary driver of global warming (Zakariazadeh et al. 2024).

Wind power has high potential, offering zero-emission operation post-installation and significantly reducing CO_2 emissions by substituting coal or natural gas (Fischedick et al. 2011). Solar energy (both CSP and PV) can drastically reduce emissions, particularly in sunny areas. Hydropower, an established renewable source, provides considerable clean energy, though ecological impacts need careful management. Geothermal energy offers continuous, reliable, and low-emission electricity and heat, making it an effective tool for emission reduction in regions with suitable geology (Weinberg and Stenzel 2022).

The cost of green energy is a crucial component in its acceptance, and these costs have been reducing over time (Verbruggen et al. 2010), making RE more competitive with conventional fossil fuels. While the initial investment in RE projects, including turbines, solar panels, hydropower plants, and geothermal drilling, may be significant, these costs are frequently offset by lower operational and maintenance costs because renewable systems do not require fuel and have fewer components that wear out. Renewables' Levelized Cost of Energy (LCOE) has been falling, particularly for wind and solar, because of technology advancements and economies of scale. Government regulations, subsidies, and incentives like tax feed-in tariffs (FITs), grants, and credits are critical in making RE projects financially viable (Osman et al. 2023).

5.6 POLICY, FINANCING, AND IMPLEMENTATION FOR GREEN ENERGY

5.6.1 Green Energy Policy and Implementation

Climate action is facilitated by policies that enhance access to green energy. To achieve the Paris Agreements and SDGs, countries double the green energy policy through the communication of its nationally determined contribution (Shirizadeh 2023). Greener energy policies reduce GHG emissions by shifting financing away from energy-intensive and polluting industries and towards the recovery of an environmentally friendly economy (Okumuş et al. 2021). Regulation, encouraging private investment, and public investment initiatives are the primary policy facets underpinning most green energy policies (UNCTAD 2023).

Policies may incorporate green energy frameworks like FITs, net metering auctions, and direct-use regulations for heating, cooling, and transportation. These policies may involve grants, subsidies, and refunds for infrastructure investment, as well as tax breaks for green technology development. The following are the green energy policies being adopted in various areas.

I. **Tax incentives**: A tax incentive policy in the implementation of green energy refers to various tax benefits granted by the government to encourage the development, adoption, and usage of RE technology. These measures are intended to lessen the financial burden on firms and individuals engaging in green energy initiatives.

   These are the instruments that are used for boosting investments across the world. In addition, profit-driven tax breaks, including lower corporate income taxes and tax holidays, are the most frequent tools used by policymakers in developing countries, i.e., 70% of least developed countries and 64% of developing nations. However, developed nations often encourage the application of production-based tax credits and expenditure-based incentives (UNCTAD 2023).

   Furthermore, Egypt's new investment law provides financial incentives for RE projects. These incentives include a 30% deduction from net taxable earnings over the first seven years of a project's existence and a 2%

reduction in customs charges on machinery and equipment. These incentives can include lower corporate tax rates, tax rebates for R&D, or exemptions from customs duties.

II. **Non-fiscal incentives**: These tools promote investment in green energy using established policy tools that are already in place to promote investment in other sectors. Grants and subsidies are used more often than loans to promote green energy projects in developed countries. Furthermore, 16% used loans for developing countries, and only 13% used by developed countries (UNCTAD 2023).

III. **Feed-in tariff** is a framework that governments established in place to encourage the use of green energy. Their basis lies in a guarantee that they will purchase electricity from RE providers for extended periods at a fixed price. For example, for over 20 years, the government pledged to purchase all wind electricity generated in the nation at $0.15/kWh (UNCTAD 2023). These policies are the major policy instruments that are most efficient in developed countries and less so in developing countries. In addition, a change from a fixed price model to one that incorporates full or partial inflation adjustment, considering the unstable macroeconomic environment and the country's high inflation rate, may result in greater private investment in the energy industry (Qudrat-Ullah 2024). FIT has been implemented by many developing countries. However, they differ in terms of existing policy and implementation processes.

Germany enacted the Renewable Energy Act in 2000. This is a FIT arrangement that provides all producers of RE with a fixed price above the market for 20 years. Moreover, despite obstacles, this policy showed success, for example, the percentage of electricity consumed from renewable sources increased by 35% between 2000 and 2021(UNCTAD 2023). On the other hand, the Philippines used FITs in 2012 and achieved remarkable outcomes. After five years of the programme's inception, the nation's solar, biomass, and wind energy capacities had increased eight times. Furthermore, some North African nations are using structured procurement products like FITs and competitive auctions in their expanding power sector policies (Renewable Energy Agency 2023).

IV. **Net metering** regulations aim to encourage people to create their green energy generating system by providing safeguards. Households can draw electricity from the grid when they are unable to generate enough of it, and they can upload any excess electricity they generate into the grid. For example, Small Island Developing States (SIDS) have prioritized net-metering and net-billing over FITs as methods of accelerating the growth of RE over the previous 10 years. In 2016, Morocco enacted Law No. 58-15, which amended the law on RE and instituted a net metering system for onshore wind farms and solar PV systems. In addition, the legislation covers both private generators that may sell the grid up to 20% of their output and power plants that are linked to the high-voltage grid (Renewable Energy Agency 2023).

However, not all policies have been shown to be successful at promptly or significantly expanding the deployment of green energy (Song et al. 2022). For example, in China, the level of energy-efficient utilization is greatly increased by the green credit scheme, while the effects of this improvement vary by location. On the other hand, low-carbon city pilot (LCCP) programmes have the potential to significantly increase energy efficiency by stimulating technical innovation and transformation in businesses (Song et al. 2022). Combining policies across all sectors is crucial for effectively deploying green energy. The execution of policies can be done at both local and international levels (Qudrat-Ullah 2024).

5.6.2 Green Energy Financing

Members of the Kyoto Protocol declared to address climate change and achieve sustainable development goals in addition to strengthening green finance through investment, financing, and financial services that prioritize environmental sustainability. The most appropriate and helpful finance instrument for energy efficiency is green financing, among others (Liu et al. 2018).

Green energy would be possible with better access and availability of financing for climate action. It is estimated to cost between US$5 trillion and US$7 trillion annually (Shirizadeh 2023). Both government finance and private finance are crucial for the deployment of green technologies (IPCC 2023). Finance is essential for the current energy revolution. However, public funding has only contributed 24% of direct investments in RE, most of the financing for RE has come from the private sector (Renewable Energy Agency 2023). For example, public banks provide funding to project developers, alleviate investment risks, foster financial learning, and increase trust in low-carbon technologies. Furthermore, for residential installations up to 200 kW, Egypt offers soft loans at a 4% interest rate for projects between 200 and 500 kW; the interest rate can go up to 8% (Renewable Energy Agency 2023).

On the other hand, bilateral agencies and multilateral development banks (MDBs) are the biggest investors in green energy (Isah et al. 2023). For example, the International Fund Corporation and the state-owned phosphate and fertilizer manufacturer agreed upon a US$ 110 million loan to fund four solar plants to power the nation's industrial operations in April 2023 (Renewable Energy Agency 2023). Scaling up financial flow mechanisms for climate action is crucial, particularly in developing countries (IPCC 2023). To foster investment in low-carbon technology, sustainable energy storage, and industrial decarbonization, the Climate Investment Funds (CIF) have pledged $7.5 billion in blended finance products to developing nations. Furthermore, technology diffusion and low-emission innovation have been successfully supported by policy packages that are adapted to national contexts and technological features.

5.7 CHALLENGES AND BARRIERS

Financing green energy will require a large investment. Nevertheless, the annual investments needed to accelerate this transformation are currently less than US$2 trillion (Shirizadeh 2023). Providing cleaner energy in developing countries remains a challenge that is particularly affordable, dependable, and environmentally friendly (Edenhofer et al. 2011). Furthermore, in developing countries, limited financing options, weak public-private partnerships among financial stakeholders, substantial policy risks, and a lack of political will are the major constraints. In addition, insufficient cooperation between bilateral agencies results in unintentionally raised risk profiles and transaction costs associated with green technology (Isah et al. 2023).

Political barriers lead to unclear political pathways; lack of transparent and unambiguous regulatory systems affects the initiation and implementation of green energy. Moreover, investment in energy projects must be contingent upon long-term policy and political approval from the central government agencies involved in energy (Senpong and Wiwattanadate 2023). For example, Libya has not released new energy transition strategies since 2012, and its political leadership is perceived as less stable, also the nation has not joined the Paris Agreement. Furthermore, the lack of transparency in the EU's green hydrogen regulatory framework was perceived as a significant deterrent to investments in this sector.

Market barrier: There is no perfect market for green energy which means that investing in green energy is based on a local perspective, not on a global one. At the macro level, capital flows may be restricted by global inflationary shocks such as the energy crisis in 2022 brought on by the conflict between Russia and Ukraine, increasing the cost of funding green initiatives. For example, in Sierra Leone, where inflation increased by 40% in 2022–2023 the country's currency lost 40% of its value of the US dollar. In addition, typically, green initiatives last more than 25 years.

Transitional obstacles: In COP 28 parties discussed phasing out fossil fuels which remains a challenge. Therefore, abruptly shutting down fossil fuel plants to green energy without a quick and sustainable replacement might harm companies, financial systems, and jobs in places where stranded asset risks are high, such as Kazakhstan which is the largest exporter of fossil fuel. On the other hand, the capacity to adopt green technology is often challenging, especially in developing countries. The community opposes both biomass and waste-to-energy power plants because they must burn solid to produce heat and release combustion pollutants, particularly fly dust. This difficulty leads to interruptions in power supply. For example, Nigeria stands out as one of the worst-affected nations, experiencing 4,600 h of power outages in 2018. This is primarily attributed to inadequate grid management, infrastructure, and experts (Senpong and Wiwattanadate 2023).

5.8 FUTURE OUTLOOKS

According to the IEA, all fossil fuel subsidies must be phased out during the next few years to reach net zero emissions by 2050. Thus, green energy and low-carbon technology will dominate. Furthermore, extending green finance options, strengthening global collaboration, and improving green recovery strategies, including coordinating financial flows for climate action in a way that is consistent with funding requirements and ambition levels and mobilizing and expanding access to capital, especially for developing nations, vulnerable areas, sectors, and organizations, are essential. Future policy improvements like net metering, FITs, and priority grid access are necessary for investment development, particularly in Africa (Renewable Energy Agency 2023).

5.9 CONCLUSION

To summarize, green energy is critical for lowering carbon footprints and mitigating climate change. Utilizing renewable resources such as hydro, solar, wind, and geothermal energy allows us to reduce reliance on fossil and natural fuels, cutting emissions of GHGs and limiting the negative effects of climate change, ultimately sustaining development. Rapid advancements in green energy technologies, supporting regulations, and lower costs have considerably increased global usage of RE. Making the transition to green energy not only solves environmental issues but also fosters economic expansion and energy security. Infrastructure investments for RE provide jobs, encourage innovation, and advance sustainable development.

REFERENCES

Arvizu, D., P. Balaya, L. F. Cabeza, K. G. T. Hollands, A. Jäger-Waldau, M. Kondo, C. Konseibo, et al. 2011. "Direct Solar Energy." In *Renewable Energy Sources and Climate Change Mitigation*, edited by O. Edenhofer, R. Pichs-Madruga, Y. Sokona, K. Seyboth, S. Kadner, T. Zwickel, P. Eickemeier, et al., 333–400. Cambridge University Press. doi:10.1017/CBO9781139151153.007

Behar, O., A. Khellaf, and K. Mohammedi. 2013. "A Review of Studies on Central Receiver Solar Thermal Power Plants." *Renewable and Sustainable Energy Reviews* 23: 12–39. doi:10.1016/j.rser.2013.02.017

Bundschuh, J., N. Ghaffour, and B. Tomaszewska. 2021. "State-of-the-Art of Renewable Energy Sources Used in Water Desalination: Present and Future Prospects." *Desalination* 508 (March): 115035. doi:10.1016/j.desal.2021.115035

Čotar, A. F. 2012. *Andrej Photovoltaic Systems*. Darko Jardas, Dipl. Ing. REA Kvarner Ltd. https://www.irena-istra.hr/uploads/media/Photovoltaic_systems.pdf

Daikin. 2000. "Ground-Source Heat Pump Systems". 23 (4): 204.

Demarco, P. M. 2017. *Transforming the Energy System. Pathways to Our Sustainable Future*. doi:10.2307/j.ctt1wf4d66.9

Demirkaya, E., O. Dal, and A. Yüksel. 2019. "Liquefaction of Waste Hazelnut Shell by Using Sub- and Supercritical Solvents as a Reaction Medium." *Journal of Supercritical Fluids* 150: 11–20. doi:10.1016/j.supflu.2019.03.019

Dubie, R. 2019. "Paris Agreement (United Nations Framework Convention on Climate Change)," edited by Ian Muehlenhaus, *Geography Today: An Encyclopedia of Concepts, Issues, and Technology*, Vol. 1, 326–327. ABCCLIO / Bloomsbury publishing

Edenhofer, O., R. P. Madruga, Y. Sokona, K. Seyboth, P. Matschoss, S. Kadner, T. Zwickel, et al. 2011. *Renewable Energy Sources and Climate Change Mitigation: Special Report of the Intergovernmental Panel on Climate Change*. Cambridge University Press. doi:10.1017/CBO9781139151153

Eslamian, S., K. Ostad, A. Askari, and M. Soltani. 2017. "Climate Variability : Integration of Renewable Energy into Present and Future Energy Systems in Designing Residential Buildings." *International Journal of Rural Development, Environment and Health Research (IJREH)* 2: 18–30.

Fischedick, M., R. Schaeffer, A. Adedoyin, M. Akai, T. Bruckner, L. Clarke, V. Krey, et al. 2011. "Mitigation Potential and Costs." *Renewable Energy Sources and Climate Change Mitigation*, 791–864. doi:10.1017/cbo9781139151153.014

Garcia-Maraver, A. 2015. "Biomass Pelletization Process." *Developments in Combustion and Pyrolysis Research* 85: 53–66. doi:10.2495/978-1-84566-062-8/004

Gleick, P. H. 2009. "Three Gorges Dam Project, Yangtze River, China." *The Worlds Water 2008–2009* 11 (1): 139–150.

Gowrisankaran, G., Reynolds, S.S. and Samano, M., 2016. "Intermittency and the value of renewable energy. *Journal of Political Economy*." 124(4): 1187–1234. doi:10.3386/w17086

Hevia-Koch, P., and H. K. Jacobsen. 2019. "Comparing Offshore and Onshore Wind Development Considering Acceptance Costs." *Energy Policy* 125 (February 2018): 9–19. doi:10.1016/j.enpol.2018.10.019

Hoogwijk, M., A. Faaij, B. Eickhout, B. De Vries, and W. Turkenburg. 2005. "Potential of Biomass Energy out to 2100, for Four IPCC SRES Land-Use Scenarios." *Biomass and Bioenergy* 29 (4): 225–257. doi:10.1016/j.biombioe.2005.05.002

IEA. 2021. *World Energy Outlook 2021*. IEA Publications, 15.

IEA. 2023. *Renewable Energy Market Update: Outlook for 2023 and 2024 - International Energy Agency*, 87.

Investing in Energy. 2012. "Hydroelectric Power," 145–152. doi:10.1002/9781119204442.ch16

IPCC. 2022. *Mitigation Pathways Compatible with 1.5°C in the Context of Sustainable Development. Global Warming of 1.5°C*. doi:10.1017/9781009157940.004

IPCC. 2023. "Summary for Policymakers: Synthesis Report." *Climate Change 2023: Synthesis Report. Contribution of Working Groups I, II and III to the Sixth Assessment Report of the Intergovernmental Panel on Climate Change*, 1–34.

IPCC, H. Chum, A. Faaij, J. Moreira, G. Berndes, P. Dhamija, H. Dong, et al. 2012. "SRREN - Chapter 2 - Bioenergy." *Bioenergy, no. In IPCC Special Report on Renewable Energy Sources and Climate Change Mitigation: 209–332.*

IRENA. 2019. *Future of Wind: Deployment, Investment, Technology, Grid Integration and Socio-Economic Aspects(A Global Energy Transformation Paper)*. International Renewable Energy Agency.

IRENA. 2022. *Renewable Energy Statistics 2022 Statistiques D'Energie Renouvelable 2022 Estadisticas De Energia Renovablle 2022.*

IRENA. 2023. *Renewable Capacity Statistiques 2023.*

IRENA, IEA, and REN21. 2018. "Renewable Energy Policies in a Time of Transition." *International Journal of Production Research* 53: 28–31.

Isah, A., M. O. Dioha, R. Debnath, M. C. Abraham-Dukuma, and H. M. Butu. 2023. "Financing Renewable Energy: Policy Insights from Brazil and Nigeria." *Energy, Sustainability and Society* 13 (1): 1–16. doi:10.1186/s13705-022-00379-9

Jacobson, M. Z., and M. A. Delucchi. 2011. "Providing All Global Energy with Wind, Water, and Solar Power, Part I: Technologies, Energy Resources, Quantities and Areas of Infrastructure, and Materials." *Energy Policy* 39 (3): 1154–1169. doi:10.1016/j.enpol.2010.11.040

Kamran, M.. 2023. "Chapter 6 - Planning and Modeling of Wind Energy Systems." In edited by Muhammad B.T., *Fundamentals of Smart Grid Systems Kamran*, 271–298. Academic Press. https://doi.org/10.1016/B978-0-323-99560-3.00003-X

Kanoğlu, M., Y. Çengel, and I. Dincer. 2012. *Efficiency Evaluation of Energy Systems, Springer Briefs in Energy*. ISBN 978-1-4614-2241-9. *Mehmet Kanoğlu, Yunus A.* Çengel, *İbrahim Dinçer, 2012*. doi:10.1007/978-1-4614-2242-6

Kaunda, C. S., C. Z. Kimambo, and T. K. Nielsen. 2012. "Hydropower in the Context of Sustainable Energy Supply: A Review of Technologies and Challenges." *ISRN Renewable Energy* 2012 (December 2012): 1–15. doi:10.5402/2012/730631

Kumar, A. 2022. "Challenge of Climate Change." In *Ecosystem-Based Adaptation*, 39–104. Elsevier. doi:10.1016/B978-0-12-815025-2.00002-2

Kumar, A., T. Schei, A. Ahenkorah, R. Caceres Rodriguez, J.-M. Devernay, M. Freitas, D. Hall, Å. Killingtveit, Z. Liu. 2011. "Hydropower." In *IPCC Special Report on Renewable Energy Sources and Climate Change Mitigation*, edited by O. Edenhofer, R. Pichs-Madruga, Y. Sokona, K. Seyboth, P. Matschoss, S. Kadner, T. Zwickel, P. Eickemeier, G. Hansen, S. Schlömer, C. von Stechow, 437–496. Cambridge University Press, Cambridge, United Kingdom and New York, NY, USA. doi:10.1017/cbo9781139151153.009

Ladanai, S., and J. Vinterbäck. 2009. *Global Potential of Sustainable Biomass for Energy*. SLU, Institutionen För Energi Och Teknik Swedish University of Agricultural Sciences,

Department of Energy and Technology, Uppsala, Sweden. https://pub.epsilon.slu.se/4523/1/ladanai_et_al_100211.pdf

Lang, A., D. Bradley, and G. Gauthier. 2016. *Global Bioenergy Statistics*." World Bioenergy Association.

Letcher, T. M. 2022. "1 - Global Warming, Greenhouse Gases, Renewable Energy, and Storing Energy." In edited by Trevor M B T - Storing Energy (Second Edition) Letcher, 3–12. Elsevier. https://doi.org/10.1016/B978-0-12-824510-1.00011-8

Liu, Y., Q. Chen, Q. Ge, J. Dai, Y. Qin, L. Dai, X. Zou, and J. Chen. 2018. "Modelling the Impacts of Climate Change and Crop Management on Phenological Trends of Spring and Winter Wheat in China." *Agricultural and Forest Meteorology* 248: 518–526.

Lund, J. W., and A. N. Toth. 2021. "Direct Utilization of Geothermal Energy 2020 Worldwide Review." *Geothermics* 90: 25–29. doi:10.1016/j.geothermics.2020.101915

Mathew, S., and G. S. Philip. 2012. "2.05 - Wind Turbines: Evolution, Basic Principles, and Classifications." In *Comprehensive Renewable Energy*. Vol. 2. Elsevier Ltd. doi:10.1016/B978-0-08-087872-0.00205-5

Mdee, O. J., T. K. Nielsen, C. Z. Kimambo, and J. Kihedu. 2018. "Assessment of Hydropower Resources in Tanzania. A Review Article." *Renewable Energy and Environmental Sustainability* 3: 4. doi:10.1051/rees/2018004

Meena, M., S. Shubham, K. Paritosh, N. Pareek, and V. Vivekanand. 2021. "Production of Biofuels from Biomass: Predicting the Energy Employing Artificial Intelligence Modelling." *Bioresource Technology* 340 (July): 125642. doi:10.1016/j.biortech.2021.125642

Mock, J. E., J. W. Tester, and P. M. Wright. 1997. "Geothermal Energy from the Earth: Its Potential Impact as an Environmentally Sustainable Resource." *Annual Review of Energy and the Environment* 22 (1): 305–356. doi:10.1146/annurev.energy.22.1.305

Moore, B., C. Verfuerth, A. M. Minas, C. Tipping, S. Mander, I. Lorenzoni, C. Hoolohan, A. J. Jordan, and L. Whitmarsh. 2021. "Transformations for Climate Change Mitigation: A Systematic Review of Terminology, Concepts, and Characteristics." *WIREs Climate Change* 12 (6): e738. doi:10.1002/wcc.738

Ndaki, P. M., L. Kyaruzi, and R. Y. M. Kangalawe. 2022. "The Role of Renewable Energy Policies for Effective Climate Change Mitigation Actions in Tanzania - A Systematic Review." *Journal of the Geographical Association of Tanzania* 42 (2): 23–52.

Okumuş, İ., A. Güzel, and M. Destek. 2021. "Renewable, Non-Renewable Energy Consumption and Economic Growth Nexus in G7: Fresh Evidence from CS-ARDL." *Environmental Science and Pollution Research* 28 (October). doi:10.1007/s11356-021-14618-7

Osman, A. I., L. Chen, M. Yang, G. Msigwa, M. Farghali, S. Fawzy, D. W. Rooney, and P. Seng Yap. 2023. "Cost, Environmental Impact, and Resilience of Renewable Energy Under a Changing Climate: A Review." *Environmental Chemistry Letters* 21 (2): 741–764. doi:10.1007/s10311-022-01532-8

Pamba, S. 2019. *Stiegler's Gorge Hydropower Project: The Reasons behind the Project and Its Far Reaching Benefits. The Report on Responses and Mitigations: Beyond the Trues Cost of Power. Tanzania High Commission, New Delhi-India,* March, 3–63. Tanzania High Commission. https://www.in.tzembassy.go.tz

Pavolová, H., K. Čulková, and T. Bakalár. 2015. "Utilization of Geothermal Energy in Slovakia." *Acta Montanistica Slovaca* 20 (4): 326–333.

Pitz-Paal, R. 2020. "Concentrating Solar Power." In *Future Energy: Improved, Sustainable and Clean Options for Our Planet*. Elsevier Ltd. doi:10.1016/B978-0-08-102886-5.00019-0

Prakash, A., and K. Lal. 2023. "A Comprehensive Analysis of Biomass Energy in India for Agricultural and Domestic Use." *Journal of Agriculture and Ecology* 17: 26–29. doi:10.58628/jae-2317-303

Pruess, K., N. Spycher, and L. Berkeley. 2010. "Enhanced Geothermal Systems (EGS) with CO_2 as Heat Transmission Fluid - A Scheme for Combining Recovery of Renewable Energy with Geologic Storage of CO_2." *Proceedings of the World Geothermal Congress 2010*, Bali, Indonesia, 25–29 April 2010.

Qudrat-Ullah, H. 2024. "A Framework for Developing and Implementing FIT Policies for Renewable Energy Based on Local Economic Conditions." *Sustainable Futures* 7: 100170. doi:10.1016/j.sftr.2024.100170

Raghuvanshi, S. P., A. K. Raghav, and A. Chandra. 2008. "Renewable Energy Resources for Climate Change Mitigation." *Applied Ecology and Environmental Research* 6 (4): 15–27. doi:10.15666/aeer/0604_015027

Renewable Energy Agency, International. 2023. *North Africa Policies and Finance for Renewable Energy Deployment 2.*

Ruhul, M., A. Bhuiyan, H. Mamur, and J. Begum. 2021. "A Brief Review on Renewable and Sustainable Energy Resources in Bangladesh." *Cleaner Engineering and Technology* 4: 100208. doi:10.1016/j.clet.2021.100208

Sadhwani, N., Z. liu, M. Eden, and S. A. 2013. "Simulation, Analysis, and Assessment of CO_2 Enhanced Biomass Gasification." *Elsevier* 3: 3–8.

Sawant, M., S. Thakare, A. P. Rao, A. E. Feijóo-Lorenzo, and N. D. Bokde. 2021. "A Review on State-of-the-Art Reviews in Wind-Turbine-and Wind-Farm-Related Topics." *Energies* 14 (8): 1–30. doi:10.3390/en14082041

Senpong, C., and D. Wiwattanadate. 2023. "Krabi's Renewable Energy Transition towards Sustainable Energy: Drivers, Barriers, and Challenges." *Environment, Development and Sustainability*, 0123456789. doi:10.1007/s10668-023-02986-8

Shirizadeh, B.. 2023. *Financing the Green Energy Transition A US$50 Trillion Catch. In Deloitte*. doi:10.13140/RG.2.2.18130.94406

Sims, R., P. Mercado, W. Krewitt, G. Bhuyan, D. Flynn, H. Holttinen, G. Jannuzzi, S. Khennas, Y. Liu, M. O'Malley, L. J. Nilsson, J. Ogden, K. Ogimoto, H. Outhred, Ø. Ulleberg, F. van Hulle. 2011. "Integration of Renewable Energy into Present and Future Energy Systems." In *IPCC Special Report on Renewable Energy Sources and Climate Change Mitigation*, edited by O. Edenhofer, R. Pichs-Madruga, Y. Sokona, K. Seyboth, P. Matschoss, S. Kadner, T. Zwickel, et al., 609–706. Cambridge University, Cambridge, United Kingdom and New York, NY, USA.

Song, J., J. Wang, and Z. Chen. 2022. "How Low-Carbon Pilots Affect Chinese Urban Energy Efficiency: An Explanation from Technological Progress." *International Journal of Environmental Research and Public Health* 19 (23). doi:10.3390/ijerph192315563

UNCTAD. 2023. "Investment Policies for the Energy Transition," no. 26.

Verbruggen, A., M. Fischedick, W. Moomaw, T. Weir, A. Nadaï, L. J. Nilsson, J. Nyboer, and J. Sathaye. 2010. "Renewable Energy Costs, Potentials, Barriers: Conceptual Issues." *Energy Policy* 38 (2): 850–861. doi:10.1016/j.enpol.2009.10.036

Viebahn, P., Y. Lechon, and F. Trieb. 2011. "The Potential Role of Concentrated Solar Power (CSP) in Africa and Europe-A Dynamic Assessment of Technology Development, Cost Development and Life Cycle Inventories Until 2050." *Energy Policy* 39 (8): 4420–4430. doi:10.1016/j.enpol.2010.09.026

Weinberg, J., and F. Stenzel. 2022. *Mitigation Measures in Energy Systems*. Stockholm International Water Institute.

Wiles, J. 2008. *Photovoltaic Power Systems and the 2005 National Electrical Code: Suggested Practices. Report SAND2001–0674.*

Wiser, R., Z. Yang, C. Kjaer, F. Rahimzadeh, R. Wiser, Z. Yang, M. Hand, et al. 2011. "Wind Energy." In *IPCC Special Report on Renewable Energy Sources and Climate Change Mitigation*, edited by O. Edenhofer, R. Pichs-Madruga, Y. Sokona, K. Seyboth, P. Matschoss, S. Kadner, T. Zwickel, P. Eickemeier, G. Hansen, S. Schlömer, C. von Stechow, 1–74. Cambridge University Press, Cambridge, United Kingdom and New York, NY, USA.

Zakariazadeh, A., R. Ahshan, R. Al, and Mohammed Al-abri. 2024. "Renewable Energy Integration in Sustainable Water Systems : A Review." *Cleaner Engineering and Technology* 18 (December 2023): 100722. doi:10.1016/j.clet.2024.100722

6 Managing Forests for Climate Change Mitigation

Hassan Mhanna and Fatima Sami Khalifa

6.1 INTRODUCTION

Forests, covering about 42 million km² across tropical, temperate, and boreal regions, make up roughly 30% of the Earth's land area. These forests provide a vast range of ecological, economic, social, and aesthetic benefits. They act as habitats for diverse species, supply food and medicinal resources, regulate the hydrological cycle, protect soil, and offer spaces for recreation and spiritual activities (Bonan 2008). Moreover, forests play a significant role in climate regulation by exchanging energy, water, carbon dioxide, and other chemicals with the atmosphere. They store about 45% of terrestrial carbon and account for around 50% of terrestrial net primary production, offsetting about 33% of human-caused carbon emissions in the 1990s by absorbing roughly 2.6 petagrams of carbon annually (Pan et al. 2011). This carbon uptake helps mitigate climate change by reducing atmospheric carbon dioxide levels. However, forests can also contribute to planetary warming due to their low surface albedo, which can enhance solar heating (Bonan 2008; Denman et al. 2007; MacKenzie 2012). Forests support the hydrological cycle through evapotranspiration (ET), cooling the climate by affecting cloud formation and precipitation patterns. This process is generally less efficient in forests compared to some crops and varies between coniferous and deciduous broadleaf forests. The efficiency of ET depends on tree species, forest structure, and climatic conditions, influencing the overall water and energy exchange with the atmosphere (Ellison et al. 2017).

6.2 FOREST CARBON CYCLE

Carbon continuously circulates between the atmosphere and forest trees in a process known as the forest carbon cycle. Forests alternate between phases of growth and decay, during which they absorb and release carbon, respectively. During their growth phase, trees accumulate carbon through photosynthesis, converting carbon dioxide, water, and nutrients into biomass, which is stored in leaves, branches, stems, roots, and the surrounding soil. This process is crucial for carbon sequestration, and is essential for carbon sequestration, which helps mitigate climate change by lowering atmospheric carbon dioxide levels (Bonan 2008). Forests act as carbon sinks when they absorb more carbon dioxide than they emit. This occurs during periods of rapid

DOI: 10.1201/9781003492733-6

growth, when photosynthesis exceeds respiration and decomposition. Conversely, forests can become carbon sources when they emit more carbon dioxide than they absorb, often due to disturbances such as deforestation, wildfires, storms, pest infestations, or human activities (Pan et al. 2011). The balance between these phases is critical for maintaining the atmospheric carbon levels. Natural carbon sinks, such as forests, soils, and oceans, play a vital role in regulating atmospheric carbon dioxide levels. Artificial sinks, including technologies like carbon capture and storage, are also being developed to help manage carbon emissions. Forests, as one of the primary natural carbon sinks, are integral to the global carbon cycle, sequestering about 2.4 billion tons of carbon annually, which is approximately one-third of the carbon dioxide emitted from fossil fuels each year (FAO 2020). The efficiency of the forest carbon cycle is influenced by various factors, including tree species, age, growth patterns, and environmental conditions such as sunlight, water, and nutrient availability (Pan et al. 2011). For example, tropical forests have a higher carbon sequestration capacity compared to boreal forests due to their faster growth rates and longer growing seasons. However, they are also more vulnerable to deforestation and land-use changes, which can significantly reduce their carbon storage capacity. Soil, closely linked to vegetation, is another vital carbon sink. Carbon enters the soil through plant roots and decaying biomass, where it is stored as soil organic matter. Microorganisms in the soil slowly release carbon dioxide through biodegradation. The efficiency of soil carbon storage depends on factors such as biomass type, soil composition, moisture, and temperature variations (Lal et al. 2007). For instance, soils in tropical forests tend to have higher organic carbon content compared to those in temperate regions due to the continuous input of plant litter and favorable conditions for microbial activity. Forest management practices can significantly impact the forest carbon cycle. Practices such as reforestation, afforestation, and sustainable forest management (SFM) can enhance carbon sequestration. For example, reforestation of degraded lands can restore their carbon sequestration potential, while implementing SFM practices can preserve forest health and productivity, ensuring continued carbon uptake. Conversely, deforestation and land-use changes can transform forests from carbon sinks to carbon sources, worsening climate change. Modern scientific tools and methodologies, such as eddy covariance flux towers, satellite sensors, and climate models, have enhanced our understanding of forest-atmosphere interactions. These tools help measure carbon fluxes and evaluate the effects of various forest management practices on carbon sequestration. For instance, satellite data has been instrumental in monitoring deforestation rates and assessing the effectiveness of conservation efforts.

6.3 IMPORTANCE OF FORESTS IN SEQUESTERING CARBON DIOXIDE AND MITIGATING CLIMATE CHANGE

6.3.1 Carbon Sequestration Mechanisms

- **Biomass carbon storage**: Forests serve as substantial carbon sinks, storing large quantities of carbon in both living biomass and soil. Mature forests can sequester large quantities of carbon, which helps offset CO_2 emissions from human activities. According to research, forests constitute approximately 68% of US terrestrial carbon stocks, making them crucial for carbon storage and sequestration efforts (Gao et al. 2020; Lloyd and Farquhar 2008).

- **Soil carbon storage**: Forest soils are crucial for carbon sequestration, playing a significant role in storing carbon. The organic matter in forest soils, including decomposed plant and animal material, stores carbon over long periods. This storage is influenced by forest management practices and natural disturbances (Change 2014).
- **Long-term carbon storage**: Wood products derived from forests can continue to store carbon over extended periods. SFM practices encourage the use of wood in construction and other durable products contributing to long-term carbon sequestration (Change 2014; Raison et al. 1985).

6.3.2 Climate Change Mitigation

- **Offsetting emissions**: Forests help mitigate climate change by absorbing a portion of the CO_2 emitted by fossil fuels. For instance, in the United States, forestland sequesters nearly 16% of the country's annual CO_2 emissions, highlighting the importance of forest conservation and reforestation efforts (Pan et al. 2011; Domke et al. 2020).
- **Biodiversity and ecosystem services**: Beyond carbon storage, forests offer vital ecosystem services, including biodiversity conservation, water regulation, and soil protection. These services are essential for maintaining ecological balance and enhancing the resilience of ecosystems to climate change impacts (Ellison et al. 2017; Chazdon et al. 2016).
- **Climate-smart forestry**: This approach integrates climate change mitigation and adaptation strategies into forest management. Climate-smart forestry practices aim to increase forest carbon storage, enhance ecosystem resilience, and sustainably manage wood resources. This method underscores the need to adapt traditional forestry practices to address climate change more effectively (Change 2014).

6.4 DEFORESTATION, ITS CAUSES, AND THE IMPACT ON CLIMATE

Due to human activities, approximately 47% of Earth's forests have disappeared over the last 8,000 years. Despite this, forests still cover about one-third of the world's land area, totaling around 4.06 billion hectares. Over half of these forests are concentrated in just five countries: the Russian Federation, Brazil, Canada, the USA, and China (Bodo et al. 2021).

However, the rate of destruction of tropical rainforests continues to increase around the world, causing the area of tropical rainforests to decrease despite increased activity and environmental awareness (Okia 2012).

Forests are considered the lifeblood of the Earth's surface, harboring over 75% of the planet's biodiversity and providing essential resources for humanity including food, timber, fibers, fuel wood, and medicine. Additionally, forests play a pivotal role in climate regulation, water conservation, and purification (Cook et al. 2013;

Derbyshire 1984; Ojha et al. 2019; Sheram 1993), as well as contributing to flood and disease control. The importance of forests extends beyond these aspects alone, as they form an integral part of soil nutrient cycling and formation processes, and through their ability to sequester carbon, they contribute to the sustainability of primary production through photosynthetic processes (Assessment 2005; Khan et al. 2019; Melnykovych et al. 2018).

By preserving the forest and its biodiversity, we ensure the long-term survival of humanity. We can gain food security, improved agriculture, recreational activities, and other benefits that can be derived through the various species of life that the forest harbors (Meyer et al. 2003).

6.4.1 Deforestation

Deforestation involves transforming natural forested regions into areas devoid of trees. This process includes the removal of tree cover and the clearing of existing vegetation to make way for permanent non-forested uses like agriculture or urbanization (Chakravarty et al. 2012; Van Kooten and Bulte 2000). According to the Food and Agriculture Organization (FAO), deforestation is defined as the transformation of forest land to other land uses, whether driven by human activity or not (FAO 2020).

'Forest area net change': The term 'forest area net change' refers to the overall balance of forest losses (deforestation) and gains (forest expansion) over a certain period. Thus, net change can be positive or negative, depending on whether forest expansions exceed deforestation or vice versa (FAO 2020).

$$\text{Forest area net change} = \sum \text{gains}\left(\text{forest expansion}\right) - \sum \text{losses}\left(\text{deforestation}\right)$$

According to the Global Forest Resources Assessment 2020, Africa experienced the highest annual net forest loss between 2010 and 2020, losing 3.9 million hectares per year. South America also saw significant losses, with an annual net forest loss of 2.6 million hectares during the same period. Conversely, Asia saw the largest net gain in forest area from 2010 to 2020. Oceania, however, experienced net forest losses during the decades of 1990–2000 and 2000–2010 (FAO 2020).

6.4.2 Causes of Deforestation

Deforestation is caused by several interconnected factors. Understanding these causes is essential for developing effective conservation strategies (Okia 2012).

6.4.2.1 Agricultural Land Expansion

A major reason for deforestation is the expansion of agricultural land. This alone accounts for about 60% of tropical deforestation (Okia 2012).

- **Shifting agriculture**: Shifting agriculture, also known as slash-and-burn farming, involves clearing forested land to grow crops. Farmers use the land until the soil nutrients are depleted, then move on to clear new forest areas, repeating the cycle. This practice is prevalent in tropical regions and is responsible for nearly half of tropical deforestation, with some estimates suggesting it could account for up to two-thirds (Angelsen 1995; Geist and Lambin 2002).

- **Permanent agriculture**: There's a shift toward more permanent agriculture in tropical areas. Farmers stay on the same land by using fertilizers and other inputs to keep the soil fertile. This method can slow down deforestation rates after the initial clearing, but the first impact is still significant (Geist and Lambin 2002). Over time, permanent farming can also lead to soil degradation and loss of biodiversity, so sustainable practices are needed (Foley et al. 2011).

6.4.2.2 Timber Harvesting and Wood Extraction

Logging doesn't always cause deforestation directly, but it often leads to serious forest degradation. Logging creates roads that allow more people to move in and clear additional land, speeding up deforestation (Buys 2007; Putz et al. 2001).

- **Commercial logging**: Commercial logging is a major cause of deforestation, especially in boreal and temperate forests. While some areas are replanted after logging, this doesn't always happen, leading to long-term forest loss (Jardine 1994; Plantinga et al. 2007). In tropical forests, selective logging for valuable trees can lead to further exploitation and agricultural use, worsening the problem (Asner et al. 2016; Geist and Lambin 2002).
- **Fuel wood harvesting**: In tropical dry forests, the collection of fuel wood significantly contributes to deforestation and forest degradation. While it is not the primary cause of deforestation in humid tropics, it can be in densely populated areas with less forest cover, such as the Philippines, Thailand, and parts of Central America (Repetto 1988, 1990; Rowe et al. 1992). Finding alternative energy sources can help reduce the need for fuelwood (Arnold et al. 2006).

6.4.2.3 Overgrazing

Overgrazing is particularly harmful to forests, especially in dry tropical regions, degrading pastures and causing soil erosion. Livestock eat tree leaves and trample small plants, contributing to forest loss. The link between deforestation and cattle ranching is significant, with forested areas shrinking by almost 40% in the past four decades as grazing lands and cattle numbers have expanded (Hays et al. 2005; Steinfeld et al. 2006). Overgrazing also reduces biodiversity by shrinking the habitat available for wildlife (Laurance et al. 2014).

6.4.2.4 Mining

Mining is a destructive activity that contributes significantly to deforestation. Roads built for mining open up forested areas to more farming, ranching, and development. In the Philippines, mining activities, alongside logging, have dramatically reduced forest cover, decreasing from 17 million hectares in 1934 to merely 3 million in 2003, an 82% reduction (Okia 2012). Mining also often leads to pollution, further damaging forest ecosystems (Sonter et al. 2017). Over a 34-year period, the annual rate of deforestation in the southeastern Peruvian Amazon has been significantly impacted by

Artisanal-scale gold mining (ASGM). Notably, there was a sharp increase in mining activity in 2010, which aligns with a rise in gold prices. This trend was followed by sustained high levels of mining activity despite fluctuations in gold prices. The period is further distinguished by stages related to the construction of infrastructure: Pre-Interoceanic Highway (PIOH), Construction Phase of the Interoceanic Highway (CIOH), and the Operational Phase (IOH).

6.4.2.5 Urbanization, Industrialization, and Infrastructure

Growing populations, especially in tropical regions, increase the demand for agricultural land, leading to more forest being cleared (Geist and Lambin 2002). Infrastructure projects like oil extraction, logging concessions, and hydropower dams result in expanded road networks, which cut into forests (Eliasch 2012). These activities not only remove forests but also break up habitats, making it harder for wildlife to survive (Haddad et al. 2015). In temperate forests, while natural disasters and industrial activities contribute to deforestation, replanting efforts often help reduce long-term forest loss (Plantinga et al. 2007).

Since human colonization in 1638, Mauritius has experienced notable population growth and significant deforestation. The upper part of the analysis highlights the relationship between increasing population numbers and the extent of forest loss over time. The lower part provides a visual representation of six time periods, illustrating areas of native forest removal since the initial colonization (Norder et al. 2017).

6.4.2.6 Overpopulation and Poverty

Organizations like the FAO and international bodies highlight poverty and overpopulation as major causes of forest loss. Rapid population growth increases the need for food and space, leading to more land being cleared for farming and living. This results in more deforestation. Poverty worsens the problem as poor communities depend heavily on forest resources for survival, such as timber, fuelwood, and land for farming (Geist and Lambin 2002). Addressing these social and economic issues is key to managing forests sustainably (Wunder 2001).

6.4.2.7 Fire

Fire is often used to clear land for agriculture and plantations. Along with drought, it is a major cause of deforestation (Carmenta et al. 2011). Fires destroy large areas of forest and release significant amounts of carbon dioxide, contributing to climate change (Van der Werf et al. 2009). Managing fire use and promoting fire-resistant farming techniques are important to reduce forest loss (Bowman et al. 2009). In 2022, the Amazon biome experienced significant forest loss, with deforestation and fire hotspots distributed across the region. This spatial data, compiled by UMD/GLAD and ACA/MAAP, provides a comprehensive overview of the areas most affected by these disturbances (Finer and Mamani 2020).

6.4.2.8 Wars

Military operations have historically caused deforestation, such as during the Vietnam War. Wars can lead to large-scale forest clearing for military infrastructure, disrupting ecosystems and displacing wildlife (Gorsevski et al. 2012).

6.4.3 Impact of Deforestation on Climate

Deforestation significantly impacts both local and global climates by altering the balance of greenhouse gases in the atmosphere and affecting temperatures.

6.4.3.1 Greenhouse Gas Emissions

Deforestation is a key factor in climate change, accounting for 12%–20% of global greenhouse gas emissions.

Forests are vital as carbon sinks, absorbing carbon dioxide from the atmosphere and storing it. When deforestation occurs, the carbon stored in trees, soil, and other vegetation is released into the atmosphere as carbon dioxide (CO_2) through several mechanisms:

- **Biomass decomposition**: When trees are cut down, microorganisms such as bacteria and fungi break down the organic material in the trees, releasing carbon dioxide (CO_2) as a byproduct of respiration. This decomposition process contributes to the immediate release of carbon stored in the biomass of the trees (Chambers et al. 2000). The decomposition rate and subsequent CO_2 release depend on the type of biomass, climate conditions, and microbial activity (Davidson and Janssens 2006).
- **Burning (combustion)**: As mentioned previously, deforestation often involves the practice of slash-and-burn agriculture, where forests are cleared by cutting and burning trees and plants. Burning biomass releases substantial amounts of carbon dioxide and other greenhouse gases, including methane (CH_4) and nitrous oxide (N_2O) into the atmosphere, contributing to rising atmospheric carbon dioxide levels (Van der Werf et al. 2009).
- **Soil carbon loss**: Forest soils are rich in organic matter, which contains carbon. When forests are cleared, or burned, the soil will be disturbed and the organic matter is exposed to decomposition processes, further increasing carbon dioxide emission (Davidson and Janssens 2006).

6.4.3.2 Biophysical Effects

Deforestation impacts the climate through several biophysical mechanisms, including changes in albedo, ET, and surface roughness:

- **Albedo**: Albedo refers to the reflectivity of a surface, indicating the portion of solar energy (shortwave radiation) reflected from the Earth back into space. Forests generally have a low albedo, which means they absorb more solar radiation compared to agricultural lands or grasslands, which have a higher albedo and reflect more sunlight, the albedo increases, leading to more reflected sunlight and potentially cooler local temperatures, outweighing the warming effects of increased CO_2 (Davin and Noblet-Ducoudré 2010).
- **Evapotranspiration**: ET is vital for Earth's water cycle and climate. In forests, plants release water vapor into the air through transpiration. This process influences humidity, cloud formation, and rainfall. When forests

are removed, ET decreases, leading to higher local temperatures; this effect is particularly significant in tropical regions where the loss of ET can cause substantial warming (Luyssaert et al. 2014).

- **Surface roughness**: Forests have tall trees and dense vegetation, creating a rough surface with various obstacles like branches and leaves. On the other hand, cleared or deforested lands have a smoother surface with fewer obstacles. Forests create a rougher surface compared to deforested lands, affecting the transfer of heat and moisture between the land and atmosphere. Deforestation reduces surface roughness, potentially altering local wind patterns and affecting climate dynamics (Luyssaert et al. 2014).

6.4.3.3 Biodiversity Loss

Deforestation negatively impacts biodiversity through several mechanisms, leading to ecosystem disturbance, habitat degradation, and species loss. Below is an overview of each mechanism of deforestation's impact on biodiversity.

- **Habitat loss**: The most immediate and direct effect of deforestation is the loss of natural habitats. Forests host a wide and diverse range of species, many are highly specialized and entirely dependent on their native environment for survival. When forests are cut down or cleared, these creatures lose their natural habitats, leading to a significant decline in their numbers, and in extreme cases, they may even become completely extinct. The Amazon rainforest is a prime example of this. Deforestation in the Amazon region has led to significant destruction of natural habitats, posing a serious threat to many species that live there. As a result, this region has witnessed a significant decline in biodiversity, as animal and plant species have lost habitats without which they cannot survive, leading to a sharp decline in their numbers, and increasing the risk of extinction of some of them (WWF 2019).
- **Fragmentation**: Large continuous forests are broken into smaller, isolated patches. Fragmentation greatly reduces biodiversity, with losses ranging from 13% to 75%. It also disrupts ecosystem functions by lowering biomass and changing nutrient cycles. These negative effects are strongest in the smallest and most isolated habitat fragments, and they get worse over time. Fragmentation consistently weakens species survival, species diversity, nutrient retention, and overall ecosystem health (Haddad et al. 2015).
- **Impact on ecosystem functions**: Deforestation leads to the loss of animals, which impacts many ecological processes, such as pollination, pest control, nutrient cycling, and water quality.

  For instance, when pollinators like bees and butterflies decline, there will be less seed production, which makes it harder for plant populations to regenerate. This can result in fewer plants overall, which affects the entire ecosystem. Similarly, when the number of vertebrates, such as birds and mammals, decreases, there can be an increase in the population of herbivores like insects and small mammals. Without predators to keep their numbers in check, these herbivores can over-consume plants, leading to a reduction in plant biomass. This imbalance can disrupt the ecosystem, affecting everything from soil health to the availability of food for other species (Dirzo et al. 2014).

6.4.3.4 Feedback Loops

Cutting down trees can start a cycle that makes climate change worse. When trees are removed, carbon stored in the soil is released into the air, and the local climate changes, which can cause more trees to be cut down. The carbon released from cutting down trees and the fires that often follow adds to climate change. This makes droughts worse and makes forests more likely to catch fire. So, warmer temperatures lead to drier conditions, which then make fires more likely to happen, creating a dangerous cycle (Woodwell Climate Research Center 2022).

6.4.3.5 Impact on Hydrological Cycle

Deforestation significantly affects the hydrological cycle. It changes rainfall patterns, decreases groundwater levels, and raises the chances of floods and droughts.

- **Reduction in ET altered precipitation patterns**: Water is transferred from soil and plant surfaces to the atmosphere through the ET process. Deforestation reduces the number of trees, thereby decreasing ET, which can result in less rain. This alteration in precipitation patterns could eventually lead to regions that were once covered in forests becoming drier over time (Ellison et al. 2017).
- **Increased surface runoff, floods, and soil erosion and decline in groundwater levels**: Trees play a role as a natural barrier to slow surface runoff and facilitate water infiltration into the soil and replenish groundwater. When this natural barrier is removed, surface runoff and peak flows increase during rainy seasons, exacerbating the potential for flooding, leading to higher rates of soil erosion and lower groundwater levels (Bruijnzeel 2004).

6.5 SUSTAINABLE FOREST MANAGEMENT: A COMPREHENSIVE DEFINITION

SFM involves the stewardship and use of forest lands in a manner that maintains their ecological processes, biodiversity, and productivity over the long term. It encompasses practices that ensure the conservation of soil, air, water, and wildlife habitats, while simultaneously providing economic, social, and cultural benefits to present and future generations. This approach integrates ecological constraints with the need to enhance net social benefits from forest use, thereby ensuring that forest ecosystems remain viable and resilient, without depleting resources or causing harm to other ecosystems. SFM aims to achieve a balance between ecological, economic, and social objectives, promoting a stable relationship between human activities and the natural environment to support ongoing development and well-being (Finer and Mamani 2020).

6.5.1 A Historical Overview of SFM

Before the 20th century, indigenous communities worldwide practiced SFM, employing techniques like controlled burns, selective logging, and biodiversity conservation, grounded in deep ecological knowledge (Berkes 2012). In medieval Europe, forest laws aimed to regulate activities like logging and hunting to prevent overexploitation and foster forest regeneration (Rackham 1980). Japan's Satoyama system integrated sustainable forestry with agriculture, maintaining biodiversity while meeting human needs (Takeuchi 2010). Early colonial America, influenced by indigenous practices, implemented regulations to ensure timber supply (Cronon 2011). However, the 19th-century Industrial Era saw rampant deforestation globally due to the unchecked expansion of industries like logging and agriculture, prioritizing short-term gains over sustainability (Williams 2003). Clear-cutting in North America and intensive logging in Europe led to significant environmental degradation (Bonnicksen 2000; Williams 2003), while colonial powers exploited African forests without regard for long-term consequences (Fairhead and Leach 1996). This era exemplified the prioritization of economic profit at the expense of forest ecosystems' health and sustainability.

6.5.2 Sustainable Forest Management Current Challenges

6.5.2.1 Natural Disturbance Regimes and Climate Change

SFM faces significant challenges due to changes in natural disturbance regimes. Wildfires, essential for maintaining biodiversity and forest dynamics, are expected to become more frequent and severe due to climate change, leading to shifts in forest composition, increased carbon emissions, and reduced old-growth forests, thereby impacting ecosystem resilience (Groot et al. 2013; Girona et al. 2023; Robinne et al. 2018). Weather-related disturbances like heat, drought, wind, floods, and snow also affect tree mortality and ecosystem stability, with climate change exacerbating these impacts (Girona et al. 2023; Gregow et al. 2011; Price et al. 2013). Additionally, biotic disturbances such as insect outbreaks and diseases, influenced by climate change, contribute to forest degradation and biodiversity loss. Interactions between these disturbances, like wildfires and insect outbreaks, further amplify their effects, complicating forest management and resilience (Girona et al. 2023; Seidl et al. 2017).

6.5.2.2 Biodiversity Loss

Biodiversity loss poses a major challenge for SFM. The rapid transformation of boreal forest landscapes, largely due to mechanized forestry practices like clear-cutting, has led to a decrease in natural disturbances and an increase in human-induced disturbances (Mori et al. 2021). This shift has resulted in managed forests lacking essential structural elements such as dead wood and old trees, which are crucial for biodiversity (Bonnicksen 2000). Simplistic forest management approaches, particularly in Fennoscandia, have led to a decline in species richness and put numerous species at risk (Hyvärinen et al. 2019). For example, in Finland and Sweden, many threatened species rely on forest habitats, and habitat change driven by forest management practices is a primary driver of their population decline (Hyvärinen et al. 2019;

Artdatabanken 2020). The loss of biodiversity not only affects species survival but also impacts the functioning and health of forest ecosystems, making them more susceptible to disturbances like pest outbreaks (Bohn and Huth 2017). Restoring habitat complexity by reintroducing natural elements like deadwood is essential for sustaining biodiversity. However, managed forests often lack sufficient deadwood, which is vital for various species including beetles and birds (Culbert et al. 2013). Habitat fragmentation and the absence of old-growth forests further exacerbate biodiversity loss, underscoring the need for landscape-level conservation efforts (Junninen and Komonen 2011). Maintaining a mix of habitats at different successional stages within managed forest landscapes is crucial for conserving biodiversity (Mielikäinen and Hynynen 2003).

6.5.2.3 Loss of Old-Growth Forests

Losing old-growth forests presents major challenges for SFM. These forests are complex and diverse, with varying tree ages and types shaped by environmental conditions and natural disturbances (Kulha et al. 2020). Sustaining this diversity is tough with traditional forest management, which often aims for uniformity. In boreal regions, old-growth forests may seem similar due to fewer tree species and smaller trees (Martin et al. 2019). This can lead to management practices overlooking their unique roles and habitats. Large-scale logging since the mid-20th century has drastically reduced old-growth areas, altering tree types and landscape uniformity (Dupuis et al. 2020). This threatens natural forest structure and function, making sustainable management challenging amidst economic pressures. Old-growth forests harbor significant deadwood crucial for many species (Moussaoui et al. 2016). Logging reduces this deadwood, jeopardizing dependent species. Sustainable management requires strategies to retain deadwood and support biodiversity. Climate change adds uncertainty and severity, with more frequent and intense natural disturbances (Bergeron et al. 2017). Balancing immediate forest management with long-term climate threats complicates sustainability efforts, necessitating adaptability to both current and future changes. Conservation often prioritizes quantity over quality of protected old-growth forest (Martin et al. 2019). Remnant patches in managed landscapes may differ in productivity and structure, potentially losing ecological value. Effective management should prioritize both size and quality of conserved areas. Fragmentation disrupts ecological processes and species movement in old-growth forests. Sustainable management should enhance landscape connectivity to bolster ecosystem resilience and species survival.

6.5.2.4 Biotic Stress Factors

Biotic stress factors, like insects, pathogens, and invasive species, are major challenges for SFM. They play a crucial role in driving ecological changes and can sometimes even counteract natural disturbances and climate change effects (Nuttle et al. 2013). Integrating these factors into ecosystem-based management strategies is tricky because their impacts can accumulate over time or strike suddenly. For instance, in boreal forests, rising moose populations have reshaped forest composition and structure in North America and Fennoscandia, causing significant ecological shifts (Vriendt et al. 2021). Similarly, the northward migration of white-tailed deer in Canada, spurred by climate change, has also influenced forest composition (Dawe and Boutin 2016). Moreover, the global movement of goods and people has increased the spread of invasive species, posing a substantial threat to biodiversity and forest dynamics (Bellard et al. 2013). These factors highlight the complexity of managing forests sustainably amidst evolving biotic stressors and changing environmental conditions.

6.5.2.5 Carbon Sequestration

Carbon sequestration in boreal forests poses a challenge for sustainable management. These forests play a crucial role in storing carbon, but their capacity to sequester carbon varies widely across regions. Hof et al. (2021) note that predicting how boreal forests will respond to global change is uncertain due to complex factors like tree growth, respiration, and disturbances. Disturbances, both natural and human-caused, such as fires and insect outbreaks, can disrupt the carbon balance by releasing stored carbon back into the atmosphere. This makes it difficult to maintain and enhance the forests' carbon sequestration potential (Pappas et al. 2020). Managing these disturbances while preserving carbon sinks remains a significant challenge in SFM efforts.

6.5.2.6 Historical Silvicultural Practices

One of the key challenges in SFM is the impact of historical silvicultural practices. Traditional methods like clear-cutting and even-aged approaches have simplified forest landscapes, reducing biodiversity and resilience (Girona et al. 2023). This simplification negatively affects species dependent on diverse forest structures, hindering ecosystem recovery and adaptation to climate change (Splawinski et al. 2019); silvicultural practices are crucial. While alternatives like partial harvesting show promise, their applicability to North American boreal forests is hindered by limited experimentation (Montoro et al. 2018). Moreover, forest fragmentation resulting from extensive clear-cutting exacerbates these challenges, increasing edge.

6.5.3 Sustainable Forest Management Practices

- SFM aims to ensure that forests continue to provide environmental, economic, social, and cultural benefits for present and future generations. Here are some key practices associated with SFM:
- **Selective logging**: This involves removing certain trees while leaving most of the forest intact. It helps reduce the impact on the forest compared to clear-cutting. According to the FAO (2021), selective logging is essential for maintaining biodiversity and forest health.
- **Reforestation and afforestation**: Reforestation is planting trees in deforested areas, while afforestation is planting trees in areas that were not previously forested. These practices help restore forest cover, improve carbon sequestration, and restore habitats. The World Bank (2020) stresses that reforestation and afforestation are crucial for fighting climate change and maintaining ecosystem services.
- **Reduced impact logging (RIL)**: RIL involves planning and controlling logging operations to minimize environmental damage. Techniques include

careful planning of logging roads, directional felling, and minimizing soil disturbance.

- **Community-based forest management (CBFM)**: This empowers local communities to manage forests sustainably, integrating their needs and knowledge into forest management.
- **Certification schemes**: Forest certification schemes, like those by the Forest Stewardship Council (FSC) and the Programme for the Endorsement of Forest Certification (PEFC), set standards for responsible forest management. Certified forests are managed to preserve biodiversity and benefit local communities. FSC and PEFC standards are globally recognized for promoting SFM.
- **Agroforestry**: This combines agricultural and forestry techniques to create productive and sustainable land-use systems. It can improve soil quality, increase biodiversity, and provide economic benefits to farmers.

6.6 ROLE OF TECHNOLOGY IN MONITORING AND MANAGING FORESTS

The integration of advanced technologies such as satellite monitoring and Geographic Information Systems (GIS) has significantly enhanced the ability to monitor and manage forests effectively. These technologies provide comprehensive tools for data collection, analysis, and application in forest management.

6.6.1 Satellite Monitoring

Satellite remote sensing technology provides crucial data on forest cover, health, and biomass. This technology enables continuous and large-scale monitoring of forests, which is essential for detecting deforestation, forest degradation, and changes in land use. High-resolution satellite images from sources like Landsat and Sentinel offer detailed insights into the structural characteristics and health of forest ecosystems (Asner et al. 2016). Satellites equipped with LiDAR (Light Detection and Ranging) technology help estimate forest carbon stocks by measuring tree height and density. This data is crucial for comprehending the role of forests in carbon sequestration and for monitoring carbon storage changes over time (Dubayah et al. 2020; Jiang et al. 2023).

6.6.2 Geographic Information Systems (GIS)

GIS integrates various spatial data to produce detailed maps of forest resources. These maps facilitate the planning and implementation of SFM practices by providing information on forest types, species distribution, and land-use patterns (Vijitharan et al. 2022). GIS tools are essential for tracking biodiversity within forest ecosystems. By mapping habitats and monitoring changes, GIS supports conservation efforts aimed at preserving species diversity and maintaining ecosystem services (Bicknell et al. 2015).

6.6.3 Unmanned Aerial Vehicles and Drones

Unmanned aerial vehicles (UAVs), commonly referred to as drones, provide high-resolution imagery for precise forest monitoring. Drones are particularly useful in areas with difficult terrain, allowing for accurate mapping of tree species, canopy gaps, and monitoring of tree health and pest infestations. Drones enable rapid assessments of forest conditions, which is crucial during events such as forest fires or storms. They help in evaluating damage and planning recovery efforts efficiently (Jiang et al. 2023).

6.6.4 Early Warning Systems

Advanced remote sensing technologies facilitate the development of early warning systems for forest fires, illegal logging, and other threats. These systems enhance the capacity to respond promptly and mitigate potential damage (Jiang et al. 2023). Combining data from satellites, UAVs, and field surveys within GIS platforms allows for comprehensive analysis and better-informed decision-making in forest management. However, challenges remain in ensuring data accuracy, consistency, and accessibility.

6.7 COMMUNITY INVOLVEMENT AND INDIGENOUS KNOWLEDGE IN FOREST CONSERVATION EFFORTS

Community involvement and the integration of indigenous knowledge are crucial for effective forest conservation efforts. These approaches are rooted in the understanding that local and indigenous communities possess deep, place-based ecological knowledge and have historically managed their environments sustainably.

6.7.1 Importance of Indigenous Knowledge

Indigenous and local knowledge (ILK) encompasses the understanding, practices, and beliefs developed by indigenous and local communities over centuries of interaction with their environment. This knowledge is invaluable for biodiversity conservation and sustainable resource management. It is often based on a holistic view of ecosystems, recognizing the interconnections between different species, natural resources, and human activities.

For instance, a study on the Bulang people in Yunnan Province, China, highlighted how indigenous practices, such as moral constraints and worship rituals, effectively manage forest and wildlife resources without the need for formal enforcement measures (Su et al. 2020). These practices include community-enforced restrictions and traditional ceremonies that reinforce conservation behaviors.

6.7.2 Case Studies

- **Bulang people, China**: The Bulang community in Mangba Village utilizes indigenous knowledge to protect their local forests and wildlife. Their conservation system revolves around the sacred banyan trees and the Derbyan

Parakeet, demonstrating a symbiotic relationship between cultural beliefs and biodiversity conservation. Despite these efforts, the study found that the integration of scientific knowledge is necessary to address gaps and prevent practices that might unintentionally harm the ecosystem, such as artificial feeding of birds for tourism (Su et al. 2020).

- **Global systematic review**: A systematic review highlighted that integrating ILK into landscape-based studies and conservation projects has been successful in areas like agricultural systems, social-ecological systems, and climate change adaptation. These projects often involve multi-stakeholder collaborations that include local communities in the planning and decision-making processes, which enhance the sustainability and resilience of landscapes (Goglio et al. 2020).
- **Global conservation efforts**: Research published in *Ecology and Society* highlighted that involving indigenous peoples and local communities is essential for achieving effective and equitable conservation outcomes. This study underlined that such communities contribute significantly to biodiversity conservation through their traditional management practices, which are often more effective than conventional conservation strategies alone (Dawson et al. 2021).

6.8 CONCLUSION

Forests are essential for mitigating climate change by capturing carbon dioxide and maintaining ecological balance. Effective forest management is critical in addressing the multifaceted challenges posed by deforestation, which significantly contributes to global carbon emissions and climate instability. SFM strategies, including reforestation, afforestation, and sustainable logging, are essential for preserving forest ecosystems and enhancing their carbon sequestration capacity. Technological advancements, such as satellite monitoring and GIS, have revolutionized our ability to monitor and manage forests, providing invaluable data for informed decision-making and policy formulation. These technologies enable precise tracking of deforestation rates, forest health, and biodiversity, facilitating more effective conservation efforts. Community involvement and the integration of indigenous knowledge are paramount in forest conservation. Local communities and indigenous peoples have an in-depth knowledge of their natural environments, and their active participation ensures that conservation strategies are culturally appropriate, sustainable, and more likely to succeed. Case studies from various regions underscore the effectiveness of community-led conservation initiatives and highlight the importance of collaborative approaches in forest management. Ultimately, the successful management of forests for climate change mitigation requires a holistic and integrated approach, combining technological innovation, sustainable practices, and the empowerment of local communities. By prioritizing these elements, we can enhance the resilience of forest ecosystems, reduce carbon emissions, and contribute significantly to global climate change mitigation efforts. Through concerted global action and commitment, we can safeguard our forests for future generations and ensure their ongoing contribution to maintaining the health and stability of our planet.

REFERENCES

Angelsen, A. 1995. "Shifting Cultivation and 'Deforestation': A Study from Indonesia." *World Development* 23: 1713–1729.

Arnold, J. M., G. Köhlin, and R. Persson. 2006. "Woodfuels, Livelihoods, and Policy Interventions: Changing Perspectives." *World Development* 34: 596–611.

Artdatabanken, S. L. U. 2020. "Tillstånd Och Trender För Arter Och Deras Livsmiljöer–Rödlistade Arter i Sverige 2020." *SLU Artdatabanken* 24: 571–584.

Asner, G. P., P. G. Brodrick, C. B. Anderson, N. Vaughn, D. E. Knapp, R. E. Martin, and others. 2016. "Progressive Forest Canopy Water Loss during the 2012–2015 California Drought." *Proceedings of the National Academy of Sciences* 113: E249–E255.

Assessment, M. E. 2005. *Ecosystems and Human Well-Being: Wetlands and Water*. World Resources Institute, Washington, DC.

Bellard, C., W. Thuiller, B. Leroy, P. Genovesi, M. Bakkenes, and F. Courchamp. 2013. "Will Climate Change Promote Future Invasions?" *Global Change Biology* 19: 3740–3748.

Bergeron, Y., D. B. I. P. Vijayakumar, H. Ouzennou, F. Raulier, A. Leduc, S. Gauthier, and others. 2017. "Projections of Future Forest Age Class Structure under the Influence of Fire and Harvesting: Implications for Forest Management in the Boreal Forest of Eastern Canada." *Forestry: An International Journal of Forest Research* 90: 485–495.

Berkes, F. 2012. *Sacred Ecology*. New York & London: Routledge.

Bicknell, J. E., M. J. Struebig, and Z. G. Davies. 2015. "Reconciling Timber Extraction with Biodiversity Conservation in Tropical Forests Using Reduced-impact Logging." *Journal of Applied Ecology* 52: 379–388.

Bodo, T., B. G. Gimah, and K. J. Seomoni. 2021. "Deforestation and Habitat Loss: Human Causes, Consequences and Possible Solutions." *Journal of Geographical Research* 4: 22–30.

Bohn, F. J., and A. Huth. 2017. "The Importance of Forest Structure to Biodiversity–Productivity Relationships." *Royal Society Open Science* 4: 160521.

Bonan, G. B. 2008. "Forests and Climate Change: Forcings, Feedbacks, and the Climate Benefits of Forests." *Science* 320: 1444–1449.

Bonnicksen, T. M. 2000. *America's Ancient Forests: From the Ice Age to the Age of Discovery*. John Wiley & Sons, USA.

Bowman, D. M., J. K. Balch, P. Artaxo, W. J. Bond, J. M. Carlson, M. A. Cochrane, and S. J. Pyne. 2009. "Fire in the Earth System." *Science* 324: 481–484.

Bruijnzeel, L. A. 2004. "Hydrological Functions of Tropical Forests: Not Seeing the Soil for the Trees?" *Agriculture, Ecosystems & Environment* 104: 185–228.

Buys, P. 2007. *At Loggerheads?: Agricultural Expansion, Poverty Reduction, and Environment in the Tropical Forests*. World Bank Publications, Washington, DC.

Carmenta, R., L. Parry, A. Blackburn, S. Vermeylen, and J. Barlow. 2011. "Understanding Human-Fire Interactions in Tropical Forest Regions: A Case for Interdisciplinary Research across the Natural and Social Sciences." *Ecology and Society* 16: 53.

Chakravarty, S., S. K. Ghosh, C. P. Suresh, A. N. Dey, and G. Shukla. 2012. "Deforestation: Causes, Effects and Control Strategies." *Global Perspectives on Sustainable Forest Management* 1: 1–26.

Chambers, J. Q., N. Higuchi, J. P. Schimel, L. V. Ferreira, and J. M. Melack. 2000. "Decomposition and Carbon Cycling of Dead Trees in Tropical Forests of the Central Amazon." *Journal of Ecology* 122: 380–388.

Change, I. C. 2014. "Mitigation of Climate Change." *Contribution of Working Group III to the Fifth Assessment Report of the Intergovernmental Panel on Climate Change* 1454: 147.

Chazdon, R. L., P. H. Brancalion, L. Laestadius, A. Bennett-Curry, K. Buckingham, C. Kumar, and S. J. Wilson. 2016. "When Is a Forest a Forest? Forest Concepts and Definitions in the Era of Forest and Landscape Restoration." *Ambio* 45: 538–550.

Cook, E. R., J. G. Palmer, M. Ahmed, C. A. Woodhouse, P. Fenwick, M. U. Zafar, N. Khan, and others. 2013. "Five Centuries of Upper Indus River Flow from Tree Rings." *Journal of Hydrology* 486: 365–375.

Cronon, W. 2011. *Changes in the Land: Indians, Colonists, and the Ecology of New England.* Hill and Wang, New York.

Culbert, P. G., V. C. Radeloff, C. H. Flather, J. M. Kellndorfer, C. D. Rittenhouse, A. M. Pidgeon, and others. 2013. "The Influence of Vertical and Horizontal Habitat Structure on Nationwide Patterns of Avian Biodiversity." *The Auk* 130: 656–665.

Davidson, E. A., and I. A. Janssens. 2006. "Temperature Sensitivity of Soil Carbon Decomposition and Feedbacks to Climate Change." *Nature* 440: 165–173.

Davin, E. L., and N. de Noblet-Ducoudré. 2010. "Climatic Impact of Global-Scale Deforestation: Radiative versus Nonradiative Processes." *Journal of Climate* 23 (1): 97–112.

Dawe, K. L., and S. Boutin. 2016. "Climate Change Is the Primary Driver of White-tailed Deer (Odocoileus Virginianus) Range Expansion at the Northern Extent of Its Range; Land Use is Secondary." *Ecology and Evolution* 6: 6435–6451.

Dawson, N. M., B. Coolsaet, E. J. Sterling, R. Loveridge, N. D. Gross-Camp, S. Wongbusarakum, F. J. Rosado-May, and others. 2021. "The Role of Indigenous Peoples and Local Communities in Effective and Equitable Conservation." *Ecology and Society* 26(3): 19.

Denman, K.L., G. Brasseur, A. Chidthaisong, P. Ciais, P.M. Cox, R.E. Dickinson, D. Hauglustaine, C. Heinze, E. Holland, D. Jacob, U. Lohmann, S Ramachandran, P.L. da Silva Dias, S.C. Wofsy and X. Zhang, 2007: Couplings Between Changes in the Climate System and Biogeochemistry. In: Climate Change 2007: The Physical Science Basis. Contribution of Working Group I to the Fourth Assessment Report of the Intergovernmental Panel on Climate Change [Solomon, S., D. Qin, M. Manning, Z. Chen, M. Marquis, K.B. Averyt, M.Tignor and H.L. Miller (eds.)]. Cambridge University Press, Cambridge, United Kingdom and New York, NY, USA.Derbyshire, E. 1984. "Quaternary Glacial History of the Hunza Valley, Karakoram Mountains, Pakistan." *The International Karakoram Project* 2: 456–495.

Dirzo, R., H. S. Young, M. Galetti, G. Ceballos, N. J. Isaac, B. Collen, and others. 2014. "Defaunation in the Anthropocene." *Science* 345: 401–406.

Domke, G. M., S. N. Oswalt, B. F. Walters, and R. S. Morin. 2020. "Tree Planting Has the Potential to Increase Carbon Sequestration Capacity of Forests in the United States." *Proceedings of the National Academy of Sciences* 117: 24649–24651.

Dubayah, R., J. B. Blair, S. Goetz, L. Fatoyinbo, M. Hansen, S. Healey, C. Silva, and others. 2020. "The Global Ecosystem Dynamics Investigation: High-Resolution Laser Ranging of the Earth's Forests and Topography." *Science of Remote Sensing* 1: 100002.

Dupuis, S., V. Danneyrolles, J. Laflamme, Y. Boucher, and D. Arseneault. 2020. "Forest Transformation Following European Settlement in the Saguenay-Lac-St-Jean Valley in Eastern Quèbec, Canada." *Frontiers in Ecology and Evolution* 8: 257.

Eliasch, J. 2012. *Climate Change: Financing Global Forests: The Eliasch Review.* Routledge, London.

Ellison, D., C. E. Morris, B. Locatelli, D. Sheil, J. Cohen, D. Murdiyarso, C. A. Sullivan, and others. 2017. "Trees, Forests and Water: Cool Insights for a Hot World." *Global Environmental Change* 43: 51–61.

Fairhead, J., and M. Leach. 1996. *Misreading the African Landscape: Society and Ecology in a Forest-Savanna Mosaic.* Cambridge University Press, United Kingdom.

FAO. 2020. *Global Forest Resources Assessment 2020: Main Report.* Rome.

FAO. 2021. *FAO Statistics, Emissions-Total.* Rome, Italy: Food and Agriculture Organization of the United Nations (FAO). https://www.fao.org/faostat/en/#data/GT

Finer, M., and N. Mamani. 2020. "MAAP Synthesis: 2019 Amazon Deforestation Trends and Hotspots." *MAAP Synthesis* 4: 8.

Foley, J. A., N. Ramankutty, K. A. Brauman, E. S. Cassidy, J. S. Gerber, M. Johnston, Zaks, and others. 2011. "Solutions for a Cultivated Planet." *Nature* 478: 337–342.

Gao, T., F. Liu, Y. Wang, S. Mu, and L. Qiu. 2020. "Reduction of Atmospheric Suspended Particulate Matter Concentration and Influencing Factors of Green Space in Urban Forest Park." *Forests* 11 (9): 950.

Geist, H. J., and E. F. Lambin. 2002. "Proximate Causes and Underlying Driving Forces of Tropical Deforestation: Tropical Forests Are Disappearing as the Result of Many Pressures, Both Local and Regional, Acting in Various Combinations in Different Geographical Locations." *BioScience* 52 (2): 143–150.

Girona, M.-A., T. Aakala, N. Aquilué, A. C. Bélisle, E. Chaste, V. Danneyrolles, and S. Gauthier. 2023. "Challenges for the Sustainable Management of the Boreal Forest under Climate Change." *Boreal Forests in the Face of Climate Change: Sustainable Management*. Cham: Springer International Publishing, 773–837.

Goglio, P., A. G. Williams, N. Balta-Ozkan, N. R. P. Harris, P. Williamson, D. Huisingh, and others. 2020. "Advances and Challenges of Life Cycle Assessment (LCA) of Greenhouse Gas Removal Technologies to Fight Climate Changes." *Journal of Cleaner Production* 244: 118896. doi:10.1016/j.jclepro.2019.118896

Gorsevski, V., E. Kasischke, J. Dempewolf, T. Loboda, and F. Grossmann. 2012. "Analysis of the Impacts of Armed Conflict on the Eastern Afromontane Forest Region on the South Sudan–Uganda Border Using Multitemporal Landsat Imagery." *Remote Sensing of Environment* 118: 10–20.

Gregow, H., H. Peltola, M. Laapas, S. Saku, and A. Venäläinen. 2011. "Combined Occurrence of Wind, Snow Loading and Soil Frost with Implications for Risks to Forestry in Finland under the Current and Changing Climatic Conditions." *Silva Fennica* 45: 35–54.

Groot, W. J. de, M. D. Flannigan, and A. S. Cantin. 2013. "Climate Change Impacts on Future Boreal Fire Regimes." *Forest Ecology and Management* 294: 35–44.

Haddad, N. M., L. A. Brudvig, J. Clobert, K. F. Davies, A. Gonzalez, R. D. Holt, T. E. Lovejoy, et al. 2015. "Habitat Fragmentation and its Lasting Impact on Earth's Ecosystems." *Science Advances* 1 (10): e1500052.

Hays, G. C., A. J. Richardson, and C. Robinson. 2005. "Climate Change and Marine Plankton." *Trends in Ecology & Evolution* 20 (6): 337–344.

Hof, A. R., M.-A. M. Girona, M.-J. Fortin, and J.-P. A. Tremblay. 2021. "Editorial: Using Landscape Simulation Models to Help Balance Conflicting Goals in Changing Forests [Editorial]." *Frontiers in Ecology and Evolution* 9.1–3.

Hyvärinen, E., A. Juslén, E. Kemppainen, A. Uddström, and U. M. Liukko. 2019. *Suomen Lajien Uhanalaisuus 2019-Punainen Kirja: The 2019 Red List of Finnish Species*. Helsinki: Ympäristöministeriö & Suomen ympäristökeskus.

Jardine, K. 1994. *The Carbon Bomb: Climate Change and the Fate of the Northern Boreal Forests*. Amsterdam: Stichting Greenpeace Council.

Jiang, M., J. Kong, Z. Zhang, J. Hu, Y. Qin, K. Shang, and J. Zhang. 2023. "Seeing Trees from Drones: The Role of Leaf Phenology Transition in Mapping Species Distribution in Species-Rich Montane Forests." *Forests* 14 (5): 908.

Junninen, K., and A. Komonen. 2011. "Conservation Ecology of Boreal Polypores: A Review." *Biological Conservation* 144 (1): 11–20.

Khan, A. S., H. Yi, L. Zhang, X. Yu, E. Mbanzamihigo, G. Umuhumuza, S. I. A. Yevide, et al. 2019. "An Integrated Social-Ecological Assessment of Ecosystem Service Benefits in the Kagera River Basin in Eastern Africa." *Regional Environmental Change* 19 (1): 39–53.

Kulha, N., L. Pasanen, L. Holmström, L. De Grandpré, S. Gauthier, T. Kuuluvainen, and T. Aakala. 2020. "The Structure of Boreal Old-Growth Forests Changes at Multiple Spatial Scales over Decades." *Landscape Ecology* 35 (4): 843–858.

Lal, R., R. F. Follett, B. A. Stewart, and J. M. Kimble. 2007. "Soil Carbon Sequestration to Mitigate Climate Change and Advance Food Security." *Soil Science* 172 (12): 943–956.

Laurance, W. F., J. Sayer, and K. G .Cassman. 2014. "Agricultural Expansion and Its Impacts on Tropical Nature." *Trends in Ecology & Evolution* 29 (2): 107–116.

Lloyd, J., and G. D. Farquhar. 2008. "Effects of Rising Temperatures and [CO2] on the Physiology of Tropical Forest Trees." *Philosophical Transactions of the Royal Society B: Biological Sciences* 363 (1498): 1811–1817.

Luyssaert, S., M. -F. Jammet, P. C. Stoy, S. Estel, J. Pongratz, E. Ceschia, L. P. Chini, et al. 2014. "Land Management and Land-Cover Change Have Impacts of Similar Magnitude on Surface Temperature." *Nature Climate Change* 4 (5): 389–393.

MacKenzie, S. H. 2012. *The Global Carbon Cycle: Integrating Humans, Climate, and the Natural World*. Vol. 62. Island Press,Washington, DC.

Martin, M., N. J. Fenton, and H. Morin. 2019. "Boreal Old-Growth Forest Structural Diversity Challenges Aerial Photographic Survey Accuracy." *Canadian Journal of Forest Research* 50 (2): 155–169.

Melnykovych, M., M. Nijnik, I. Soloviy, A. Nijnik, S. Sarkki, Y. Bihun, O. Vyshnevska, I. Vasylyna, N. Zadorozhna, and R. Zubyk. 2018. "Social-Ecological Innovation in Remote Mountain Areas: Adaptive Responses of Forest-Dependent Communities to the Challenges of a Changing World." *Science of the Total Environment* 613: 894–906.

Meyer, A. L., G. C. Van Kooten, and S. Wang. 2003. "Institutional, Social and Economic Roots of Deforestation: A Cross-Country Comparison." *International Forestry Review* 5 (1): 29–37.

Mielikäinen, K., and J. Hynynen. 2003. "Silvicultural Management in Maintaining Biodiversity and Resistance of Forests in Europe–Boreal Zone: Case Finland." *Journal of Environmental Management* 67 (1): 47–54.

Montoro, G. M.-A., L. Navarro, and H. Morin. 2018. "A Secret Hidden in the Sediments: Lepidoptera Scales." *Frontiers in Ecology and Evolution* 6: 1–5.

Mori, A. S., L. E. Dee, A. Gonzalez, S. Ohashi, J. Cowles, A. J. Wright, and F. Isbell. 2021. "Biodiversity–Productivity Relationships Are Key to Nature-Based Climate Solutions." *Nature Climate Change* 11 (7): 543–550.

Moussaoui, L., N. J. Fenton, A. Leduc, and Y. Bergeron. 2016. "Can Retention Harvest Maintain Natural Structural Complexity? A Comparison of Post-Harvest and Post-Fire Residual Patches in Boreal Forest." *Forests* 7 (11): 243.

Norder, S. J., A. C. Seijmonsbergen, S. D. Rughooputh, E. E. Van Loon, V. Tatayah, A. T. Kamminga, and K. F. Rijsdijk. 2017. "Assessing Temporal Couplings in Social–Ecological Island Systems: Historical Deforestation and Soil Loss on Mauritius (Indian Ocean)." *Ecology and Society* 22 (2): 29.

Nuttle, T., A. A. Royo, M. B. Adams, and W. P. Carson. 2013. "Historic Disturbance Regimes Promote Tree Diversity Only under Low Browsing Regimes in Eastern Deciduous Forest." *Ecological Monographs* 83 (1): 3–17.

Ojha, H., T. Maraseni, A. Nightingale, and B. Bhattarai. 2019. "Rescuing Forests from the Carbon Trap." *Forest Policy and Economics* 101: 15–18.

Okia, C. A. 2012. *Global Perspectives on Sustainable Forest Management*. BoD–Books on Demand, Rijeka, Croatia.

Pan, Y., R. A. Birdsey, J. Fang, R. A. Houghton, P. E. Kauppi, W. A. Kurz, O. L. Phillips, et al. 2011. "A Large and Persistent Carbon Sink in the World's Forests." *Science* 333 (6045): 988–993.

Pappas, C., J. Maillet, S. Rakowski, J L. Baltzer, A. G. Barr, T. A. Black, and T. Zha. 2020. "Aboveground Tree Growth Is a Minor and Decoupled Fraction of Boreal Forest Carbon Input." *Agricultural and Forest Meteorology* 290: 108030.

Plantinga, A. J. Alig, R. J. Eichman, H. Lewis, D. J. 2007. "Linking land-use projections and forest fragmentation analysis". Res. Pap. PNWRP-570. Portland, OR: U.S. Department of Agriculture, Forest Service, Pacific Northwest Station. 41 p.

Price, D. T., R. I. Alfaro, K. J. Brown, M. D. Flannigan, R. A. Fleming, E. H. Hogg, M. P. Girardin, et al. 2013. "Anticipating the Consequences of Climate Change for Canada's Boreal Forest Ecosystems." *Environmental Reviews* 21 (4): 322–365.

Putz, F. E., G. M. Blate, K. H. Redford, R. Fimbel, and J. Robinson. 2001. "Tropical Forest Management and Conservation of Biodiversity: An Overview." *Conservation Biology* 15 (1): 7–20.

Rackham, O. 1980. *Ancient Woodland, Its History, Vegetation and Uses in England*. Edward Arnold, London, UK.

Raison, R. J., P. K. Khanna, and P. V. Woods. 1985. "Mechanisms of Element Transfer to the Atmosphere during Vegetation Fires." *Canadian Journal of Forest Research* 15 (1): 132–140.

Repetto, R.. 1988. *Forest for the Trees? Government Policies and the Misuse of Forest Resources*. World Resources Institute,Washington, DC.

Repetto, R. 1990. "Deforestation in the Tropics." *Scientific American* 262 (4): 36–45.

Robinne, F.-N., J. Burns, P. Kant, M. Flannigan, M. Kleine, W. de Groot, D. M. Wotton, A. Held editor, P. Hörling editor, and C. Wildburger editor. 2018. *Global Fire Challenges in a Warming World*. IUFRO , Vienna.

Rowe, R., N. P. Sharma, and J. Browder. 1992. "Deforestation: Problems, Causes and Concerns." *Managing the World's Forests*, 33–46.

Seidl, R., D. Thom, M. Kautz, D. Martin-Benito, M. Peltoniemi, G. Vacchiano, J. Wild, et al. 2017. "Forest Disturbances under Climate Change." *Nature Climate Change* 7 (6): 395–402.

Sheram, K. 1993. *The Environmental Data Book: A Guide to Statistics on the Environment and Development*. International Bank for Reconstruction and Development, The World Bank, 1818 H Street, NW, Washington, DC 20433.

Sonter, L. J., D. Herrera, D. J. Barrett, G. L. Galford, C. J. Moran, and B. S. Soares-Filho. 2017. "Mining Drives Extensive Deforestation in the Brazilian Amazon." *Nature Communications* 8 (1): 1–9.

Splawinski, T. B., D. Cyr, S. Gauthier, J. P. Jetté, and Y. Bergeron. 2019. "Analyzing Risk of Regeneration Failure in the Managed Boreal Forest of Northwestern Quebec." *Canadian Journal of Forest Research* 49 (7): 680–691.

Steinfeld, H., P. Gerber, T. Wassenaar, V. Castel, M. Rosales, and C. De Haan. 2006. *Livestock's Long Shadow: Environmental Issues and Options*. Food & Agriculture Org.

Su, K., J. Ren, Y. Qin, Y. Hou, and Y. Wen. 2020. "Efforts of Indigenous Knowledge in Forest and Wildlife Conservation: A Case Study on Bulang People in Mangba Village in Yunnan Province, China." *Forests* 11 (11): 1178.

Takeuchi, K. 2010. "Rebuilding the Relationship between People and Nature: The Satoyama Initiative." *Ecological Research* 25 (6): 891–897.

Van der Werf, G. R, D. C. Morton, R. S. DeFries, J. G. J. Olivier, P. S. Kasibhatla, R. B. Jackson, G. J. Collatz, and J. T. Randerson. 2009. "CO_2 Emissions from Forest Loss." *Nature Geoscience* 2 (11): 737–738.

Van Kooten, G. C., and E. H. Bulte. 2000. *The Economics of Nature: Managing Biological Assets*. Wiley-Blackwell Oxford, UK.

Vijitharan, S., N. Sasaki, M. Venkatappa, N. K. Tripathi, I. Abe, and T. W. Tsusaka. 2022. "Assessment of Forest Cover Changes in Vavuniya District, Sri Lanka: Implications for the Establishment of Subnational Forest Reference Emission Level." *Land* 11 (7): 1061.

Vriendt, L. De, S. Lavoie, M. Barrette, and J. P. Tremblay. 2021. "From Delayed Succession to Alternative Successional Trajectory: How Different Moose Browsing Pressures Contribute to Forest Dynamics Following Clear-cutting." *Journal of Vegetation Science* 32: e12945.

Williams, M. 2003. *Deforesting the Earth: From Prehistory to Global Crisis*. University of Chicago Press Chicago and London.

Woodwell Climate Research Center. 2022. "The Vicious Cycle of Climate Change, Deforestation, and Fire in the Amazon." https://www.woodwellclimate.org/climate-change-deforestation-and-fire-in-the-amazon/

Wunder, S.. 2001. "Poverty Alleviation and Tropical Forests—What Scope for Synergies?" *World Development* 27 (11): 1817–1833.

WWF. 2019. "Climate, Nature and Our 1.5°C Future: A Synthesis of IPCC and IPBES Reports." Gland,Switzerland. https://wwfint.awsassets.panda.org/downloads/wwf_climate__nature_and_our_1_5c_future_report.pdf

7 Life Cycle Assessment
A Tool for Climate Change Mitigation

Shweta Singh and Satvasheel Powar

7.1 INTRODUCTION

Climate change, resulting from the accumulation of greenhouse gases (GHGs) in the atmosphere, presents a complex and multifaceted issue. The Earth's atmosphere is experiencing heat retention due to an unparalleled rise in carbon dioxide, methane, and other GHGs stemming from the combustion of fossil fuels, deforestation, and other industrial activity. The greenhouse effect is elevating global temperatures, leading to various environmental consequences (Roux et al. 2016). The thawing of polar ice sheets and glaciers contributes to rising sea levels, threatening coastal communities and ecosystems. The intensification of weather patterns results from increasing global temperatures, potentially causing more frequent and severe cyclones, droughts, and heatwaves. These occurrences may profoundly affect human health, infrastructure, and agriculture. Climate change is not merely an environmental issue; it also possesses significant economic and societal ramifications. The most susceptible populations in poor nations are often the most impacted by climate-related disasters and possess the fewest means of adaptation, thereby intensifying disparities. Health concerns related to air pollution and vector-borne diseases are increasing, while food and water security are becoming increasingly precarious. The financial repercussions of climate change effects, encompassing infrastructure deterioration and declines in productivity, are significant and escalating. Consequently, tackling climate change necessitates a holistic approach.

This method takes into account the interplay of numerous environmental, economic, and social factors. Traditional environmental management approaches frequently focus on specific stages of a product's life cycle or isolated processes, sometimes overlooking broader systemic impacts. Consequently, a comprehensive framework is required for the creation and execution of successful climate change mitigation plans that encompass the interrelated nature of environmental impacts. Life cycle assessment (LCA) is a vital tool for formulating strategies that foster long-term sustainability and tackle the root causes of climate change (Goglio et al. 2020). LCA provides an extensive assessment of environmental impacts, including energy consumption, GHG emissions, water usage, and waste creation, by examining each phase of a product's lifespan, from raw material extraction to production, utilization, and disposal or recycling. This tool offers a comprehensive analysis, enabling stakeholders to pinpoint crucial stages where actions are most impactful.

DOI: 10.1201/9781003492733-7

The ability of LCA to highlight the primary sources of GHG emissions and other environmental consequences over the product life cycle is particularly beneficial. In the manufacturing sector, LCA might reveal that a substantial portion of a product's carbon footprint arises from the extraction and refinement of raw materials. This information allows companies to decrease emissions at this stage by pursuing alternate materials, enhancing efficiency, or developing new processes (Vieira et al. 2016). In the energy industry, LCA may demonstrate the enduring benefits of shifting from fossil fuels to renewable energy sources by evaluating the comprehensive life cycle consequences of different energy systems (Laurent et al. 2017). Moreover, LCA enhances policy development by providing a robust, evidence-based framework for decision-making. LCA data can help policymakers make rules and offer incentives that encourage environmentally friendly practices in many different areas. For example, it can help them make rules that support low-impact materials, energy-efficient technologies, or recycling and reducing waste, making sure that these steps are effective and targeted (Bento and Klotz 2014). Technological innovation is further enhanced by the insights derived from LCA. Companies can utilize LCA to guide their research and development initiatives toward creating products that meet consumer demands while minimizing environmental effects. This may involve enhancing product durability to prolong its lifespan, creating innovative materials that need fewer resources, or designing items for easier disassembly and recycling. These technologies are crucial for establishing a circular economy, characterized by the perpetual reprocessing, remanufacturing, and recycling of products and materials, thus minimizing waste and resource utilization. Furthermore, LCA enhances resource efficiency across supply chains. Businesses can alleviate environmental impacts, save operational costs, and decrease energy and material usage by identifying inefficiencies and inefficient practices. The optimization of transportation routes and procedures can diminish fuel consumption and emissions, while the refinement of production processes can decrease energy consumption and material waste. LCA enables customers to make educated decisions by offering openness and accountability for consumer behavior. The publication of LCA results by corporations enhances consumer confidence and encourages the acquisition of environmentally beneficial products. This transparency can encourage market rivalry, compelling enterprises to continually improve their environmental performance to satisfy consumer expectations and differentiate themselves in the marketplace. Overall, LCA has demonstrated its utility as a strategy for mitigating climate change in several ways. This chapter stresses the importance of LCA as a tool for reducing climate change. It also shows a case study of the environmental effects of the cradle-to-gate life cycle phase of monocrystalline silicon modules (mc-Si), because solar energy plays a big part in the global effort to reduce GHG emissions and fight climate change. The geographical area chosen for the study is the country of India. This study facilitates a thorough environmental impact analysis of the life cycle phases of mc-Si, commencing from raw material extraction to the installation phase at the production facility. This initiative will assist us in improving the sustainability of the mc-Si module by optimizing energy efficiency, resource management, and waste reduction.

7.1.1 Case Study: Cradle-to-Gate Life Cycle Assessment of mc-Si Technology in India

The solar energy sector in India is experiencing significant growth due to the country's ambitious renewable energy objectives and commitment to reducing GHG emissions. India has established itself as a global leader in solar energy deployment by aiming to achieve 500 GW of renewable energy capacity by 2030 (Dawn et al. 2016). India's geographic location renders it very suitable for solar energy, making it an optimal choice. Because of this, the solar energy industry's rapid growth makes it even more important to fully understand the environmental effects of the different solar photovoltaic (PV) technologies. Only then can this growth be sustained. mc-Si technology is widely employed for solar PV applications in India. mc-Si PV modules are renowned for their superior efficiency and long-lasting reliability, rendering them a favored choice for residential and large-scale solar applications. In comparison to alternative PV technologies, these modules exhibit a reduced footprint and enhanced power output, rendering them particularly suitable for applications in space-constrained and densely populated areas. Nonetheless, the production process of mc-Si PV modules necessitates considerable resources and comprises multiple stages, each contributing to the overall environmental impact. Each stage of the production process, from the extraction of natural resources to the fabrication of silicon wafers, solar cells, and PV modules, entails environmental consequences. These variables encompass significant energy consumption, GHG emissions, water usage, and trash generation (Soares et al. 2018). To address these issues and make solar PV technology more environmentally friendly, this chapter uses LCA as a key tool to look at the environmental effects of making mc-Si PV modules.

7.2 METHODOLOGY

7.2.1 Goal and Scope

The study performed the 'cradle-to-gate' assessment for manufacturing 1 m^2 of mc-Si module with all the inputs collected and outputs generated in primary functional unit (FU) of 'per square meter' of module production which will later change to secondary FU 'per kWh' of electricity production using equation (7.1) (Laurent et al. 2017).

$$\text{Impact}_{\text{kwh}} = \frac{\text{Impact}_{\text{m}^2}}{\text{Insolation} \times \eta \times A \times \text{PR} \times \text{LT}} \tag{7.1}$$

where PR is performance ratio of 80%, LT is lifetime, i.e., 25 years, A is active area of 75%, and η is the efficiency of about 24%. Efficiency is assumed as per Ameur et al. (2021). The system boundary includes only the raw material, energy consumption, and waste disposal and excludes transportation, use, and disposal for simplicity of study. The lifetime for mc-Si is assumed to be 25 years for this study, but in reality, it is more than 25 years.

TABLE 7.1
Life Cycle Inventory for Manufacturing of 1 m² of mc-Si Solar Module

Material/Energy	Quantity	Unit
mc-Si cell	73.4	p
Aluminum sheet, primary ingot	3.8	kg
Solar glass	11.4	kg
Ethyl vinyl acetate	1.3	kg
Electricity, medium voltage	8.33	kWh
Foil	0.32	kg
Copper, primary	0.14	kg
Tin	0.007	kg
Flux	0.01	kg
Lead	0.004	kg
Nickel	0.0002	kg
Cell waste	0.015	kg
Methanol	0.0162	kg
Cardboard	1.37	kg
Tap water	27	kg
Silicone	0.0029	kg

7.2.2 Life Cycle Inventory

Life cycle inventory (LCI) consists of all the data inputs in terms of raw material, energy flows, waste, and direct emission. The data inventory for the study is from Ecoinvent database and the existing literature, and for the final module of size 1 m^2, it is as shown in Table 7.1.

7.2.3 Life Impact Assessment

For this study, Life Cycle Impact Assessment (LCIA) is calculated with Simapro V9.0 software and Ecoinvent databases using CML method. The method utilized gives results in 11 environmental impact categories.

7.2.4 Interpretation

The last phase in the LCA process is the interpretation stage. It is a methodical approach in which the results of the LCI and the LCIA are categorized, measured, verified, and assessed. Interpretations will lead to suitable deductions and suggestions.

7.3 RESULTS AND DISCUSSION

According to the LCA, the environmental impacts produced by the 1 m^2 mc-Si modules for the 11 impact categories are presented in Table 7.2, and the same is represented graphically by indicating the contribution of each material and electricity used in manufacturing separately in Figure 7.1.

TABLE 7.2
Life Cycle Inventory for Manufacturing of 1 m² of mc-Si Solar Module

Impact Category	Unit	Total
Global warming (GWP)	kg CO_2 eq/m^2	348.1316
Ozone layer depletion (ODP)	kg CFC-11 eq/m^2	1.74×10^{-5}
Abiotic depletion	kg Sb eq/m^2	0.0005
Abiotic depletion (fossil fuels)	MJ/m^2	4.45×10^3
Human toxicity	kg 1,4-DB eq/m^2	231.445
Freshwater aquatic ecotoxicity	kg 1,4-DB eq/m^2	144.574
Marine aquatic ecotoxicity	kg 1,4-DB eq/m^2	5.56×10^5
Eutrophication	Kg PO_4 eq/m^2	0.744
Terrestrial ecotoxicity	kg 1,4-DB eq/m^2	2.041
Photochemical oxidation	kg C_2H_4 eq/m^2	0.065
Acidification	kg SO_2 eq/m^2	1.246

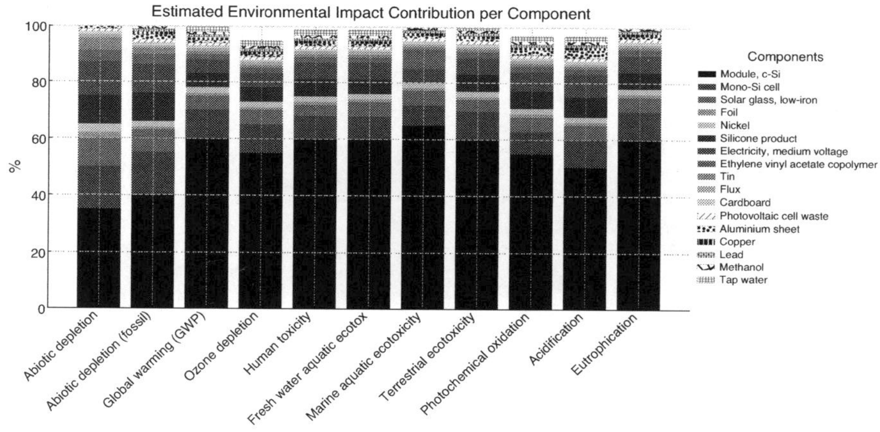

FIGURE 7.1 Environmental impact assessment for 1 m² mc-Si module.

According to the data, mc-Si cells exceed 80% for each effect indicator, with the exception of abiotic depletion, where the cell accounted for only 31.6%. Copper and zinc accounted for 35.2% and 22.4% of abiotic depletion, respectively. In all other effect categories, aside from the significant contribution of silicon cells, a little contribution has been noted from heavy metals like copper—primarily in human toxicity and ecotoxicity, as well as from solar glass, aluminum sheets, and electricity.

The primary influences on mc-Si cells stem from the silicon wafer and the electricity consumed during production. Upon further analysis of the silicon wafer's environmental impact, it is discovered that electricity usage during production has a more significant influence on the environment than the material itself, accounting for over 50% in nearly every impact category. mc-Si significantly contributes 62.9% to Ozone Depletion Potential (ODP) and 27.9% to Global Warming Potential (GWP) in materials. The utilization of surfactants contributes 82.1% to eco toxicity. After looking at the effects on the environment of a mc-Si module that covers 1 m^2, the effects are converted to a secondary unit called 'per kWh' of electricity production. This is done to make sure that the results are consistent with what has already been written about important impact categories like GWP, ODP, and human toxicity. It was found that mc-Si has a GWP of 0.048 kg CO_2 equivalent, an ODP of 2.41×10^{-9}, and a human toxicity level of 0.032 kg 1,4-dichlorobenzene equivalent. All of these values are within the range of impact indicators described in the literature (Müller et al. 2021).

7.4 CONCLUSIONS

This study analyzes the environmental impact of 1 m^2 of mc-Si PV technology through a cradle-to-gate LCA. The report presents impact values across 11 essential impact categories, including GWP, ODP, and human toxicity as significant categories. The study aims to analyze the implications of manufacturing mc-Si in India utilizing the Indian energy mix. The LCA analyzed the matter and determined that the effect value fell within the range of GWP as specified by the United Nations and documented in existing literature. Figure 7.1 for the other effect categories is similarly corroborated with the literature. Consequently, the study identified India as an advantageous site for the manufacturing of mc-Si PV technology, which will reduce imports and promote the 'Make in India' effort, thereby facilitating the expansion of economies of scale.

REFERENCES

Ameur, A., A. Berrada, K. Loudiyi, and R. Adomatis. 2021. "Performance and Energetic Modeling of Hybrid PV Systems Coupled with Battery Energy Storage." *Hybrid Energy System Models*. Elsevier, 195–238. doi:10.1016/b978-0-12-821403-9.00008-1

Bento, A. M., and R. Klotz. 2014. "Climate Policy Decisions Require Policy-Based Lifecycle Analysis." *Environmental Science & Technology* 48: 5379–5387. doi:10.1021/es405164g

Dawn, S., P. K. Tiwari, A. K. Goswami, and M. K. Mishra. 2016. "Recent Developments of Solar Energy in India: Perspectives, Strategies and Future Goals." *Renewable and Sustainable Energy Reviews* 62: 215–235. doi:10.1016/j.rser.2016.04.040

Goglio, P., A. G. Williams, N. Balta-Ozkan, N. R. P. Harris, P. Williamson, D. Huisingh, and others. 2020. "Advances and Challenges of Life Cycle Assessment (LCA) of Greenhouse Gas Removal Technologies to Fight Climate Changes." *Journal of Cleaner Production* 244: 118896. doi:10.1016/j.jclepro.2019.118896

Laurent, A., N. Espinosa, and M. Z. Hauschild. 2017. *LCA of Energy Systems. Life Cycle Assessment: Theory and Practice*. Springer International Publishing. doi:10.1007/978-3-319-56475-3_26.

Müller, A., L. Friedrich, C. Reichel, S. Herceg, M. Mittag, and D. H. Neuhaus. 2021. "A Comparative Life Cycle Assessment of Silicon PV Modules: Impact of Module Design, Manufacturing Location and Inventory." *Solar Energy Materials and Solar Cells* 230: 111277. doi:10.1016/j.solmat.2021.111277

Roux, C., P. Schalbart, E. Assoumou, and B. Peuportier. 2016. "Integrating Climate Change and Energy Mix Scenarios in LCA of Buildings and Districts." *Applied Energy* 184: 619–629. doi:10.1016/j.apenergy.2016.10.043

Soares, W. M., D. D. Athayde, and E. H. M. Nunes. 2018. "LCA Study of Photovoltaic Systems Based on Different Technologies." *International Journal of Green Energy* 15: 577–583. doi:10.1080/15435075.2018.1510408

Vieira, D. R., J. L. Calmon, and F. Z. Coelho. 2016. "Life Cycle Assessment (LCA) Applied to the Manufacturing of Common and Ecological Concrete: A Review." *Construction and Building Materials* 124: 656–666. doi:10.1016/j.conbuildmat.2016.07.125

8 Low-Cost Sustainable Practices for Climate Change Mitigation

Fatima Sami Khalifa and Alaa Ayman Hasan

8.1 INTRODUCTION

In today's world, the imperative to mitigate climate change and promote sustainability has intensified the focus on energy efficiency, renewable energy sources, and innovative waste management practices. Clean energy solutions require both enhancing energy efficiency and integrating renewable energy sources. Measures such as upgrading building insulation and using energy-efficient appliances reduce energy consumption and emissions. Additionally, renewable energy technologies, such as solar and wind power, offer sustainable solution alternatives to fossil fuels, contributing to lower greenhouse gas emissions and energy security (Situmeang et al. 2024).

Electricity consumption is projected to constitute an increasing share of global energy demand over the next 20 years, with rising fossil fuel prices and environmental concerns renewing interest in alternative energy resources. The Fukushima Daiichi accident was a pivotal moment, underscoring the necessity for safer energy sources. As a result, renewable energy is now seen as a more preferable choice compared to nuclear power because it does not carry the same risks and potential for disasters (Abolhosseini et al. 2014).

This chapter delves into low-cost sustainable practices for climate change mitigation, focusing on initiatives such as composting, rainwater harvesting, energy-efficient technologies, agroecology, and waste management. We explore the economic and environmental benefits of these practices, including cost savings, job creation, environmental protection, and resource conservation. Through successful case studies, we highlight practical implementations: small-scale biogas technology in Nepal, advancing second-generation biofuels in India, agroforestry and conservation in Indonesia, sustainable resource management and energy efficiency in the Nordic sugar industry, and sustainable waste management practices at Sathaye College in India.

Moreover, we address the challenges in mitigating climate change, including informational and awareness obstacles, sociocultural challenges, institutional and policy hindrances, economic barriers, and resource and technical issues. The chapter also examines the environmental and health impacts, water resource impacts, community and infrastructure challenges, and physical factors associated with these efforts.

DOI: 10.1201/9781003492733-8

Government policies and regulations are essential in encouraging low-cost sustainable practices. We emphasize the significance of having detailed action plans with clear accountability, incentivizing markets and enforcing change, enhancing innovation through increased funding, fostering conditions for private sector investment, improving the design and implementation of green initiatives, acting as role models, advocating for a whole-of-society, people-centered approach, and ensuring a fair transition. By integrating these diverse strategies, this chapter presents a comprehensive approach to achieving sustainability and combating climate change through practical, low-cost initiatives.

8.2 LOW-COST INITIATIVES

8.2.1 Composting

Due to ongoing population growth, agricultural expansion, and urbanization, there has been a notable rise in the production of organic waste worldwide, including livestock manure, municipal solid waste, sewage sludge, wheat straw, and other types of waste (Wang et al. 2016). Composting is a low-cost initiative that contributes to both waste reduction and renewable energy generation. Composting organic waste reduces landfill use. Composting not only reduces methane emissions but also generates nutrient-rich compost for soil improvement, thereby decreasing reliance on chemical fertilizers (Okedu et al. 2022). Moreover, anaerobic digestion, a process within composting, produces biogas, a renewable energy source suitable for generating electricity or providing heat (Al-Rumaihi et al. 2020). In another study on compost fertilizers, employing both pre-composting and vermicomposting led to a notable reduction in CH_4 and N_2O emissions by 84.2% and 80.9%, respectively (Wang et al. 2014).

Many recent research studies have proven the necessity of replacing chemical fertilizers with organic ones due to their health and environmental risks, in addition to their high cost. Conversely, organic fertilizers contribute to enhancing crop yields while mitigating soil and water pollution, thus ensuring healthy plant and animal production.

Hasan's (2024) research has shown that using spent coffee grounds as fertilizer resulted in elevated seed protein content in vetch species, which in turn economically restricts it as one of the most important crop candidates for increasing sustainability and the global supply of plant protein. It also reduces the harmful effects of coffee waste on the environment due to its high caffeine and tannin content, especially if it is disposed of without proper reuse. Therefore, many solutions have been proposed for reuse in many industries due to its richness in phenolic compounds, such as: ceramic industry, flavorings, solid fuel, ethanol production, animal food, fertilizer, and many other industries (Soares et al. 2000; Nyangito 2000; Saenger et al. 2001; Somashekar and Anu Appaiah 2013; Didanna 2014; Hoseini et al. 2021; Acchar et al. 2013).

8.2.2 Rainwater Harvesting

Global pressure on water resources is increasing due to population growth, rising per capita consumption, depletion of fossil water sources, and climate change. Domestic water needs constitute a substantial portion of this demand, which, under favorable conditions, can be partially or fully met through rainwater harvesting (Thomas 1998). Rainwater harvesting presents another opportunity for harnessing renewable resources. According to research by Arora (2021), rainwater harvesting systems can supplement conventional water sources, particularly in regions facing water scarcity. Collecting and storing rainwater for irrigation and other uses decreases dependence on municipal water supplies, conserves water, and reduces storm water runoff (Umar 2020). These systems reduce reliance on energy-intensive water pumping and treatment processes.

Ensuring a sustainable future hinges on addressing critical water supply issues globally. Urban cities are expanding rapidly, with infrastructure evolving to meet the growing needs of residents. Numerous studies emphasize the importance of developing greener and smarter urban areas to meet current social and environmental challenges. As a result, the management of global water resources has become an international priority, underscored by the rapid growth in global population (Ghisi et al. 2009). UNEP has highlighted that less than 1% (2.5% including frozen and snow-covered) of all freshwater resources are accessible for human consumption. Despite this limited availability, even these freshwater reserves are increasingly threatened by incremental pollution (UNEP 2008).

Rainwater harvesting has demonstrated significant benefits in urban areas. However, in numerous rural regions worldwide where access to safe drinking water is either absent or limited, implementing rainwater harvesting systems plays a crucial role in ensuring the sustainable future of these populations (Farreny et al. 2011).

8.2.3 Energy-Efficient Technologies

Energy efficiency and the use of renewable energy sources are essential foundations for transitioning to a sustainable future. According to Lu et al. (2020), progress in energy-efficient technologies presents considerable potential for decreasing energy consumption and reducing greenhouse gas emissions. For instance, the widespread adoption of LED lighting systems and smart thermostats has been shown to substantially decrease electricity usage in residential, commercial, and industrial settings (Conti et al. 2016), making energy efficiency accessible and affordable (Yang and Song 2023).

Future sustainable energy and technology are vital for advancing energy efficiency to reduce consumption, secure and replenish energy supplies, expand and reuse alternative energy sources, and address the damage caused by past energy practices. For instance, a solar-powered eco-friendly dehumidifier and a farming robot offer smart solutions to agricultural challenges (Singh et al. 2022).

8.3 Agroecology and Waste Management: Sustainable Farming and Recycling Initiatives

8.3.1 Agroecology

Agroecology integrates research, education, action, and change to bring sustainability to all aspects of the food system, including ecological, economic, and social dimensions (Gliessman 2018).

Agroecology emphasizes sustainable agricultural practices that prioritize environmental health, biodiversity, and social equity. One of the key principles of agroecology is the integration of natural processes to enhance productivity while minimizing environmental impact (Altieri 2018). This approach includes techniques such as crop diversification, agroforestry, and organic farming, which promote soil fertility, pest control, and resilience to climate change.

When humans are seen as integral components of the ecosystem, subject to its natural laws and the consequences of its success, it becomes imperative to make agroecosystems as sustainable as possible for the long term. Despite our current disruptive impact on the ecosystem, we have the capability to design systems that close nutrient cycles, rely more on renewable energy, minimize production inefficiencies, and promote environmental health. By applying system design principles and properties that mimic natural ecosystems, we can achieve these goals (Francis et al. 2003).

8.3.2 Waste Management

In waste management, recycling initiatives play a pivotal role in reducing resource depletion and environmental pollution. According to Alkahtani et al. (2021), recycling conserves energy, reduces greenhouse gas emissions, and mitigates the need for landfill space. By reprocessing materials such as paper, plastic, glass, and metals into new products, recycling closes the loop on resource utilization and contributes to a circular economy.

Understanding waste composition is crucial for selecting appropriate waste treatment and disposal methods. Significant investments in waste management facilities, training, and education are necessary to mitigate the health impacts of improper waste disposal. However, addressing health risks associated with waste management requires action on multiple fronts, including:

i. Implementing measures and incentives for waste minimization, prevention, recycling, and composting.
ii. Incorporating waste management costs into consumer products.
iii. Increasing public participation in deciding local and regional waste management practices.
iv. Enhancing public health surveillance.
v. Utilizing biomarker epidemiology techniques in future studies.

These efforts are essential for improving waste management practices and safeguarding public health (Giusti 2009).

8.4 ECONOMIC AND ENVIRONMENTAL BENEFITS

Adopting low-cost sustainable practices offers significant economic and environmental benefits:

8.4.1 Cost Savings

Energy-efficient technologies and renewable energy sources reduce energy bills and operating costs. For example, businesses and households can save money by using LED lighting and solar panels (Situmeang et al. 2024; Allcott and Greenstone 2012), renewable energy sources offer long-term economic benefits by ensuring stable and predictable energy prices. In contrast, fossil fuels are susceptible to price volatility (Hirth 2013).

8.4.2 Job Creation

The renewable energy sector generates jobs in manufacturing, installation, and maintenance, thereby contributing to economic growth (Yang and Song 2023; Gillingham et al. 2009).

8.4.3 Environmental Protection

Lowering energy consumption and adopting renewable energy sources help reduce greenhouse gas emissions, thereby mitigating climate change and decreasing air pollution (Umar 2020). Furthermore, the deployment of renewable energy infrastructure reduces habitat destruction and land degradation associated with fossil fuel extraction and mining activities (Tong et al. 2019).

8.4.4 Resource Conservation

Practices like composting and rainwater harvesting conserve natural resources, enhance soil health, and promote sustainable water use (Okedu et al. 2022).

8.5 SUCCESSFUL CASE STUDIES

Case Study 8.1: Small-Scale Biogas Technology in Nepal

In Nepal, small-scale biogas technology has played a crucial role in providing clean cooking fuel and supporting the achievement of Sustainable Development Goal (SDG) 7.1, which aims to increase the population's access to modern cooking solutions. With over 60 years of successful biogas development, Nepal has installed more than 431,000 household biogas plants through collaboration among the government, international development agencies, and private companies. Despite these efforts, only a small fraction of the biogas potential has been utilized, with significant untapped capacity remaining. The study highlights that biogas production from livestock manure alone could produce about 3043.58 million cubic meters of biogas annually, potentially avoiding 4.35 million tons of CO_2 emissions each year. Additionally, biogas technology offers benefits beyond clean cooking, including the production of nutrient-rich biofertilizer, improved agricultural productivity, and enhanced rural livelihoods. While progress in biogas system technology has been slow, there is an urgent need to develop native technologies capable of processing varied feedstock and coping with fluctuating temperatures. The Nepal case study underscores the importance of robust policies and promotion strategies to fully harness biogas potential and align it with other SDGs such as health, gender equality, climate action, and reduced deforestation. This holistic approach can serve as a model for other low-income countries aiming to deploy local renewable energy technologies in synergy with broader development goals (Lohani et al. 2021).

Case Study 8.2: Advancing Second-Generation Biofuels in India

In India, public sector undertaking (PSU) companies, working with established private technology providers, have set up 12 second-generation ethanol biorefineries in 11 states. This project follows the 2009 National Policy on Biofuels, which aims to meet India's bioethanol and biodiesel targets that cannot be achieved with first-generation biofuels. Praj Industries Limited is a key technology partner in this initiative, providing advanced technology for several of the biorefineries. Their 'Enfinity' biorefineries use cellulosic feedstock, which is plant material not used for food, to produce ethanol. These biorefineries feature Praj's specially developed microorganisms and patented pre-treatment processes created at their research center, Praj Matrix.

These new biorefineries can be added to existing ethanol plants, sugar factories, or co-generation plants, making it easier to produce biofuels, biochemicals, and biofertilizers. This integration not only supports the bioeconomy by providing products like animal feed and healthy foods but also addresses the challenge of collecting and using agricultural residues. By converting these residues into valuable biomass, the project creates local jobs, increases farmers' incomes, and reduces greenhouse gas emissions and air pollution from burning residues.

Started in 2016, this government program involves PSU companies and Praj Industries, benefiting farmers and the wider community. It supports sustainable agriculture, bio-based chemicals, bioenergy, and waste management. The initiative aligns with several SDGs and promotes a circular economy, especially through efficient water use and recycling organic fertilizers in local farms. Success factors include effective use of biomass residues, creating markets for bioproducts, and strategic planning that considers local landscapes (Gomez San Juan et al. 2019).

Case Study 8.3: Agroforestry and Conservation in Indonesia

Kutai Timber, a forestry company in East Java, Indonesia, has moved from cutting down trees in natural forests to managing planted forests sustainably, combined with agroforestry. This change supports the national goal of achieving biomass sustainability certifications for export markets. In 2017, Kutai Timber set up a fully integrated mill that uses waste to create various bioproducts through a cascading approach. This means they turn primary wood processing waste into products like furniture parts, doors, and musical instruments, and secondary processing waste into items like pellets and particleboard. This process not only increases revenue but also cuts down transport costs.

The company uses clustering to involve local producers and reduce costs, which helps develop small and medium-sized businesses. Small-scale farmers are organized into cooperatives, encouraging them to plant fast-growing trees and food crops using eco-friendly practices. This approach boosts food security and supports environmental health.

Since 2008, this private sector initiative led by Kutai Timber has benefited cooperatives and related industries in agriculture, bio-based construction materials, pulp and paper, bioenergy, and waste management. The project focuses on making sustainable products to replace those sourced from fossil fuels or unsustainable methods. It aims to protect biodiversity, increase the value of biomass, revitalize rural areas, and support vulnerable communities.

Key factors for success include using all by-products of biomass, applying innovative practices, developing markets for bioproducts, integrating different sectors, collaborating with stakeholders, and ensuring sustainability certifications. This approach helps ensure that all relevant parties are involved in the shift toward a sustainable bio-economy (Gomez San Juan et al. 2019).

Case Study 8.4: Sustainable Resource Management and Energy Efficiency in the Nordic Sugar Industry

Danish organizations have effectively reduced their climate and environmental impact through the use of standards, particularly focusing on resource and energy efficiency. The Energy Management standard, EN/ISO 50001:2018, plays a crucial role in assisting companies to optimize energy use and reduce resource consumption, in alignment with the UN's SDG 7, which focuses on affordable and clean energy. A study of nearly 600 Danish companies revealed that standards not only reduce environmental impact but also enhance the use of climate-friendly materials and the dissemination of green solutions. About 45% of these companies utilize standards to implement green solutions, which builds customer trust and simplifies communication about their sustainability efforts. Notably, Nordic Sugar, a major sugar producer, has significantly improved its energy efficiency, halving its energy usage since 1990 by integrating the standard into its operations. This includes utilizing water from sugar beets in production and repurposing by-products. The ISO 50001 standard ensures top management involvement and systematic energy management, leading to approximately 10% energy savings in the first year of implementation. The standard's comprehensive approach not only aids in meeting political climate goals but also supports companies in demonstrating their commitment to sustainability and improving their operational efficiency (UNECE 2024).

Case Study 8.5: Sustainable Waste Management Practices in Sathaye College, India

Over the span of seven years, the Microbiology Department at Sathaye College in Vile Parle has undertaken a commendable endeavor: managing their biodegradable waste. Led by Prof. Kudva, the department embarked on an experimental journey to establish a vermi-composting system for the campus's leaf litter. Initially armed with scant information, Prof. Kudva and the team ventured into a realm of trial and error, ingeniously crafting a 6×3 pit using reclaimed materials, with a makeshift cover fashioned from old windows to aerate the composting process. By purchasing worms from a nearby nursery for a reasonable sum of INR 300, they initiated the composting process by mixing horticultural waste with vegetable peels from the canteen. The construction of the pit incurred an additional expense of INR 30,000. The resulting compost now serves as invaluable nourishment for the college's greenery. Concurrently, the challenge of paper waste, particularly from the storage of examination answer sheets for up to three years, prompted collaboration with Recyclekaro.com, an initiative spearheaded by alumnus Laxmikanth Deshpande. Recyclekaro.com handles the collection of paper and e-waste, while local waste-pickers manage high-value recyclables such as bottles, plastics, and cardboard. Science labs further contribute to the recycling effort by entrusting their equipment to local vendors. Additionally, the college actively promotes sustainability by discouraging the use of bottled water, instead opting for water filters installed on every floor. Principal Smt. Kavita Rege champions a culture of environmental stewardship, emphasizing the pivotal role of education in fostering eco-consciousness among students, thus ensuring the enduring success of the college's green initiatives (Iyer 2016).

8.6 CHALLENGES IN MITIGATING CLIMATE CHANGE

8.6.1 Informational and Awareness Barriers

The implementation of sustainable practices is significantly hindered by critical informational gaps and limited awareness. Developing countries, particularly in rural regions, frequently struggle with a deficiency of precise data on natural resources and a shortage of skilled human resources (Edenhofer et al. 2011). There is a notable lack of awareness about renewable energy among the public and institutions (Edenhofer et al. 2011). Environmental awareness is crucial for addressing ecological problems and encouraging individual sustainability actions (Earth5R 2024). Low literacy levels and limited knowledge about agroforestry significantly reduce technology adoption rates, with farmers having higher education levels being better equipped to understand and implement these technologies (Gello 2021).

8.6.2 Sociocultural Barriers

Societal and individual values and norms can significantly influence the perception and acceptance of renewable energy, with these perspectives often evolving slowly (Edenhofer et al. 2011). Secure land rights are crucial for farmers to invest in agroforestry, but gender disparities, particularly under patrilineal land tenure systems, create substantial barriers. Interestingly, research indicates that households led by women are more inclined to adopt agroforestry practices compared to those led by men (Gello 2021).

8.6.3 Institutional and Policy Barriers

Existing industry structures and energy market regulations are typically designed around monopolistic providers, making it difficult for renewable energy technologies to penetrate the market (Edenhofer et al. 2011). Technical regulations and standards are predominantly based on assumptions of large, centralized energy systems (Edenhofer et al. 2011). Effective policies and financing mechanisms are essential to overcome these barriers and promote renewable energy technologies (Edenhofer et al. 2011). In many regions, the lack of extension workers and technical support means farmers remain unaware of agroforestry benefits and lack the necessary knowledge for effective implementation (Gello 2021). The global economy's dependence on fossil fuels and the difficulty of eliminating fossil fuel subsidies further hinder the transition to sustainable energy sources (UNDP 2023).

8.6.4 Economic Barriers

Major challenges include high initial investment costs and financial risks, particularly for technologies not yet fully developed (Edenhofer et al. 2011). Market failures, such as unpriced environmental impacts and monopolistic powers in energy markets, hinder renewable energy deployment (Edenhofer et al. 2011). Waste management, especially for biomedical waste, poses significant financial and logistical challenges (Earth5R 2024). In agricultural contexts, high costs associated with acquiring seeds, seedlings, and nursery equipment prevent low-income farmers from adopting agroforestry, with wealthier farmers more likely to implement these practices due to better resource access (Gello 2021). The increasing demand for critical minerals essential to decarbonization presents additional challenges, with lengthy mining project timelines and supply chain vulnerabilities potentially slowing decarbonization efforts (UNDP 2023).

8.6.5 Resource and Technical Issues

Renewable resources like wind and solar may be too small or variable in certain locations, with their low energy density potentially insufficient for large-scale industrial facilities (Edenhofer et al. 2011). In agricultural contexts, inconsistent or unavailable sources of quality seeds and seedlings significantly hinder agroforestry adoption (Gello 2021). The lack of essential nursery materials and equipment further poses a substantial barrier to establishing and maintaining agroforestry practices (Gello 2021).

8.6.6 Environmental and Health Impacts

Traditional biomass use leads to significant health impacts due to high pollutant concentrations (Edenhofer et al. 2011). Cities like Indore face challenges with untreated sewage and biomedical waste, resulting in pollution and health risks (Earth5R 2024). The rush to invest in the mining sector to meet critical mineral demand risks exacerbating environmental degradation, economic and governance issues, and social inequalities (UNDP 2023).

8.6.7 Water Resource Impacts

The availability of cooling water for thermal power plants can be limited, affecting their efficiency (Edenhofer et al. 2011). Indore experiences significant water scarcity due to over-reliance on river water and insufficient groundwater recharge (Earth5R 2024).

8.6.8 Community and Infrastructure Challenges

Rapid urbanization in Indore has led to deforestation and loss of green cover, affecting the local ecosystem (Earth5R 2024). Promoting renewable energy sources like solar and implementing energy-efficient technologies is critical (Earth5R 2024). Agriculture, recognized as the leading cause of global deforestation, requires transformation in food systems. Shifting to regenerative agricultural practices could ensure a sustainable food supply and significantly reduce greenhouse gas emissions (UNDP 2023).

8.6.9 Physical Factors

Topographical challenges, including steep slopes and river course changes, affect land availability and suitability for agroforestry. Extreme weather conditions such as droughts, floods, and high temperatures discourage farmers from planting trees (Gello 2021).

These challenges highlight the complexity of transitioning to renewable energy and managing environmental sustainability. Addressing these barriers requires comprehensive strategies and coordinated efforts across various sectors (Gello 2021; World Bank 2024).

8.7 GOVERNMENT POLICIES AND REGULATIONS' ROLE IN PROMOTING LOW-COST SUSTAINABLE PRACTICES

Government policies and regulations play a pivotal role in promoting low-cost sustainable practices by providing structured action plans, incentivizing market behavior, boosting innovation through increased funding, and setting a precedent for other sectors to follow. A sustainable society requires a strong partnership between the government, businesses, and individuals. While government policies can encourage sustainable practices, it is ultimately up to everyone to make a difference. Here are the key ways in which these policies can drive the green transition:

8.7.1 Detailed Action Plans with Clear Accountability

Governments need to develop detailed, sector-specific road maps in collaboration with industries to bridge the gap between long-term environmental pledges and short-term actions. These plans should specify policy measures, desired outcomes, timelines, and necessary resources, thereby making it clear how each sector will contribute to achieving sustainability goals (Atalla et al. 2022; Rovers 2003). Ambitious climate policies are crucial for reducing greenhouse gas emissions and promoting sustainable practices.

8.7.2 Incentivizing Market and Mandating Change

Bold policies such as green taxes on harmful environmental activities, subsidies for green investments, and stricter regulations can significantly drive the transition to a green economy. For instance, Germany's €2.5 billion investment in EV infrastructure and subsidies per vehicle demonstrate how financial incentives can accelerate the adoption of sustainable technologies (Atalla et al. 2022; Rovers 2003). Additionally, successful policies like the Netherlands' recycling and reuse initiatives, achieved through a mix of regulations, taxes, and bans on dumping, showcase the effectiveness of integrated policies (Rovers 2003). Governments can provide tax incentives for businesses that adopt sustainable practices, such as tax credits for renewable energy investments, energy-efficient buildings, and sustainable transportation.

8.7.3 Boosting Innovation through Increased Funding

Governments must increase funding for research and development to stimulate innovation in sustainable technologies. For example, the Pakistani government's significant investment in afforestation programs not only captures carbon but also creates job opportunities, illustrating a multifaceted approach to sustainability (Atalla et al. 2022; Rovers 2003). Government funding for research into renewable energy technologies such as solar and wind power can create new industries and jobs while reducing environmental impact.

8.7.4 Creating Conditions for Private Sector Investment

Governments must foster environments conducive to private investment in areas where financial organizations are hesitant. This involves developing structured financial solutions specific to various sectors and implementing policies to mitigate risks. Public-private partnerships, such as the Breakthrough Energy Catalyst, demonstrate how collaboration can drive the development of climate technologies (Rovers 2003).

8.7.5 Enhancing the Design and Execution of Green Initiatives

Governments need to carefully plan and implement green initiatives to minimize the risk of failure. This involves having a clear vision, setting realistic timelines, securing adequate funding, and establishing a supportive regulatory framework. Adopting an integrated approach across government and sectors helps in understanding the complex factors that can influence the outcome of programs and devising strategies to address potential issues (Rovers 2003). Ongoing monitoring and evaluation of climate policies are crucial to ensure their effectiveness and to make necessary adjustments.

8.7.6 Acting as Role Models

By embedding sustainability into their core functions and policies, governments can set a standard for other sectors to follow. This includes creating green infrastructure, implementing energy-efficient practices, and fostering a culture of sustainability across all government activities (Rovers 2003).

8.7.7 Promoting a Whole-of-Society, People-Centered Approach

Engaging all sectors of society, including local and regional governments, businesses, and citizens, is crucial. For instance, the Green City Action Plan (GCAP) in Yerevan, Armenia, involved extensive stakeholder consultation and collaboration, leading to the successful adoption of a comprehensive environmental plan. Engaging the public through participatory forms, like citizens' assemblies, can generate ambitious climate policies and encourage sustainable behavior. Educational and awareness campaigns can help people understand the environmental impact of their actions and promote sustainable practices.

8.7.8 Ensuring a Just Transition

A fair distribution of costs and benefits is critical for a just transition. This involves targeted government support, subsidies, and interventions to avoid negative impacts on vulnerable populations and ensure new green employment opportunities in affected regions. Successful examples, such as the Humber region's transformation into an offshore wind hub, demonstrate how targeted support can revitalize economies and create new jobs.

By addressing these areas, governments can effectively overcome economic, political, regulatory, and financial obstacles, setting the global stage for a sustainable future. The commitment to these priorities will ensure that environmental, economic, and social outcomes are balanced, driving the green transition forward.

8.8 CONCLUSION

In this chapter, we explored the vital role of low-cost sustainable practices in mitigating climate change. Through case studies from Nepal, India, Indonesia, the Nordic region, and other contexts, we demonstrated the effectiveness of practical initiatives such as composting, rainwater harvesting, energy-efficient technologies, agroecology, and waste management. These methods address environmental challenges while providing significant economic benefits, including cost savings, job creation, and resource conservation.

Despite the challenges of implementation—including barriers related to awareness, sociocultural factors, institutional limitations, and economic constraints—the potential for sustainable practices remains promising. Government policies and regulations are crucial in this transition, requiring comprehensive action plans, market incentives, increased innovation funding, and supportive conditions for private sector investment. By fostering collaboration between governments, businesses, and individuals, we can create a path toward a sustainable and resilient future that effectively addresses climate change challenges.

REFERENCES

Abolhosseini, S., A. Heshmati, and J. Altmann. 2014. "A Review of Renewable Energy Supply and Energy Efficiency Technologies." IZA Discussion Papers, No. 8145, Institute for the Study of Labor (IZA), Bonn.

Acchar, W., E. J. V. Dultra, and A. M. Segadães. 2013. "Untreated Coffee Husk Ashes Used as Flux in Ceramic Tiles." *Applied Clay Science* 75: 141–147.

Alkahtani, M., A. Ziout, B. Salah, M. Alatefi, A. E. E. Abd Elgawad, A. Badwelan, and U. Syarif. 2021. "An Insight into Reverse Logistics with a Focus on Collection Systems." *Sustainability* 13: 548.

Allcott, H., and M. Greenstone. 2012. "Is There an Energy Efficiency Gap?" *Journal of Economic Perspectives* 26: 3–28.

Al-Rumaihi, A., G. McKay, H. R. Mackey, and T. Al-Ansari. 2020. "Environmental Impact Assessment of Food Waste Management Using Two Composting Techniques." *Sustainability* 12: 1595.

Altieri, M. A. 2018. *Agroecology: The Science of Sustainable Agriculture*. CRC Press.

Arora, G. 2021. "A Review on Rain Water Harvesting Techniques." *ACADEMICIA: An International Multidisciplinary Research Journal* 11: 1343–1349.

Atalla, G., M. Mills, and J. McQueen. 2022. "Six Ways That Governments Can Drive the Green Transition." Ernst & Young Global Limited 13.

Conti, J., P. Holtberg, J. Diefenderfer, A. LaRose, J. T. Turnure, and L. Westfall. 2016. *International Energy Outlook 2016 with Projections to* 2040. DOE/EIA-0484 (2016). USDOE Energy Information Administration (EIA), Washington, DC (United States). Office of Energy Analysis.

Didanna, H. L. 2014. "A Critical Review on Feed Value of Coffee Waste for Livestock Feeding." *World Journal of Biology and Biological Sciences* 2: 72–86.

Earth5R. 2024. "Sustainable Waste Management in Indore: A Case Study." https://earth5r.org/sustainable-waste-management-in-indore-a-case-study/

Edenhofer, O., R. P. Madruga, Y. Sokona, K. Seyboth, P. Matschoss, S. Kadner, T. Zwickel, et al. 2011. *Renewable Energy Sources and Climate Change Mitigation: Special Report of the Intergovernmental Panel on Climate Change*. Cambridge University Press. doi:10.1017/CBO9781139151153

Farreny, R., T. Morales-Pinzón, A. Guisasola, C. Tayà, J. Rieradevall, and X. Gabarrell. 2011. "Roof Selection for Rainwater Harvesting: Quantity and Quality Assessments in Spain." *Water Research* 45: 3245–3254.

Francis, C., G. Lieblein, S. Gliessman, T. A. Breland, N. Creamer, R. Harwood, and R. Poincelot. 2003. "Agroecology: The Ecology of Food Systems." *Journal of Sustainable Agriculture* 22: 99–118.

Gello, J. 2021. "Adoption of Agroforestry Technology in Malawi: Hindrances Experienced by Farmers in Nsanje District." GRÓ Land Restoration Training Programme Final project 2021, Árleynir 22, 112 Reykjavik, Iceland.

Ghisi, E., D. da Fonseca Tavares, and V. L. Rocha. 2009. "Rainwater Harvesting in Petrol Stations in Brasília: Potential for Potable Water Savings and Investment Feasibility Analysis." *Resources, Conservation and Recycling* 54: 79–85.

Gillingham, K., R. G. Newell, and K. Palmer. 2009. "Energy Efficiency Economics and Policy." *Annual Review of Resource Economics* 1: 597–620.

Giusti, L. 2009. "A Review of Waste Management Practices and Their Impact on Human Health." *Waste Management* 29: 2227–2239.

Gliessman, S. 2018. "Defining Agroecology." *Agroecology and Sustainable Food Systems* 42: 599–600.

Gomez San Juan, M., A. Bogdanski, and O. Dubois. 2019. "Towards Sustainable Bioeconomy-Lessons Learned from Case Studies." Rome, FAO. 132 pp. Licence: CC BY-NC-SA 3.0 IGO.

Hasan, A. 2024. *Identifying the Morphological Characteristics and the Seed Proteins' Properties of Vicia L. Species and Studying the Effect of Coffee Fertilizer on Them.* Tishreen University, Syrian Arab Republic.

Hirth, L. 2013. "The Market Value of Variable Renewables: The Effect of Solar Wind Power Variability on Their Relative Price." *Energy Economics* 38: 218–236.

Hoseini, M., S. Cocco, C. Casucci, V. Cardelli, and G. Corti. 2021. "Coffee By-Products Derived Resources. A Review." *Biomass and Bioenergy* 148: 106009.

Iyer, H. 2016. "Case Study of Mumbai: Decentralised Solid Waste Management." *Procedia Environmental Sciences* 35: 101–109. doi:10.1016/j.proenv.2016.07.054

Lohani, S. P., B. Dhungana, H. Horn, and D. Khatiwada. 2021. "Small-Scale Biogas Technology and Clean Cooking Fuel: Assessing the Potential and Links with SDGs in Low-Income Countries – A Case Study of Nepal." *Sustainable Energy Technologies and Assessments* 46: 101301. doi:10.1016/j.seta.2021.101301.

Lu, P., G. Xia, Q. Zhao, R. Xu, S. Li, and Y. Guo. 2020. "Temporal Trends of the Association between Ambient Temperature and Hospitalisations for Cardiovascular Diseases in Queensland, Australia from 1995 to 2016: A Time-Stratified Case-Crossover Study." PLOS Medicine 17 (7). doi:10.1371/journal.pmed.1003176

Nyangito, H. 2000. *Delivery of Services to Smallholder Coffee Farmers and Impacts on Production under Liberalization in Kenya.* Kenya Institute for Public Policy Research and Analysis, Nairobi, Kenya.

Okedu, K. E., H. F. Barghash, and H. A. Al Nadabi. 2022. "Sustainable Waste Management Strategies for Effective Energy Utilization in Oman: A Review." *Frontiers in Bioengineering and Biotechnology* 10: 825728.

Rovers, R. 2003. "The Role of Policies in Promoting Sustainable Practices." *Environmental Science & Policy* 26: 29–32.

Saenger, M., E. U. Hartge, J. Werther, T. Ogada, and Z. Siagi. 2001. "Combustion of Coffee Husks." *Renewable Energy* 23: 103–121.

Singh, R., A. Gehlot, and D. Sharma, eds. 2022. *Futuristic Sustainable Energy and Technology.* CRC Press, London.

Situmeang, N., S. Y. Putri, S. Q. Desbrantoro, Y. S. M. Gultom, and S. A. Putri. 2024. "The International Energy Agency's Role in Supporting the Sustainability of Renewable Energy Supply in Indonesia." *SOSHUM: JurnalSosial Dan Humaniora* 14: 1–9.

Soares, M., P. Christen, A. Pandey, and C. R. Soccol. 2000. "Fruity Flavour Production by Ceratocystis Fimbriata Grown on Coffee Husk in Solid-State Fermentation." *Process Biochemistry* 35: 857–861.

Somashekar, K. L., and K. A. Anu Appaiah. 2013. "Coffee Cherry Husk–a Potential Feed Stock for Alcohol Production." *International Journal of Environment and Waste Management* 11: 410–419.

Thomas, T. 1998. "Domestic Water Supply Using Rainwater Harvesting." *Building Research & Information* 26: 94–101.

Tong, D., Q. Zhang, Y. Zheng, K. Caldeira, C. Shearer, C. Hong, and S. J. Davis. 2019. "Committed Emissions from Existing Energy Infrastructure Jeopardize 1.5 C Climate Target." *Nature* 572: 373–377.

Umar, T. 2020. "Frameworks for Reducing Greenhouse Gas (GHG) Emissions from Municipal Solid Waste in Oman." *Management of Environmental Quality: An International Journal* 31: 945–960.

UNECE. 2024. *SDG 7 Denmark.*

UNEP. 2008. "*An Overview of the State of the World's Fresh and Marine Waters.* World Bank.

United Nations Development Programme. 2023. *What is Climate Change Mitigation and Why Is It Urgent?* https://climatepromise.undp.org/news-and-stories/what-climate-change-mitigation-and-why-it-urgent

World Bank. (2024, April 9). *Climate Change Overview*. https://www.worldbank.org/en/topic/climatechange/overview3

Wang, J., Z. Hu, X. Xu, X. Jiang, B. Zheng, X. Liu, and P. Kardol. 2014. "Emissions of Ammonia and Greenhouse Gases during Combined Pre-Composting and Vermicomposting of Duck Manure." *Waste Management* 34: 1546–1552.

Wang, Q., Z. Wang, M. K. Awasthi, Y. Jiang, R. Li, X. Ren, and Z. Zhang. 2016. "Evaluation of Medical Stone Amendment for the Reduction of Nitrogen Loss and Bioavailability of Heavy Metals during Pig Manure Composting." *Bioresource Technology* 220: 297–304.

Yang, C., and X. Song. 2023. "Assessing the Determinants of Renewable Energy and Energy Efficiency on Technological Innovation: Role of Human Capital Development and Investment." *Environmental Science & Pollution Research* 30: 23301.

9 Climate Change and Human Health

Madhuri Walvekar, Abhinandan Patil, and S. H. Pawar

9.1 AN INTRODUCTION TO HUMAN HEALTH AND CLIMATE CHANGE

The long-term modification of temperature, precipitation patterns, and other climatic variables at the Earth's surface is referred to as climate change. The combustion of fossil fuels, deforestation, and industrial processes, which emit greenhouse gases into the atmosphere, are the primary causes of these changes. Consequently, there are substantial transformations occurring within the Earth's climate system, which have far-reaching consequences for weather patterns, ecosystems, and human well-being. The ramifications of climate change on human health are extensive and diverse. Elevated temperatures play a role in augmenting the occurrence and severity of severe weather phenomena, including but not limited to heat waves, hurricanes, and droughts. These events can result in health repercussions that are both immediate and long-lasting. As an illustration, heat waves present a substantial hazard to human well-being, as they give rise to heat-related ailments including heat exhaustion and heatstroke. Vulnerable groups, including the elderly, children, and individuals with pre-existing health conditions, are especially susceptible to these complications (Zhao et al. 2019; Hetherington et al. 2021).

Climate change impacts air and water quality, agricultural productivity, food security, and the prevalence of infectious diseases, in addition to extreme weather events. Variations in temperature and precipitation patterns have the potential to disrupt the distribution of vector-borne illnesses, including malaria, dengue fever, and Lyme disease, by influencing the range of mosquitoes and insects that transmit these diseases. Moreover, habitat and ecosystem disturbances can heighten the likelihood of contracting zoonotic diseases, which are transmitted from animals to humans. Poverty, inadequate infrastructure, and restricted access to healthcare services are a few of the social, economic, and environmental factors that exacerbate the adverse effects of climate change on human health. Vulnerable populations, including informal settlements, indigenous communities, and marginalized groups, experience a disproportionate impact on health hazards associated with climate change as a result of systemic inequities and discrepancies in the distribution of resources. In brief, climate change constitutes a paramount concern of our era, carrying significant ramifications for the health and welfare of humanity. It is imperative to comprehend the intricate dynamics that exist between climate change and human health in order to

DOI: 10.1201/9781003492733-9

formulate efficacious approaches for risk mitigation, resilience building, and safeguarding the well-being of present and future generations (Doocy et al. 2013).

9.2 AN EXAMINATION OF THE INTERRELATIONSHIPS BETWEEN CLIMATE CHANGE AND A RANGE OF HEALTH OUTCOMES

The intricate and multifaceted relationship between climate change and diverse health outcomes comprises an extensive array of biological, social, and environmental factors. Comprehension of this interdependence is fundamental in order to confront the health ramifications of climate change and execute efficacious measures for adaptation and mitigation. An exposition of the ways in which climate change impacts a variety of health outcomes follows (Konkel 2020; Thompson et al. 2014; Matthews and Zickfeld 2012; Chen et al. 2021).

9.2.1. Climate change is a significant factor in the escalation of the occurrence and severity of extreme weather phenomena, including but not limited to wildfires, heat waves, hurricanes, and flooding. These incidents have the potential to cause bodily harm, fatalities, and population displacement. In addition, they compromise access to pure water and healthcare, damage infrastructure, and disrupt essential services, all of which result in a variety of adverse health effects.

9.2.2. Increased temperatures, which are a consequence of climate change, intensify the vulnerability to heat-related ailments such as dehydration, heat exhaustion, and heatstroke. Vulnerable populations, including but not limited to the elderly, infants, outdoor laborers, and those with pre-existing health conditions, are especially at risk of experiencing adverse health effects associated with heat.

9.2.3. **Air character**: The character of the atmosphere is impacted by climate change via its influences on atmospheric circulation, temperature, and precipitation patterns. Elevated temperatures have the potential to generate ground-level ozone and particulate matter, both of which are known to exacerbate respiratory and cardiovascular ailments. Wildfires, which are exacerbated by elevated temperatures and arid conditions, emit air pollutants that further deteriorate respiratory health and air quality.

9.2.4. Vector-borne diseases are influenced by alterations in temperature, precipitation, and abundance, which in turn impact the behavior, distribution, and abundance of vectors including mosquitoes, fleas, and flies. Consequently, the transmission dynamics of vector-borne diseases such as dengue fever, malaria, West Nile virus, Zika virus, and Lyme disease are altered. Vector reproduction is facilitated by elevated temperatures and modified precipitation patterns, which also extend the transmission seasons and broaden the geographic distribution of vector-borne diseases.

9.2.5. **Food security and nutrition**: The adverse impacts of climate change on food production systems encompass nutritional quality, crop yields, and availability. Traverses in growing seasons, water scarcity, extreme weather phenomena, and soil degradation have the potential to induce crop failures, food shortages, and price surges. Food security and nutrition are compromised as a result of these disruptions, which worsen malnutrition, infectious illnesses, and micronutrient deficiencies.

9.2.6. **Waterborne diseases**: The quality and accessibility of freshwater resources are impacted by alterations in precipitation patterns and hydrological cycles, which in turn give rise to waterborne diseases including cholera, typhoid fever, and diarrheal maladies. Particularly in vulnerable communities with limited access to clean water and sanitation facilities, the risk of waterborne disease epidemics is heightened by flooding, water source contamination, and inadequate sanitation infrastructure.

In general, the interdependence of health outcomes and climate change highlights the critical nature of implementing coordinated and comprehensive measures to reduce greenhouse gas emissions, fortify health systems, and increase resilience to climate-related risks. Protection of human health and well-being amidst persistent environmental challenges can be achieved through the implementation of sustainable development practices and the mitigation of climate change's underlying causes.

9.3 CONSEQUENCES OF OTHER CLIMATIC EXTREMITIES AND HEAT WAVES

As a result of climate change, heat waves and other climatic extremes are becoming more frequent and severe, posing grave threats to human health and well-being. Gaining insight into the ramifications of these incidents is vital in order to formulate efficacious approaches to alleviate their consequences and safeguard susceptible demographics. The consequences of heat waves and other climatic extremes are as follows (Chen et al. 2020; Hirabayashi et al. 2013; Hartmann et al. 2013).

9.3.1 Health Consequences

Heat waves have the potential to induce various heat-related ailments, such as dehydration, heat cramping, heat exhaustion, and heatstroke. These conditions have the potential to cause severe health complications, organ impairment, and mortality, especially among susceptible demographics including the elderly, infants, outdoor laborers, and those with pre-existing medical conditions.

Cardiovascular and respiratory complications may be further intensified during heat waves due to the adverse air quality and elevated temperatures, which include chronic obstructive pulmonary disease (COPD), heart attacks, and asthma. Wildfire-caused air pollution, which is exacerbated by arid, high temperatures, adds to the burden of respiratory health issues.

Mental health consequences may arise as a result of heat waves and extended durations of extreme heat, encompassing symptoms such as melancholy, anxiety, and stress. The occurrence of heat waves can lead to mental health issues becoming more severe and an elevated susceptibility to unfavorable consequences, as well as disruptions to daily routines and social isolation.

9.3.2 Vulnerable Demographics

The elderly and small children are especially susceptible to the consequences of heat waves as a result of physiological alterations associated with aging, reduced mobility, and restricted availability of cooling resources. They are more susceptible to heat-related maladies and, during periods of extreme heat, may necessitate supplementary support and assistance.

Outdoor laborers, including those engaged in construction, agriculture, and landscaping, face an elevated susceptibility to heat-related ailments as a result of extended periods of exposure to elevated temperatures and strenuous physical activity. Occupational safety protocols, which comprise provisions for shade, hydration, and rest periods, are critical in safeguarding outdoor laborers against health hazards associated with high temperatures.

Urban regions endure elevated temperatures in comparison to their rural environs as a consequence of the heat island effect, which is caused by the accumulation of heat-absorbing and heat-retaining structures, pavement, and infrastructure. Vulnerable urban populations, including marginalized groups and low-income communities, are disproportionately impacted by heat surges and may lack access to cooling resources such as air conditioning.

9.3.3 Social and Economic Effects

Economic losses can result from heat waves and other climatic extremes; these include livestock losses, agricultural failures, decreased labor productivity, and increased healthcare expenses. Constraints on food security and livelihoods, exacerbation of poverty, and strain on local economies may result from these losses.

Heat waves have the potential to cause significant disruption to social and cultural events, encompassing community gatherings, recreational activities, and outdoor gatherings. In addition to disrupting transportation, causing power failures, and closing schools, extreme heat events have the potential to disrupt daily routines and social interactions.

In conclusion, the effects of heat waves and other climatic extremes on human health, ecosystems, and society are extensive. A multifaceted approach is required to mitigate these effects, including initiatives to address the underlying causes of climate change, urban planning and design strategies, public health interventions, and heat preparedness and response measures. We can forge more resilient communities and reduce the health risks associated with heat waves by placing emphasis on safeguarding vulnerable populations, bolstering resilience to extreme heat events, and advocating for sustainable development practices.

9.4 AN ANALYSIS OF THE HEALTH CONSEQUENCES OF HEAT WAVES: HEAT-RELATED FATALITIES AND ILLNESSES

Prolonged episodes of abnormally high temperatures, referred to as heat waves, present substantial health hazards to populations across the globe. It is imperative to comprehend the health ramifications of heat waves, encompassing heat-related fatalities and maladies, in order to effectively prepare and respond to public health emergencies. The subsequent analysis delineates the principal health ramifications linked to heat waves (Collins et al. 2013; Marsooli et al. 2019; Guo et al. 2021).

9.4.1 Heat-Related Disorders

Heat exhaustion is a prevalent heat-related ailment distinguished by symptoms including profuse perspiration, weakness, vertigo, nausea, and headache. The condition arises when the body's perspiration fails to adequately chill it, resulting in dehydration and disturbances in electrolyte levels.

Heatstroke is an extremely critical medical condition that arises when the central body temperature surges to hazardously high levels, usually surpassing 104°F or 40°C. Confusion, rapid pulse, dry, hot skin, unconsciousness, and seizures are among the symptoms. Heatstroke necessitates urgent medical intervention and, if untreated, may lead to fatal consequences such as organ impairment and coma.

Dehydration can result from prolonged exposure to high temperatures and insufficient fluid consumption; it is characterized by increased fluid loss beyond the body's intake. Symptoms of dehydration include fatigue, light-headedness, dehydration-induced parched mouth, dark urine, and tiredness. Critical dehydration poses a significant risk to life, especially among susceptible demographics like the elderly and neonates.

9.4.2 Vulnerable Demographics

The elderly constitute a particularly susceptible demographic during heat waves owing to physiological alterations associated with aging, persistent health conditions, and social isolation. These individuals may experience compromised thirst perception, impaired body temperature regulation, and restricted availability of cooling resources, all of which contribute to an elevated susceptibility to heat-related ailments and fatalities.

Young children and infants are particularly vulnerable to the consequences of heat waves owing to their diminutive stature, elevated metabolic rate, and restricted thermoregulatory capacity. They may be more susceptible to dehydration, heat exhaustion, and heatstroke due to a decreased propensity to recognize and respond to indicators of heat stress.

Individuals with chronic health conditions: During heat waves, the risk of heat-related complications is elevated for those with pre-existing medical conditions including cardiovascular disease, respiratory disorders, diabetes, and mental health disorders. Medications that induce thermoregulation disturbances or heighten susceptibility to dehydration may amplify an individual's susceptibility.

9.4.3 The Fatality Rate

Heat waves have the potential to result in excessive mortality, as fatalities can be attributed both directly and indirectly to heat-related illnesses and the worsening of pre-existing health conditions. Those who have pre-existing health conditions and the elderly are especially vulnerable to heat-related fatalities. Additionally, elevated nocturnal temperatures, referred to as "tropical nights," may contribute to fatalities caused by heat by impeding the body's ability to cool and recover during slumber.

9.4.4 Exacerbation of Chronic Conditions due to Heat

Heat waves have the potential to worsen respiratory conditions, including allergies, asthma, and COPD. Particularly predisposed individuals are more susceptible to respiratory distress, bronchoconstriction, and airway inflammation when exposed to high temperatures and poor air quality.

9.4.5 Cardiovascular Conditions

Cardiovascular conditions, including but not limited to hypertension, coronary artery disease, and heart failure, elevate the susceptibility of affected individuals to deleterious cardiovascular events that may occur during heat waves. Elevated temperatures have the potential to induce alterations in blood pressure, heart rate, and blood viscosity, which may culminate in exacerbations of heart failure, arrhythmias, angina, and myocardial infarction.

9.4.6 Thermal-Related Despairs

Heat waves can lead to extended periods of sun exposure, which can manifest as sunburn, a condition characterized by symptoms including skin redness, discomfort, and blistering. When perspiration ducts become obstructed, heat rash, also known as prickly heat, develops as small red bumps or blisters on the skin.

Enhanced risk of burns and scalds may result from increased outdoor activities and the utilization of outdoor culinary equipment during heat waves. Boiling liquids, exposed flames, and hot surfaces all present potential risks of inadvertent burns and injuries, especially when considering the elderly and children.

9.4.7 Consequences for Mental Health

Heat waves have the potential to elicit psychological tension and anxiety as a result of distress, disruptions in sleep patterns, and apprehensions regarding one's well-being and security. Extreme heat events have the potential to provoke mental health issues, including anxiety disorders and melancholy, by causing disruptions to daily routines, financial burdens, and social isolation.

Based on existing research, there is evidence to suggest that elevated temperatures might be correlated with heightened levels of aggression and violence. This could potentially result in interpersonal disputes, incidents of domestic violence, and disturbance within communities during heat waves.

9.4.8 Prolonged Health Consequences

Cumulative Effects: Prolonged exposure to elevated temperatures and frequent heat waves may result in enduring health repercussions for individuals. Extended exposure to heat has the potential to hasten the aging process, heighten the vulnerability to chronic diseases, hinder cognitive function and neurodevelopment, and exacerbate these effects in particularly susceptible populations.

9.4.9 Variations in Health Results

Socioeconomic factors, encompassing disparities in income, education, and healthcare accessibility, exert an impact on susceptibility to health hazards associated with heat. Health inequities are frequently exacerbated for low-income communities, racial and ethnic minorities, and marginalized populations due to their disproportionate exposure to extreme weather and limited access to cooling resources. Urban heat islands are a phenomenon whereby temperatures in urban areas are relatively elevated in comparison to the rural regions that surround them. Urban heat islands may exacerbate heat-related health hazards for vulnerable populations as a result of insufficient housing, restricted green space, and the absence of air conditioning, among other contributing factors.

9.5 THE IMPLICATIONS FOR PUBLIC HEALTH OF THE INCREASED FREQUENCY AND INTENSITY OF CLIMATIC EXTREMES: A DISCUSSION

The frequency and severity of climatic extremes, such as heat waves, hurricanes, droughts, flooding, and wildfires, are increasing due to climate change. These severe meteorological phenomena present substantial hazards to the health and welfare of the general public, with wide-ranging consequences for communities, individuals, and healthcare infrastructure. The following is an analysis of how heightened climatic extremes have affected public health (Guo et al. 2021; Sun et al. 2021; Ebi and Loladze 2019; Tian et al. 2019).

9.5.1 Intensity and Frequency of Extreme Weather Occurrences

Scientific evidence indicates that climate change is causing an increase in the frequency and intensity of extreme weather phenomena on a global scale. The heightened frequency of climatic extremes can be attributed to factors such as rising global temperatures, alterations in atmospheric circulation patterns, and modifications in precipitation patterns.

As a consequence of climate change, heat waves, which are characterized by extended durations of excessively high temperatures, are increasing in both frequency and intensity. Extreme heat events have the potential to cause heat-related maladies, heatstroke, and death, especially among populations that are most susceptible.

Climate change has been found to be correlated with heightened hurricane and tropical storm intensity, which manifests as increased wind velocities, heavier precipitation, and more extensive storm surges. These occurrences have the potential to induce extensive devastation, population displacement, and healthcare service disruptions, thereby elevating the susceptibility to injuries, infections, and mental health issues. Drought conditions are significantly intensified in numerous regions as a result of climate change, giving rise to food insecurity, agricultural failures, and water scarcity. Public health consequences of drought encompass malnutrition, dehydration, the transmission of waterborne illnesses, and disputes arising from the depletion of water resources. Landslides and Floods: In mountainous and coastal areas, the risk of flooding and landslides is increased due to rising sea levels and precipitation. In addition to displacing populations, infrastructure damage, and contaminating potable water sources, floods increase the likelihood of contracting waterborne illnesses, injuries, and mental health conditions.

9.5.2 Effects on the Field of Public Health

Health risks are manifold in nature and are exacerbated by climatic extremes. These risks include heat-related ailments, injuries, infectious diseases, mental health disorders, and chronic conditions. Children, the elderly, low-income communities, and individuals with pre-existing health conditions are among the vulnerable populations that experience a disproportionate impact from extreme weather phenomena. Healthcare infrastructure is significantly impacted by extreme weather events, which can lead to the overburdening of emergency departments, hospitals, and other healthcare facilities. Affecting and following climatic extremes, disruptions to transportation networks, supply chains, and healthcare services may impede the availability of lifesaving treatments and medical care. Populations impacted by extreme weather events are exposed to supplementary health risks due to climate-induced environmental alterations, including but not limited to air pollution, allergen contamination, vector-borne illnesses, and toxic algal blooms. Vector breeding grounds, poor air quality, and contaminated water sources all contribute to the transmission of infectious diseases and respiratory maladies. In conclusion, the heightened occurrence and severity of climatic extremes as a result of climate change have significant ramifications for public health. As a result, immediate measures are required to alleviate the consequences and enhance community resilience. To ensure human health is not jeopardized by the intensifying climatic challenges of the future, it is vital to implement adaptation measures, improve early warning systems, enhance disaster preparedness and response efforts, and reduce greenhouse gas emissions.

9.6 MINIMIZING THE AVAILABILITY OF FOOD AND FRESHWATER

The effects of climate change on the availability of food and freshwater are profound, posing formidable threats to global food security and water resources. An outline of the ramifications that diminishing supplies of food and freshwater have on public health is as follows (Lu et al. 2020; Wu et al. 2021).

9.6.1 Agricultural Consequences

Climate change has the potential to modify temperature and precipitation patterns, which in turn have an impact on the quality, distribution, and yields of crops. Agricultural production can be significantly disrupted by variations in water availability and temperature patterns, which can result in crop failures, diminished crop diversity, and alterations to sowing schedules. The exacerbation of water stress in agricultural regions, specifically in arid and semi-arid areas, is a consequence of rising temperatures and fluctuations in precipitation. Water scarcity compromises agricultural productivity, reduces soil moisture, and restricts irrigation capacity, thereby endangering food production and livelihoods. Extreme weather phenomena, including cyclones, floods, droughts, and heatwaves, have a detrimental impact on crops, resulting in post-harvest losses and disruptions to the food supply. Vegetables, fruits, and staple grains are vulnerable commodities that are susceptible to climate-induced stresses, invasive species, pests, and diseases.

9.6.2 Food Security Difficulties

The reduction in the availability and affordability of nutritious food as a result of climate-related factors exacerbates food insecurity and malnutrition, especially among populations that are most susceptible to such conditions. Inadequate dietary diversity, micronutrient deficiencies, and undernutrition result from restricted access to nutritious foods; these conditions heighten the likelihood of developing stunted growth, maternal and infant mortality, and developmental delays.

Climate-induced disruptions to transportation networks, storage facilities, and market access hinder the efficient distribution and accessibility of food supplies. Disruptions in the supply chain caused by severe weather conditions worsen food scarcity, price instability, and outbreaks of infectious illnesses, thereby impacting both the availability and consumption of food.

9.6.3 Water Resource Difficulties

Climate change causes a reduction in snowpack, an alteration in precipitation patterns, and acceleration in the melting of glaciers, all of which contribute to a diminished availability of freshwater in numerous regions. The adverse effects of diminishing river flows, aquifer contraction, and saline intrusion into coastal aquifers intensify issues related to water scarcity and quality, thereby affecting the provision of potable water, sanitation, and hygiene.

Insufficient availability of sanitation facilities and safe drinking water contributes to an elevated susceptibility to waterborne illnesses, including but not limited to cholera, hepatitis A, and diarrheal illnesses. The transmission dynamics of waterborne pathogens are significantly impacted by climate-related alterations in precipitation patterns and water quality. This poses a particular health hazard to vulnerable populations, including children and communities that have restricted access to clean water and sanitation facilities.

9.6.4 Effects on Livelihoods and Means of Subsistence

One of the adverse consequences of climate change is the reduction in the availability of food and freshwater, which has significant implications for the economic stability and livelihoods of rural and agriculturally reliant communities. Reduced agricultural incomes, crop failures, and livestock losses all contribute to poverty, food insecurity, and rural-urban migration, thereby aggravating economic and social disparities.

One consequence of climate-induced disturbances in the agriculture, fisheries, and forestry industries is the deprivation of income opportunities and means of subsistence for millions of individuals across the globe. Rural communities, smallholder farmers, and fishermen who depend on natural resources are especially susceptible to climate-induced disruptions that result in unemployment, food insecurity, and social unrest, as evidenced by livestock mortality, agricultural failures, and diminishing fish stocks.

9.7 VECTOR-BORNE ILLNESS TYPES

A comprehensive analysis of vector-borne illnesses, including malaria, encephalitis, dengue fever, and yellow fever, shows that vector-borne diseases encompass a group of contagious illnesses that are transmitted to animals and humans via the bite of arthropod vectors, including but not limited to sandflies, mosquitoes, and ticks (Table 9.1). Vectors function as conveyances for pathogens, encompassing bacteria, viruses, protozoa, and helminths, all of which are capable of instigating an extensive array of diseases (Table 9.2). A classification of vector-borne diseases according to the pathogens they transmit and the categories of vectors involved is as follows (Hetherington et al. 2021; Wu et al. 2021; Kim et al. 2019; Khan et al., 2021; Zhang et al., 2019; Lee et al., 2020).

9.8 AN EXAMINATION OF THE RELATIONSHIP BETWEEN FLUCTUATIONS IN WATER QUALITY AND THE INCIDENCE OF WATERBORNE DISEASES

Examination of the Impact of Water Quality and Precipitation Pattern Variations: The correlation between climatic variables and human health is exponentially apparent when it comes to waterborne illnesses. The impact of climate change on precipitation patterns and water quality has a substantial effect on the incidence and transmission of these diseases. This segment explores the mechanisms by which the dynamics of waterborne diseases are influenced by these environmental changes. It also offers a statistical analysis of the observed correlations and trends (Khan et al. 2019; Zhang et al. 2019; Lee et al. 2014; Zhao et al. 2019; Fussell and Kelly 2021; Tong et al. 2017; Ma et al. 2021).

9.8.1 The Consequences of Precipitation Patterns

9.8.1.1 Increased Flooding and Rainfall

Through the process of flooding and concurrent excessive precipitation, water treatment facilities may become inundated, resulting in the introduction of pathogens into drinking water supplies.Consequences: The introduction of sewage, feces, and agricultural effluent into water bodies by floodwaters raises the potential for the transmission of pathogens including leptospirosis, cryptosporidiosis, and cholera. In endemic regions, for instance, the incidence of cholera outbreaks increases substantially after heavy rainfall events, according to studies.

TABLE 9.1
Vector Borne Disease Datasheet

Vector-Borne Disease	Causative Agent	Vector	Transmission	Symptoms	Prevention and Control	Medication
Malaria	*Plasmodium* spp.	*Anopheles* mosquitoes	Bite of infected mosquito	Fever, chills, headache, muscle aches	Insecticide-treated bed nets, insect repellents, indoor residual spraying	Antimalarial drugs (e.g., chloroquine, artemisinin-based combination therapies)
Dengue fever	Dengue virus	*Aedes* mosquitoes	Bite of infected mosquito	Fever, severe headache, joint and muscle pain	Mosquito control measures (e.g., eliminating breeding sites), protective clothing, mosquito repellents	Supportive care, pain relievers
Zika virus infection	Zika virus	*Aedes* mosquitoes	Bite of infected mosquito	Rash, fever, joint pain, conjunctivitis	Mosquito control measures, avoiding mosquito bites	Supportive care
Chikungunya fever	Chikungunya virus	*Aedes* mosquitoes	Bite of infected mosquito	Fever, joint pain, muscle pain, rash	Mosquito control measures, avoiding mosquito bites	Supportive care, pain relievers
Yellow fever	Yellow fever virus	*Aedes* or *Haemagogus* mosquitoes	Bite of infected mosquito	Fever, headache, muscle pain, jaundice	Vaccination, mosquito control measures	Supportive care
West Nile virus	West Nile virus	*Culex* mosquitoes	Bite of infected mosquito	Fever, headache, body aches, rash	Mosquito control measures, avoiding mosquito bites	Supportive care, antiviral drugs
Lyme disease	*Borrelia burgdorferi*	Black-legged ticks (Ixodes scapularis)	Bite of infected tick	Erythema migrans (bull's-eye rash), fatigue, fever	Avoiding tick habitats, wearing protective clothing, tick checks	Antibiotics (e.g., doxycycline, amoxicillin)
Japanese encephalitis	Japanese encephalitis virus	*Culex* mosquitoes	Bite of infected mosquito	Fever, headache, neck stiffness, seizures	Vaccination, mosquito control measures	Supportive care, antiviral drugs

TABLE 9.2
Waterborne Disease Datasheet

Waterborne Disease	Causative Agent	Transmission	Symptoms	Prevention and Control	Medication
Cholera	*Vibrio cholerae*	Ingestion of contaminated water or food, particularly seafood	Watery diarrhea, vomiting, dehydration, electrolyte imbalances	Improving water quality, sanitation, hygiene practices, vaccination	Oral rehydration salts (ORS), antibiotics (e.g., doxycycline)
Typhoid fever	*Salmonella enterica Typhi*	Ingestion of water or food contaminated with fecal matter containing the bacterium	Fever, headache, malaise, abdominal pain, constipation or diarrhea	Improvement in water quality, sanitation, hygiene practices, vaccination	Antibiotics (e.g., ciprofloxacin, azithromycin)
Hepatitis A	Hepatitis A virus (HAV)	Ingestion of contaminated water or food, particularly shellfish	Fever, fatigue, nausea, vomiting, abdominal pain, jaundice	Improvement in water quality, sanitation, hygiene practices, vaccination	Supportive care, vaccination
Giardiasis	*Giardia lamblia*	Ingestion of water or food contaminated with *Giardia* cysts	Diarrhea, abdominal cramps, bloating, flatulence, weight loss	Ensuring the safety of drinking water sources, proper hygiene practices	Antiprotozoal medications (e.g., metronidazole, tinidazole)
Dysentery	Various pathogens	Ingestion of contaminated water or food	Bloody diarrhea, abdominal cramps, fever, dehydration	Improving sanitation, hygiene practices, safe food handling	Antibiotics, supportive care

(Continued)

TABLE 9.2 (*Continued*)
Waterborne Disease Datasheet

Waterborne Disease	Causative Agent	Transmission	Symptoms	Prevention and Control	Medication
Cryptosporidiosis	*Cryptosporidium* spp.	Ingestion of water or food contaminated with *Cryptosporidium* oocysts	Watery diarrhea, abdominal cramps, nausea, vomiting	Proper water treatment, boiling water, avoiding untreated water sources	Nitazoxanide
Amoebiasis	Entamoeba histolytica	Ingestion of water or food contaminated with amoebic cysts	Diarrhea, abdominal pain, bloody stools, weight loss	Improving sanitation, hygiene practices, safe food handling	Antiprotozoal medications (e.g., metronidazole, tinidazole)
Legionnaires' disease	Legionella pneumophila	Inhalation of aerosolized water contaminated with *Legionella* bacteria	Fever, cough, shortness of breath, muscle aches, headache	Proper maintenance of water systems, disinfection treatments	Antibiotics (e.g., azithromycin, levofloxacin)
Leptospirosis	*Leptospira* spp.	Contact with water or soil contaminated with urine from infected animals	Fever, chills, headache, muscle aches, vomiting, jaundice	Avoiding contact with contaminated water, protective clothing	Antibiotics (e.g., doxycycline, penicillin)
Norovirus infection	Norovirus	Ingestion of contaminated food or water, person-to-person transmission	Nausea, vomiting, diarrhea, abdominal cramps, fever	Proper hand hygiene, food safety practices, disinfection measures	Supportive care, oral rehydration therapy

9.8.2 Drought Conditions

The mechanism is that extended periods of drought diminish the accessibility of potable water and concentrate impurities in finite water reservoirs. As a consequence, populations are compelled to depend on contaminated water sources, thereby augmenting their susceptibility to pathogens. There exists a correlation between drought conditions and elevated prevalence rates of diseases such as diarrheal ailments and hepatitis A. According to statistical analyses, regions afflicted with severe drought exhibit a 30% increased susceptibility to epidemics of waterborne diseases in comparison to periods devoid of drought.

9.9 ALTERATIONS TO WATER QUALITY

9.9.1 Microbial Contamination and Climate Change

The induced fluctuations in water temperature and pH have the potential to impact the viability and proliferation of microbial pathogens residing in aquatic environments. One potential consequence of elevated water temperatures is the promotion of bacterial proliferation, including that of *Vibrio cholerae* and *Legionella*. There exists a correlation between a 1°C rise in water temperature and a 5%–10% escalation in the growth rate of these pathogens, resulting in an elevated prevalence of diseases (Ko et al. 2016; Lloyd and Farquhar 2008).

9.9.2 Chemical Contaminants

Mechanism: Variations in precipitation and runoff patterns have the potential to modify the concentration of chemical contaminants in water sources, such as heavy metals and pesticides.

Consequences: Agricultural discharge that is rich in nitrates and phosphates can facilitate the growth of algal blooms, which generate toxins that are detrimental to human health. There is a positive correlation, according to statistical models, between nutrient discharge increases and the occurrence rate of harmful algal blooms, which have the potential to cause severe health complications including gastroenteritis and liver damage.

9.9.3 Analysis of Statistics

9.9.3.1 The Correlation Analysis

Pearson correlation coefficients have been utilized by researchers to discern robust correlations that exist between the prevalence of diseases and the variability of precipitation. For example, a correlation coefficient of 0.7 signifies that heavy precipitation and cholera outbreaks are significantly and positively correlated.

9.9.3.2 Time Series Analysis

Series analyses uncover seasonal variations in the occurrence of diseases, wherein maxima frequently coincide with episodes of heavy precipitation. Illustratively, areas undergoing monsoonal precipitation observe a substantial surge in diarrheal illnesses only a few weeks after the commencement of the wet season.

9.9.3.3 Regression Models

According to the findings of linear regression models that quantify the relationship between water quality parameters and disease incidence, there exists a 15% increase in the occurrence of waterborne diseases such as cryptosporidiosis and giardiasis for every 10% increase in water turbidity.

9.9.3.4 Geographical Analysis

Spatial regression techniques and geographic information systems (GIS) facilitate the identification of waterborne disease outbreak locations in areas characterized by inadequate water infrastructure and significant fluctuations in precipitation. These analyses facilitate the identification of populations at risk and provide information for the development of targeted interventions. The importance of implementing integrated water management and climate adaptation strategies to alleviate the health consequences of climate change is emphasized by statistical evidence. Public health officials can enhance the efficacy of prevention and control measures by comprehending the interconnections that exist among waterborne diseases, drainage patterns, and precipitation quality. Adopting a comprehensive approach is critical in safeguarding vulnerable populations and establishing robust water and health systems amidst the impacts of climate change.

9.10 ECONOMIC STABILITY

Employment and income is the important base for economic stability. It basically consists of various parameters such as vulnerability, abject poverty, job security, social support systems, discrimination, and marginalization (Lloyd and Farquhar 2008; Kovats et al. 2015; Hermans and Garbe 2019).

9.10.1 Vulnerability

Due to their constrained financial resources, individuals and families with lesser income levels are more susceptible to the adverse effects of climate change. They might be incapable of adequately preparing for, responding to, and recovering from hurricanes, heat waves, and floods, among other extreme weather phenomena.

9.10.2 Job Security

Individuals who are employed in sectors that are susceptible to climate impacts, including construction, agriculture, and outdoor labor, are more susceptible to adverse effects. Employees in these sectors are exposed to increased hazards associated with heat-related ailments, physical harm, and economic volatility.

9.10.3 Abject Poverty

The lack of access to vital resources, including nourishing food, potable water, and sufficient shelter, is a consequence of poverty, which worsens the adverse health effects associated with climate change. Poverty frequently concentrates in regions that are more susceptible to environmental perils, thereby augmenting their vulnerability.

9.10.4 Social Support Systems

Community resilience is bolstered by robust social networks and support systems, which offer aid and assistance to members of the community throughout and in the aftermath of severe weather phenomena. Strong social ties enable communities to mobilize resources and provide assistance to their most vulnerable members. Furthermore, social support plays a pivotal role in safeguarding mental health by assisting individuals in managing the distress and trauma that may arise as a result of climate-related occurrences.

9.10.5 Discrimination and Marginalization

Equity: Systemic discrimination marginalizes racial and ethnic minorities, thereby intensifying their susceptibility to the adverse impacts of climate change. These demographic cohorts frequently encounter obstacles when attempting to obtain financial aid, emergency services, or healthcare. In order to effectively tackle disparities, it is imperative to implement inclusive policies that acknowledge and confront the distinct obstacles encountered by marginalized communities.

9.11 FUTURE PROSPECTS AND OBSTACLES

It is crucial to proactively anticipate the forthcoming challenges and opportunities associated with the intricate convergence of climate change and human health as we contemplate the future. The several pivotal obstacles and prospects that will influence our approach to this urgent matter are (Hasan et al. 2020; Gauderman et al. 2015) as follows:

9.11.1. Climate change mitigation is a critical issue that requires significant attention in order to prevent its detrimental effects. This requires substantial adjustments in agriculture, transportation, energy production, and other sectors. Nevertheless, it also provides a chance to shift toward more sustainable and environmentally friendly technologies and methods that are advantageous to human health and the environment.

9.11.2. The establishment of resilience to the adverse effects of climate change is critical for safeguarding human health amidst continuous environmental transformations. This entails the modification of healthcare systems, infrastructure, and communities in order to endure and recover from the detrimental effects of climate-related hazards, including heatwaves. Placing investments in disaster preparedness measures, early warning systems, and climate-resilient infrastructure is crucial for mitigating vulnerability and bolstering adaptive capacity.

9.11.3. The promotion of health equity and the mitigation of health disparities are fundamental objectives in the context of climate change adaptation initiatives. The adverse effects of climate change disproportionately affect vulnerable populations, such as indigenous peoples, low-income communities, and marginalized bands. Prioritizing the requirements of individuals and ensuring that adaptation strategies are inclusive, equitable, and culturally sensitive are of utmost importance.

9.11.4. **Interdisciplinary Collaboration:** Addressing the intricate issues of human health and climate change necessitates the establishment of collaborative partnerships that transcend sectors and disciplines. Synergy among professionals with knowledge in policy, economics, public health, environmental science, and social science can result in the development of more comprehensive and efficacious solutions. The effective management of climate-related health hazards necessitates the implementation of capacity-building initiatives, collaborative research, and knowledge sharing.

9.11.5. **Community Empowerment and Engagement:** It is crucial for the success of climate change adaptation initiatives to involve communities. Incorporating local knowledge, perspectives, and priorities into decision-making processes and adaptation strategies is essential. Enabling community engagement in the stages of planning, implementation, and monitoring adaptation measures promotes grassroots ownership and resilience.

9.11.6. Embracing innovative technologies and solutions has the potential to significantly improve our capacity to mitigate the adverse health effects associated with climate change. Technology presents prospects for enhancing healthcare provision, overseeing environmental transformations, enabling individuals to proactively address climate change, and establishing green infrastructure and telemedicine and digital health tools.

9.11.7. Effective policy frameworks and governance are critical components in promoting climate action and safeguarding human health. Enforcing comprehensive climate policies, regulations, and incentives that prioritize health, equity, and sustainability is an imperative for policymakers. In addition to promoting global health security and addressing transboundary climate impacts, international cooperation and multilateral agreements are vital.

9.12 CONCLUSION

Protection of public health depends on addressing the relationship between climate change and human health. The results emphasize the importance of all-encompassing plans to lessen the negative health impacts of growing heat waves, harsh weather, and poor air and water quality. The disproportionate impact on vulnerable people emphasizes the need for focused interventions taking social determinants of health into account. Effective combat of these issues requires a multidisciplinary strategy including community-based resilience plans, strong legislative measures, and public health programs. The problem is urgent; hence, more study and creativity are needed to create adaptable strategies that safeguard health in the face of changing climate conditions. A healthier and more resilient world population should be the main goal of future attempts to develop sustainable solutions that deal with the short- and long-term health effects of climate change.

REFERENCES

Chen, S., Y. Yong, and X. Ju. 2021. "Effect of Heat Stress on Growth and Production Performance of Livestock and Poultry: Mechanism to Prevention." *Journal of Thermal Biology* 99: 103019.

Chen, Z., T. Zhou, L. Zhang, X. Chen, W. Zhang, and J. Jiang. 2020. "Global Land Monsoon Precipitation Changes in CMIP6 Projections." *Geophysical Research Letters* 47 (14). doi:10.1029/2019gl086902

Collins, M., R. Knutti, J. Arblaster, J.-L. Dufresne, T. Fichefet, P. Friedlingstein, X. Gao, et al. 2013. "Long-Term Climate Change: Projections, Commitments and Irreversibility." In *Climate Change 2013: The Physical Science Basis. Contribution of Working Group I to the Fifth Assessment Report of the Intergovernmental Panel on Climate Change*, edited by T. F. Stocker, D. Qin, G.- K. Plattner, M. Tignor, S. K. Allen, J. Boschung, A. Nauels, Y. Xia, V. Bex, and P. M. Midgley, 1033. Cambridge, United Kingdom and New York, NY, USA: Cambridge University Press.

Doocy, S., A. Daniels, S. Murray, and T. D. Kirsch. 2013. "The Human Impact of Floods: A Historical Review of Events 1980–2009 and Systematic Literature Review." *PLoS Currents*. doi:10.1371/currents.dis.f4deb457904936b07c09daa98ee8171a

Ebi, K. L, and I. Loladze. 2019. "Elevated Atmospheric CO_2 Concentrations and Climate Change Will Affect Our Food's Quality and Quantity." *The Lancet Planetary Health* 3 (7): e283–e284.

Fussell, J. C., and F. J. Kelly. 2021. "Mechanisms Underlying the Health Effects of Desert Sand Dust." *Environment International* 157 (December): 106790. doi:10.1016/j.envint.2021.106790

Gauderman, W. J., R. McConnell, E. Avol, K. Berhane, R. Urman, E. Rappaport, R. Chang, F. Lurmann, and F. Gilliland. 2015. "Association of Improved Air Quality with Lung Development in Children." *New England Journal of Medicine* 372 (10): 905–913. doi:10.1056/nejmoa1414123

Guo, R., J. Wang, L. Bing, D. Guan, P. Ciais, S. J. Davis, R. M. Andrew, F. Xi, and Z. Liu. 2021. "Global CO_2 Uptake by Cement from 1930 to 2019." *Earth System Science Data* 13 (4): 1791–1805. doi:10.5194/essd-13-1791-2021

Hartmann, D. L., A. M. G. K. Tank, M. Rusticucci, L. V. Alexander, S. Brönnimann, Y. Charabi, F. J. Dentener, et al. 2013. "Observations: Atmosphere and Surface." In *Climate Change 2013: The Physical Science Basis: Contribution of Working Group I to the Fifth Assessment Report of the Intergovernmental Panel on Climate Change*, edited by T. F. Stocker, D. Qin, G. K. Plattner, M. Tignor, S. K. Allen, J. Boschung, A. Nauels, Y. Xia, V. Bex, and P. M. Midgley, 15–24. Cambridge: Cambridge University Press.

Hasan, M. T., G. Adhikary, S. Mahmood, N. Papri, M. H. H. Shihab, R. Kasujja, H. U. Ahmed, A. R. Azad, and M. Nasreen. 2020. "Exploring Mental Health Needs and Services among Affected Population in a Cyclone Affected Area in Costal Bangladesh: A Qualitative Case Study." *International Journal of Mental Health Systems* 14 (1). doi:10.1186/s13033-020-00351-0

Hermans, K., and L. Garbe. 2019. "Droughts, Livelihoods, and Human Migration in Northern Ethiopia." *Regional Environmental Change* 19 (4): 1101–1111. doi:10.1007/s10113-019-01473-z

Hetherington, E., K. Adhikari, L. Tomfohr-Madsen, S. Patten, and A. Metcalfe. 2021. "Birth Outcomes, Pregnancy Complications, and Postpartum Mental Health after the 2013 Calgary Flood: A Difference in Difference Analysis." *PLoS One* 16 (2). doi:10.1371/journal.pone.0246670.

Hirabayashi, Y., R. Mahendran, S. Koirala, L. Konoshima, D. Yamazaki, S. Watanabe, H. Kim, and S. Kanae. 2013. "Global Flood Risk under Climate Change." *Nature Climate Change* 3 (9): 816–821. doi:10.1038/nclimate1911.

Khan, A. S., H. Yi, L. Zhang, X. Yu, E. Mbanzamihigo, G. Umuhumuza, Simplice Innocent Asengon Yevide, et al. 2019. "An Integrated Social-Ecological Assessment of Ecosystem Service Benefits in the Kagera River Basin in Eastern Africa." *Regional Environmental Change* 19 (1): 39–53.

Ko, R., M. Hayashi, H. Hayashi, K. Hayashi, H. Kato, Y. Kurata, Y. Fuchino, et al. 2016. "Correlation between Acute Conjunctivitis and Asian Dust on Ocular Surfaces." *Journal of Toxicology and Environmental Health, Part A* 79 (8): 367–375. doi:10.1080/15287394.2016.1162248

Konkel, L. 2020. "Tropical Cyclone Exposures and Health: A New Data Set to Assess Associations over Time." *Environmental Health Perspectives* 128 (12). doi:10.1289/ehp8653

Kovats, S., S. Hales, D. Campbell-Lendrum, J. Rocklov, Y. Honda, and S. Lloyd. 2015. "Global Risk Assessment of the Effect of Climate Change on Selected Causes of Death in 2030s and 2050s." *ISEE Conference Abstracts* 2015 (1). doi:10.1289/isee.2015.2015-1204

Lee, S. Y., J. H. Primavera, F. Dahdouh-Guebas, K. McKee, J. O. Bosire, S. Cannicci, K. Diele, et al. 2014. "Ecological Role and Services of Tropical Mangrove Ecosystems: A Reassessment." *Global Ecology and Biogeography* 23 (7): 726–743.

Lloyd, J., and G. D. Farquhar. 2008. "Effects of Rising Temperatures and [CO2] on the Physiology of Tropical Forest Trees." *Philosophical Transactions of the Royal Society B: Biological Sciences* 363 (1498): 1811–1817.

Lu, P., G. Xia, Q. Zhao, R. Xu, S. Li, and Y. Guo. 2020. "Temporal Trends of the Association between Ambient Temperature and Hospitalisations for Cardiovascular Diseases in Queensland, Australia from 1995 to 2016: A Time-Stratified Case-Crossover Study." PLOS Medicine 17 (7). doi:10.1371/journal.pmed.1003176

Ma, X., W. Ma, L. Zhang, Y. Shi, Y. Shang, and H. Chen. 2021. "The Impact of Green Credit Policy on Energy Efficient Utilization in China." *Environmental Science and Pollution Research* 28 (37): 52514–52528. doi:10.1007/s11356-021-14405-4

Marsooli, R., N. Lin, K. Emanuel, and K. Feng. 2019. "Climate Change Exacerbates Hurricane Flood Hazards along US Atlantic and Gulf Coasts in Spatially Varying Patterns." *Nature Communications* 10 (1). doi:10.1038/s41467-019-11755-z

Matthews, H. D., and K. Zickfeld. 2012. "Climate Response to Zeroed Emissions of Greenhouse Gases and Aerosols." *Nature Climate Change* 2 (5): 338–341. doi:10.1038/nclimate1424

Sun, Z., Q. Wang, C. Chen, Y. Yang, M. Yan, H. Du, K. Chen, J. S. Ji, and T. Li. 2021. "Projection of Temperature-Related Excess Mortality by Integrating Population Adaptability under Changing Climate — China, 2050s and 2080s." *China CDC Weekly* 3 (33): 697–701. doi:10.46234/ccdcw2021.174

Thompson, P. R., M. A. Merrifield, J. R. Wells, and C.-M. Chang. 2014. "Wind-Driven Coastal Sea Level Variability in the Northeast Pacific." *Journal of Climate* 27 (12): 4733–4751. doi:10.1175/jcli-d-13-00225.1

Tian, B., Z. Yu, Y. Pei, Z. Zhang, E. Siemann, S. Wan, and J. Ding. 2019. "Elevated Temperature Reduces Wheat Grain Yield by Increasing Pests and Decreasing Soil Mutualists." *Pest Management Science* 75 (2): 466–475.

Tong, D. Q., J. X. Wang, T. E. Gill, H. Lei, and B. Wang. 2017. "Intensified Dust Storm Activity and Valley Fever Infection in the Southwestern United States." *Geophysical Research Letters* 44 (9): 4304–4312. doi:10.1002/2017gl073524

Wu, Y., R. Xu, B. Wen, M. de Coelho, P. H. Saldiva, S. Li, and Y. Guo. 2021. "Temperature Variability and Asthma Hospitalisation in Brazil, 2000–2015: A Nationwide Case-Crossover Study." *Thorax* 76 (10). BMJ: 962–969. doi:10.1136/thoraxjnl-2020-216549

Zhang, P., K. Wiens, R. Wang, L. Luong, D. Ansara, S. Gower, K. Bassil, and S. W. Hwang. 2019. "Cold Weather Conditions and Risk of Hypothermia among People Experiencing Homelessness: Implications for Prevention Strategies." *International Journal of Environmental Research and Public Health* 16 (18): 3259. doi:10.3390/ijerph16183259

Zhao, Q., S. Li, M. S. Coelho, P. H. Saldiva, K. Hu, R. R. Huxley, M. J. Abramson, and Y. Guo. 2019. "The Association between Heatwaves and Risk of Hospitalization in Brazil: A Nationwide Time Series Study between 2000 and 2015." *PLoS Medicine* 16 (2). doi:10.1371/journal.pmed.1002753

10 Carbon Capture Mechanisms for Industries

Amanullah Adeel, Mustafa Hasani, Sushama Tanaji Patil, Pawar Pratiksha Shivaji, and Abdulahad Ahadi

10.1 INTRODUCTION

Carbon-based greenhouse gases (GHGs) like carbon dioxide (CO_2) and methane (CH_4) resulted in climate change (Adeel and Jadhav 2024). By 2022, the industrial sector accounted for about 25% of CO_2 emissions from the worldwide energy system (Zhang et al. 2024). The International Energy Agency (IEA) attributes this significant contribution to four primary sources: the production of cement, iron and steel, petrochemical processes, and nonferrous metals, including aluminum. To keep the rise in global warming temperatures to 2°C or less, all emission-intensive industries, including the chemical, steel, and cement sectors, must attain net-zero emissions by 2080 (Åhman et al. 2017). There is growing pressure on the emission-intensive industries to lower their CO_2 emissions.

Technologies for carbon capture are necessary to reduce carbon emissions in the industrial sector (Zhang et al. 2024). Carbon dioxide (CO_2) emissions from human activity are reduced by carbon capture and storage (CCS). CO_2 is pushed into the deep subsurface for long-term storage from massive stationary sources, such as fossil fuel-fired power plants (Bandilla 2020). These technologies provide a range of techniques, including carbon mineralization, oxy-fuel combustion, pre-combustion, post-combustion, and absorption, to reduce GHG emissions and improve operational efficiency in sectors such as cement manufacturing (Shammi and Akter 2024). The economic feasibility of implementing CCS technology is a key focus, as there is a growing trend toward combining bioenergy with CCS for improved sustainability and cost-effectiveness (Ikhlas et al. 2024). Through the application of these cutting-edge technologies, enterprises can not only lower their carbon footprint but also convert waste materials into useful products, promoting economic growth and environmental preservation.

CCS technology is a feasible large-scale CO_2 emission reduction strategy. Thirty of the 196 CCS facilities that are active or in development worldwide by mid-September 2022 will have captured almost 42.5 million tons of CO_2 annually (Global CCS Institute 2022). The 0.8 Mtpa Abu Dhabi CCS project in the United Arab Emirates is the first operational facility that is economically scaled

DOI: 10.1201/9781003492733-10

for hard-to-abate industries. It collects carbon dioxide from the shaft furnace of the direct reduction of iron process using the conventional amine absorption technique (Sakaria 2017). Although the industry sector has limitedly used CCS, evaluations from academia and industry have encouraged optimistic expectations about CCS's ability to reduce emissions. According to the IEA assessment, CCS technology could contribute up to 170 Mt of CO_2 emissions to be captured by 2030 under the net-zero scenario (Zhang et al. 2024).

With an emphasis on their importance in reducing CO_2 emissions and addressing climate change, this chapter aims to give a thorough overview of the mechanisms and technologies available for carbon capture in industrial applications. The basic ideas of carbon capture, the many materials and techniques employed, and the real-world applications of these technologies in important industries will all be covered in this chapter.

10.2 FUNDAMENTALS OF CARBON CAPTURE

Carbon capture is the process of removing CO_2 from several sources by means of membrane separation, chemical looping, absorption, adsorption, and cryogenic distillation (Chai et al. 2023; Chao et al. 2021; Ikhlas et al. 2024). The performance of these technologies in terms of CO_2 capturing, energy needs, and environmental effects varies, with absorption being the most developed and extensively used approach because of its affordability and effectiveness (Chao et al. 2021). Techniques for post-combustion capture, including adsorption, membrane processes, and absorption, are widely employed. Of these, absorption is the most often utilized approach for evaluating post-combustion CCS technology (Chao et al. 2021). Strategies for improving the efficiency of carbon capture systems include the creation of composite adsorbents and the mixing of various absorbents (Ngu 2024). To achieve net-zero emissions and lessen the damaging effects of CO_2 emissions on the environment, carbon capture technologies must be used (Rutherford 2022). Furthermore, the use of CCS technologies has sparked worries about environmental impacts and global risks. Environmental concepts such as the integration principle, the precautionary principle, the rectification at source principle, the polluter pays principle, and the preventative action principle in the UK, for example, have been developed as a result of talks at international forums (Rutherford 2022).

The technique of converting fossil fuels into a hydrogen and carbon dioxide mixture prior to combustion, known as pre-combustion capture, enables CO_2 to be isolated and stored (Ibigbami et al. 2024). The goal of post-combustion capture is to extract CO_2 from flue gases through the use of different absorbents, such as ionic liquids, carbonate solutions, and amine solutions (Wang 2023). The method of capturing CO_2 is made simpler by oxygen-fuel combustion, which burns fuels in oxygen rather than air and produces flue gas that is mostly made up of CO_2 and water vapor (Ibigbami et al. 2024). These varied strategies reduce GHG emissions from sectors such as natural gas processing, power generation, and hydrogen manufacturing, which is vital in combating climate change (Allangawi et al. 2023).

The main objective of carbon capture technology is to lower CO_2 emissions through various chemical and physical processes (Figure 10.1). Chemical techniques

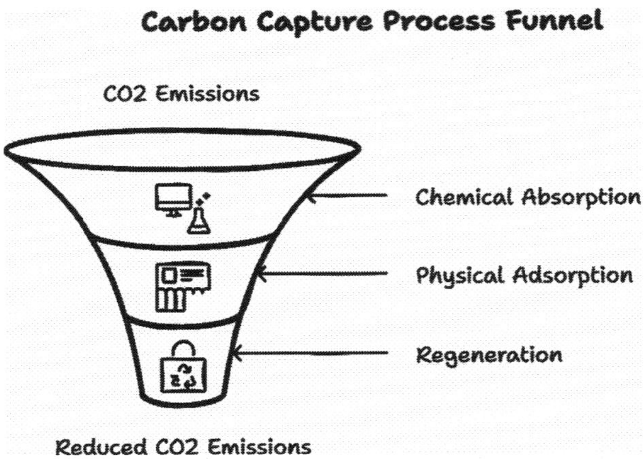

FIGURE 10.1 Carbon capture process funnel illustrates three major approaches to reduce CO_2 emissions.

include solvent-based absorption, in which CO_2 is captured by chemical and physical interactions using solvents such as mixed, chemical, or physical types (Ansarpour and Mofarahi 2023; Ngu 2024). Porous adsorbents, such as porous carbons, zeolites, activated alumina, and metal-organic frameworks (MOFs), are used in physical processes, such as physical adsorption-based separation, to adsorb CO_2 molecules based on their pore structure and surface chemistry. This allows for simple regeneration and high cycle stability (Liu et al. 2021). In order to develop effective carbon capture technologies with lower energy penalties and costs, these techniques are essential for capturing CO_2 from a variety of applications and stream conditions.

10.3 TECHNOLOGIES FOR CARBON CAPTURE

10.3.1 Pre-Combustion Capture

Pre-combustion capture technique uses advanced physical adsorbents or membranes to selectively remove CO_2 from synthesis gas before combustion, hence reducing emissions, employing a functional group-containing mesoporous carbon sorbent to remove CO_2 (Ni et al. 2023). With little energy needed for sorbent regeneration, this method, which runs at high temperatures, achieves over 90% carbon capture efficiency (Alptekin 2022). Furthermore, a two-stage procedure involving the combination of CO_2- and H_2-selective membranes with recycling has demonstrated financial benefits over conventional techniques like Selexol (Ni et al. 2023). While retaining excellent purity and recovery rates, the optimized process structure and membrane performance can dramatically lower CO_2 capture costs. Pre-combustion capture technologies present a viable way to reduce GHG emissions from power plants.

Pre-combustion carbon capture devices often employ gasification techniques to efficiently absorb carbon emissions (Figure 10.2). These approaches entail the selective removal of CO_2 from synthesis gas generated during gasification through the use of sorbents or physical adsorbents. Higher power cycle efficiencies are made possible by

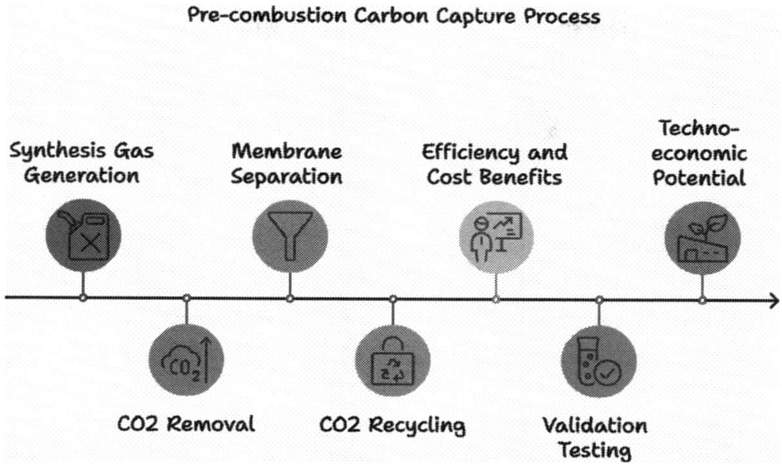

FIGURE 10.2 Pre-combustion carbon capture process initiates with synthetic gas generation and ends with economic production potential.

sorbents with grafted functional groups, such as mesoporous carbon, which have demonstrated strong CO_2 removal capabilities at temperatures above 300°C (Alptekin 2022; Alptekin et al. 2022; Jamieson 2023). Without appreciably raising the price of electricity, it has been shown that warm gas CO_2 scrubbers combined with Water-Gas-Shift (WGS) catalysts may collect over 90% of carbon emissions in Integrated Gasification Combined Cycle (IGCC) power plants (Alptekin et al. 2022). Large-scale slipstream testing and high-fidelity engineering analyses have validated these procedures, demonstrating their techno-economic potential for producing almost zero CO_2 emissions in advanced gasification-based power systems (Jamieson 2023).

10.3.2 Post-Combustion Capture

Post-combustion capture is the method of extracting CO_2 from flue gas after combustion has occurred. It is usually applied to power plants that are already in operation in order to extract CO_2 from exhaust fumes prior to their release into the atmosphere. The main objective is to capture and store CO_2 in order to lessen the negative environmental effects of burning fossil fuels. The first stage is to remove flue gas from the exhaust stack of the power plant. CO_2 is present in flue gas together with other gases like water vapor (H_2O), nitrogen (N_2), and other pollutants (Duke et al. 2010). Next, a solvent or absorbent comes into contact with the flue gas that has been caught. Mono-ethanolamine (MEA) is the most often used solvent for PCC. Because of its chemical affinity, MEA collects CO_2 from the flue gas in a selective manner. In an absorber column, the absorption process takes place (Li et al. 2022). CO_2 is absorbed by the solvent, which then gets loaded with CO_2. In a desorber column, the solvent is heated to liberate the CO_2 that has been trapped. A concentrated stream of CO_2 is produced when the heat causes CO_2 to desorb from the solvent. For storage and transportation, the desorbed CO_2 is compressed to a higher pressure. Energy is used a

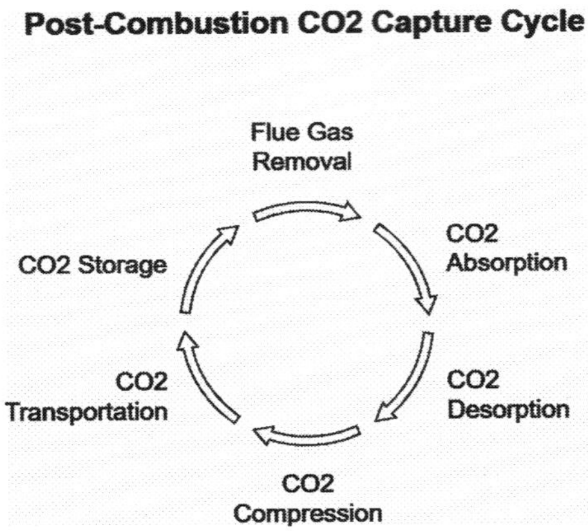

FIGURE 10.3 Shows the process of CO_2 capturing cycle in a post-combustion method. The cycle includes CO_2 absorption, desorption, compression, transportation, storage, and gas removal.

lot during the compression process (Duke et al. 2010). Pipelines can be used to move the compressed CO_2 to appropriate geological formations for storage (such as saline aquifers or depleted oil and gas reserves). By doing this action, the captured CO_2 is guaranteed to stay underground (Figure 10.3).

One way to lessen the issues with liquid amine scrubbing is to use solid adsorbents functionalized with amino acids. This can help with issues like the loss of volatile amines caused by vaporization. In order to guarantee that the amines are dispersed uniformly throughout the material, amines are utilized to functionalize high surface area and porosity support materials (Hack et al. 2022). Because of this, the functionalized materials possess the benefits of both components (an amine and a support).

10.3.3 Oxy-fuel Combustion

While oxygen-fuel technologies offer novel approaches to reduce GHG emissions, there are situations in which it can be dangerous to use pure oxygen instead of air. In the case of such air combustion processes, carbon dioxide is combined with oxygen gas to replace the inert effect of nitrogen (Talei et al. 2024). The flue gas stream will include a high concentration of CO_2 (90% on a dry basis) as a result of the recycled CO_2 replacing nitrogen, making CO_2 recovery and sequestration much simpler (Rogalev et al. 2022) (Figure 10.4). Furthermore, this method works to prevent the generation of NO_x, an unwanted byproduct of air combustion, by substituting carbon dioxide for nitrogen (Yu et al. 2021).

Oxy-fuel combustion has been extensively researched and proven in both academia and industry to be one of the most promising CCS solutions to slow down global warming. In the oxy-fuel process, air is replaced as the oxidant gas in the combustion chamber with a blend of recovered flue gas and pure O_2 from an air separation unit. As a result, flue gas might have a high concentration of CO_2, making it appropriate for later carbon storage and use (Duan and Li 2023). Other benefits include its low NO_x emissions, ease of scaling up, and suitability for use in already-existing power plants.

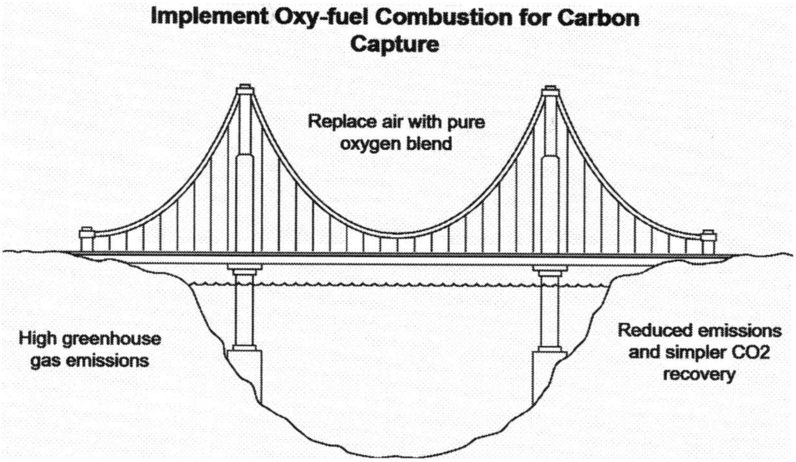

FIGURE 10.4 Bridge to cleaner combustion: The figure shows how the use of oxy-fuel combustion, by substituting air with a pure oxygen mixture, is a bridge from high greenhouse gas emissions to lower emissions and simpler CO_2 capture that aids carbon capture and climate mitigation.

10.4 CARBON CAPTURE MATERIALS AND METHODS

10.4.1 Solvent-Based Capture

Aqueous scrubbing, physical solvent scrubbing, membrane separation, cryogenic procedures, and anhydrous solvents such as ionic liquids, switchable solvents, and nanoparticle organic hybrid materials are examples of solvent-based capture techniques for carbon capture. Because of their great efficiency, amines like MEA are often employed as solvents for CO_2 capture. They are caustic and highly volatile, though, which are disadvantages. The goal of new solvents such as aqueous ammonia, carbonate systems, amino acid salts, advanced amine solvents, immiscible liquids, and ionic liquids is to increase productivity while lowering costs. These solvents increase the rates and capacities of CO_2 absorption while lowering the amount of equipment and energy needed for solvent regeneration. The versatility of amines in CO_2 removal is demonstrated by their application in industrial processes such as ammonia synthesis and biogas production. The development of amine mixtures to optimize attributes for CO_2 capture systems is the subject of ongoing research, with a primary focus on lowering the energy requirements for solvent regeneration in order to minimize process costs.

10.4.2 Solid Sorbents

According to Samanta et al. (2012), solid sorbents have great promise for carbon capture technology and may provide benefits over traditional amine-based absorption techniques. MOFs, zeolites, and activated carbons are important forms of solid sorbents (Rashid and Rafey 2023). This study also emphasized that since they are readily available and inexpensive, activated carbons are utilized extensively; nevertheless, zeolites and MOFs have greater adsorption capabilities. Because of their exceptional qualities and flexible architectures, MOFs in particular are drawing attention as the perfect carbon capture adsorbents (Tao and Xu 2024). Polymeric resins, carbonates, and microporous and mesoporous silica are other possible sorbents (Samanta et al. 2012). Equilibrium CO_2 capacities, adsorption-desorption kinetics, operating windows, stability, and renderability are among the evaluation criteria for these adsorbents (Choi et al. 2009). Many regeneration approaches, such as temperature swing adsorption (TSA), pressure swing adsorption (PSA), vacuum swing adsorption (VSA), and electric swing adsorption (ESA), are employed in carbon capture systems that use solid sorbents (Tao and Xu 2024).

The recent challenges and trends indicate that reaction mechanisms, mass transfer, material microstructure, and operating conditions are some of the factors that affect CO_2 sorption kinetics (Ji et al. 2020). For high-temperature capture, layered double hydroxide, alkali ceramic, calcium, and magnesium-based sorbents have all been thoroughly explored (Ji et al. 2020). Even with great advancements, there are still obstacles in the way of creating large-scale, economically viable sorbents that have sufficient stability and performance.

10.4.3 Membrane Technologies

When compared to more traditional techniques like amine scrubbing, membrane technology is showing promise as a carbon capture substitute because of its benefits in energy efficiency, scalability, and versatility (Lv et al. 2012). For the purpose of separating CO_2 from flue gases, several membrane types—such as gas separation and gas absorption membranes—are being developed (Hou et al. 2022). According to Ji and Zhao (2017), these technologies can be used in oxy-fuel, post-combustion, and pre-combustion processes. Membranes are also essential for the storage and exploitation of carbon, since they aid in the compression of CO_2 for subterranean storage and subsequent conversion into useful product. Notwithstanding its potential, scaling up membrane technologies for large-scale application still presents difficulties, such as material and technical issues (Hou et al. 2022).

10.4.4 Cryogenic Processes

Cryogenic carbon capture (CCC) technology is showing promise as a CO_2 removal method, with high recovery rates and purity of up to 99.99% (Aneesh and Sam 2023). Comparing the CCC technology to other major competitive carbon capture processes, it seems to use at least 30% less energy and money. A number of CCC methods, such as cryogenic packed bed and anti-sublimation, have been developed, the main obstacle being continuous dry ice removal. By 2030, Chinese CCS project cost is estimated to be 13–57$/t, declining to 3–19$/t (Shen et al. 2022).

The advantages of CCC include its ease of retrofitting into current systems, its ability to handle gas contaminants, and its potential for direct air capture and energy storage (Baxter et al. 2021). It also offers effective removal of other pollutants such as Hg, SO_x, and NO_2. Nevertheless, the accurate forecasting of frost formation and the ongoing evacuation of dry ice into storage tanks pose difficulties (Aneesh and Sam 2023). To increase efficiency, hybrid systems that combine membrane and cryogenic techniques—such as low-temperature membrane cryogenics (LTMC) and hybrid membrane cryogenics (HMC)—are being investigated (Quaranta 2023).

10.5 CASE STUDIES IN INDUSTRIAL APPLICATIONS

Several large-scale and experimental CCS projects are underway or in development globally, with differing degrees of execution in various nations (Romasheva and Ilinova 2019). For instance, the Moomba CCS project in Australia uses both new and existing infrastructure to cut emissions by 1.7 Mtpa CO_2-e (Harley 2021). Although CCS has advantages for the environment, there are possible hazards related to CO_2 migration. As such, comprehensive risk evaluations and monitoring strategies are required (Gaurina-Međimurec et al., 2018). Technological developments, policy incentives, and regulatory frameworks are some of the variables that affect the success of CCS projects (Romasheva and Ilinova 2019). Site integrity and storage complex performance are dependent on efficient monitoring systems, such as the one used in the French Lacq pilot project (Gaurina-Međimurec et al., 2018).

Several projects have been implemented, are in progress, and have been planned for CCS.

IEA reported some CCS projects (https://www.iea.org/) as follows:

Indonesia's energy industry and emissions have been significantly impacted by its amazing economic expansion over the past 50 years, with coal playing a major part in both. To achieve its goal of net-zero emissions by 2060, Indonesia will require energy efficiency measures and renewable energy options for its new power plants and industrial facilities that produce steel, iron, and cement. In order to accomplish that, carbon capture, utilization, and storage, or CCUS, can be a crucial technological advancement. Early in 2023, Indonesia will finalize the first regulatory framework for CCUS in Southeast Asia, demonstrating that it is already making significant progress. However, further actions are required to broaden the current framework outside the oil and gas industry in order for CCUS to fulfill its varied role in Indonesia's decarburization.

Opportunities for CCUS adoption in industry, power generation, and fuel transformation are examined in this policy brief. It provides important insights for enhancing Indonesia's current legislative framework in order to facilitate the decarburization of other industries and position Indonesia as a leader in the deployment of CCUS in the region.

In accordance with global trends, interest in CCUS has been rising in Southeast Asia. In just the first half of 2021, proposals for around 30 commercial CCUS sites were revealed worldwide. At least seven possible projects in Indonesia, Malaysia, Singapore, and Timor-Leste have been identified and are in the initial stages of development throughout Southeast Asia. Singapore is actively pursuing research and international partnerships, particularly with Australia, as part of its long-term emissions-reduction strategy, which has highlighted an important role for CCUS. Another important step and chance to progress CCUS in the region is the creation of the Asia CCUS Network in June 2021, which aims to facilitate cooperation and the deployment of CCUS.

10.6 CONCLUSION

Overall, “Carbon Capture Mechanisms for Industries” highlights how crucial CCS technology is to addressing industrial CO_2 emissions in order to combat climate change. It describes the different processes, applications, and limitations of the numerous technologies and methodologies, including oxy-fuel combustion, post-combustion, and pre-combustion. The chapter also explored the components and techniques of CCS, covering the advantages and disadvantages of solvent-based capture, solid sorbents, membrane technologies, and cryogenic procedures. The chapter offered insightful information on the real-world use of CCS projects through case studies from important industries like chemicals, steel, cement, and power generation. The comprehensive analysis emphasizes how crucial it is to implement CCS to achieve a sustainable industrial future, and it makes a strong case for industry executives, decision-makers, and scholars to work together and innovate in this crucial field.

REFERENCES

Adeel, A.., and A. S. Jadhav. 2024. “Identifying the Optimal Sugarcane Residue Percentage for Soil Enhancement: A Statistical and GIS Approach.” *Circular Economy and Sustainability*, November. doi:10.1007/s43615-024-00456-5

Åhman, M., L. J. Nilsson, and B. Johansson. 2017. “Global Climate Policy and Deep Decarbonization of Energy-Intensive Industries.” *Climate Policy* 17 (5): 634–649. doi:10.1080/14693062.2016.1167009

Allangawi, A., E. F. H. Alzaimoor, H. H. Shanaah, H. A. Mohammed, H. Saqer, A. A. El-Fattah, and A. H. Kamel. 2023. “Carbon Capture Materials in Post-Combustion: Adsorption and Absorption-Based Processes.” *C* 9 (1): 17. doi:10.3390/c9010017

Alptekin, G. O. 2022. *Pilot Testing of a Highly Efficient Pre-Combustion Sorbent-Based Carbon Capture System*. TDA-2942-TR. TDA Research, Inc., Wheat Ridge, CO. doi:10.2172/1906977

Alptekin, G. O., A. Jayaraman, M. Bonnema, and D. Gribble. 2022. *Integrated Water-Gas-Shift Pre-Combustion Carbon Capture Process*. TDA-R-2201-001-PhII-F. TDA Research, Inc., Golden, CO (United States). doi:10.2172/1838103

Aneesh, A. M., and A. A. Sam. 2023. “A Mini-Review on Cryogenic Carbon Capture Technology by Desublimation: Theoretical and Modeling Aspects.” *Frontiers in Energy Research* 11 (May): 1167099. doi:10.3389/fenrg.2023.1167099

Ansarpour, M., and M. Mofarahi. 2023. “Carbon Capture by Swing Adsorption Processes.” In *Reference Module in Earth Systems and Environmental Sciences*, B9780323939409000918. Elsevier. doi:10.1016/B978-0-323-93940-9.00091-8

Bandilla, K. W. 2020. “31 - Carbon Capture and Storage.” In *Future Energy* (Third Edition), edited by Trevor M. Letcher, 669–692. Elsevier. doi:10.1016/B978-0-08–102886-5.00031-1

Baxter, L., C. Hoeger, K. Stitt, S. Burt, and A. Baxter. 2021. “Cryogenic Carbon Capture™ (CCC) Status Report.” *SSRN Electronic Journal*. doi:10.2139/ssrn.3819906

Chai, Y. H., N. A. Rashidi, M. Mohamed, B. L. F. Chin, and S. Yusup. 2023. “Basic Principles of CO_2 Capture and Conversion Technologies.” In *Nanomaterials for Carbon Dioxide Capture and Conversion Technologies*, edited by Shaukat Ali Mazari, Nabisab Mujawar Mubarak, and Manoj Tripathi, 25–61. Micro and Nano Technologies. Elsevier. doi:10.1016/B978-0-323-89851-5.00006-8

Chao, C., Y. Deng, R. Dewil, J. Baeyens, and X. Fan. 2021. “Post-Combustion Carbon Capture.” *Renewable and Sustainable Energy Reviews* 138 (March): 110490. doi:10.1016/j.rser.2020.110490

Choi, S., J. H. Drese, and C. W. Jones. 2009. "Adsorbent Materials for Carbon Dioxide Capture from Large Anthropogenic Point Sources." *ChemSusChem* 2 (9): 796–854. doi:10.1002/cssc.200900036

Duan, L., and L. Li. 2023. "OCAC Technology in Oxy-Fuel Combustion for Carbon Capture." In *Oxygen-Carrier-Aided Combustion Technology for Solid-Fuel Conversion in Fluidized Bed*, edited by Lunbo Duan and Lin Li, 65–77. Singapore: Springer Nature. doi:10.1007/978-981-19-9127-1_4

Duke, M. C., B. Ladewig, S. Smart, V. Rudolph, and J. C. D. da Costa. 2010. "Assessment of Postcombustion Carbon Capture Technologies for Power Generation." *Frontiers of Chemical Engineering in China* 4 (2): 184–195. doi:10.1007/s11705-009-0234-1

Faculty of Mining, Geology and Petroleum Engineering, Pierottijeva 6, Zagreb, Croatia, Full Professor, N. Gaurina-Međimurec, K. Novak-Mavar, INA-Industry of Oil Plc., Lovinčićeva 4, Zagreb, Croatia; SD & HSE Expert, Dsc., Matej Majić, and majic90@hotmail.com. 2018. "CCS Technology: Overview of Projects, Technology and Monitoring." *Rudarsko-Geološko-Naftni Zbornik* 33 (2): 1–14. doi:10.17794/rgn.2018.2.1

Hack, J., N. Maeda, and D. M. Meier. 2022. "Review on CO_2 Capture Using Amine-Functionalized Materials." *ACS Omega* 7 (44): 39520–39530. doi:10.1021/acsomega.2c03385

Harley, N. 2021. "Moomba Carbon Capture and Storage Case Study: Material Emissions Reduction at Moomba Plant." *The APPEA Journal* 61 (2): 402. doi:10.1071/AJ20136

Hou, R., C. Fong, B. D. Freeman, M. R. Hill, and Z. Xie. 2022. "Current Status and Advances in Membrane Technology for Carbon Capture." *Separation and Purification Technology* 300 (November): 121863. doi:10.1016/j.seppur.2022.121863

Ibigbami, O. A., O. D. Onilearo, and R. O. Akinyeye. 2024. "Post-Combustion Capture and Other Carbon Capture and Sequestration (CCS) Technologies: A Review." *Environmental Quality Management* 34 (1): e22180. doi:10.1002/tqem.22180

Ikhlas, N., I. D. A. A. Warmadewanthi, N. N. Arifianingsih, I. Y. Septiariva, and M. Wulandari. 2024. "Chapter Eleven - Carbon Dioxide Capture Technologies for the Conventional Energy Sector." In *Decarbonization Strategies and Drivers to Achieve Carbon Neutrality for Sustainability*, edited by Majeti Narasimha Vara Prasad, Larry E. Erickson, Fabio Carvalho Nunes, and Bimastyaji Surya Ramadan, 203–222. Elsevier. doi:10.1016/B978-0-443-13607-8.00032-8

Jamieson, M. 2023. *Coal Gasification Operations with Carbon Capture (CCS).* National Energy Technology Laboratory (NETL), Pittsburgh, PA, Morgantown, WV, and Albany, OR (United States). doi:10.2172/1922942

Ji, G., H. Yang, M. Z. Memon, Y. Gao, B. Qu, W. Fu, G. Olguin, M. Zhao, and A. Li. 2020. "Recent Advances on Kinetics of Carbon Dioxide Capture Using Solid Sorbents at Elevated Temperatures." *Applied Energy* 267 (June): 114874. doi:10.1016/j.apenergy.2020.114874

Ji, G., and M. Zhao. 2017. "Membrane Separation Technology in Carbon Capture." In *Recent Advances in Carbon Capture and Storage*, edited by Yongseung Yun. InTech. doi:10.5772/65723

Li, J., T. Wang, P. Liu, and Z. Li. 2022. "Dynamic Modelling and Simulation of a Post-Combustion CO2 Capture Process for Coal-Fired Power Plants." *Frontiers of Chemical Science and Engineering* 16 (2): 198–209. doi:10.1007/s11705-021-2057-7

Liu, R.-S., X.-D. Shi, C.-T. Wang, Y.-Z. Gao, S. Xu, G.-P. Hao, S. Chen, and A.-H. Lu. 2021. "Advances in Post-Combustion CO_2 Capture by Physical Adsorption: From Materials Innovation to Separation Practice." *ChemSusChem* 14 (6): 1428–1471. doi:10.1002/cssc.202002677

Lv, Y. X., G. H. Yan, C. Q. Xu, M. Xu, and L. Sun. 2012. "Review on Membrane Technologies for Carbon Dioxide Capture from Power Plant Flue Gas." *Advanced Materials Research* 602–604 (December): 1140–1144. doi:10.4028/www.scientific.net/AMR.602-604.1140

Ngu, L. H. 2024. "Carbon Capture Technologies." In *Encyclopedia of Sustainable Technologies* (Second Edition), edited by Martin A. Abraham, 358–377. Oxford: Elsevier. doi:10.1016/B978-0-323-90386-8.00028-0

Ni, Z., Y. Cao, X. Zhang, N. Zhang, W. Xiao, J. Bao, and G. He. 2023. "Synchronous Design of Membrane Material and Process for Pre-Combustion CO_2 Capture: A Superstructure Method Integrating Membrane Type Selection." *Membranes* 13 (3): 318. doi:10.3390/membranes13030318

Quaranta, E. 2023. "Editorial: Reviews in Carbon Capture, Utilization and Storage: 2022." *Frontiers in Energy Research* 11 (September): 1286062. doi:10.3389/fenrg.2023.1286062

Rashid, H., and A. Rafey. 2023. "Solid Adsorbents for Carbon Dioxide Capture: A Review." *Chemistry and Ecology* 39 (7): 775–791. doi:10.1080/02757540.2023.2227622

Rogalev, N., A. Rogalev, V. Kindra, O. Zlyvko, and P. Bryzgunov. 2022. "Review of Closed SCO2 and Semi-Closed Oxy–Fuel Combustion Power Cycles for Multi-Scale Power Generation in Terms of Energy, Ecology and Economic Efficiency." Energies 15 (23): 9226. doi:10.3390/en15239226

Romasheva, N., and A. Ilinova. 2019. "CCS Projects: How Regulatory Framework Influences Their Deployment." *Resources* 8 (4): 181. doi:10.3390/resources8040181

Rutherford, A. 2022. "The Application of the Environment Act 2021 Principles to Carbon Capture and Storage." *Laws* 11 (1): 15. doi:10.3390/laws11010015

Samanta, A., A. Zhao, G. K. H. Shimizu, P. Sarkar, and R. Gupta. 2012. "Post-Combustion CO_2 Capture Using Solid Sorbents: A Review." *Industrial & Engineering Chemistry Research* 51 (4): 1438–1463. doi:10.1021/ie200686q

Shammi, M., and B. Akter. 2024. "CO_2 Sequestration for Conventional Utilization and Industrial Application." In *Decarbonization Strategies and Drivers to Achieve Carbon Neutrality for Sustainability*, edited by Majeti Narasimha Vara Prasad, Larry E. Erickson, Fabio Carvalho Nunes, and Bimastyaji Surya Ramadan, 299–319. Elsevier. doi:10.1016/B978-0-443-13607-8.00011-0

Shen, M., L. Tong, S. Yin, C. Liu, L. Wang, W. Feng, and Y. Ding. 2022. "Cryogenic Technology Progress for CO_2 Capture under Carbon Neutrality Goals: A Review." *Separation and Purification Technology* 299 (October): 121734. doi:10.1016/j.seppur.2022.121734

Talei, S., D. Fozer, P. S. Varbanov, A. Szanyi, and P. Mizsey. 2024. "Oxyfuel Combustion Makes Carbon Capture More Efficient." *ACS Omega* 9 (3): 3250–3261. doi:10.1021/acsomega.3c05034

Tao, Y. R., and H. J. Xu. 2024. "A Critical Review on Potential Applications of Metal-Organic Frameworks (MOFs) in Adsorptive Carbon Capture Technologies." *Applied Thermal Engineering* 236 (January): 121504. doi:10.1016/j.applthermaleng.2023.121504

Wang, X. 2023. "Progress in Post-Combustion Technology for Carbon Dioxide Capture." *Highlights in Science, Engineering and Technology* 73 (November): 445–451. doi:10.54097/hset.v73i.14050

Yu, D., X. Yu, J. Wu, J. Han, F. Liu, and H. Pan. 2021. "A Comprehensive Review of Ash Issues in Oxyfuel Combustion of Coal and Biomass: Mineral Matter Transformation, Ash Formation, and Deposition." *Energy & Fuels* 35 (21): 17241–17260. doi:10.1021/acs.energyfuels.1c02231

Zhang, T., M. Zhang, L. Jin, M. Xu, and J. Li. 2024. "Advancing Carbon Capture in Hard-to-Abate Industries: Technology, Cost, and Policy Insights." *Clean Technologies and Environmental Policy* 26 (7): 2077–2094. doi:10.1007/s10098-024-02810-5

11 Adaptation to Climate Change

Pooja M. Patil, Vaishnavi A. Falke, and Rasiya C. Padalkar

11.1 INTRODUCTION

The domain of atmospheric resilience has markedly enhanced awareness and understanding of concerns related to climate change (CC) in recent decades. The primary focus of these initiatives was the assessment of CC and the physiological, social, and economic impacts of climate hazards, as well as the development and implementation of methods for measuring ecosystem and community vulnerability (Du Monceau et al. 2011). Recent years have witnessed an increase in demands for enhanced decision-orientated research, as the emphasis has transitioned from identifying consequences and deficiencies to advocating for mitigation solutions and developing adaptation initiatives in response to the escalating global temperature of 20°C due to CC. This underscores the significance of focusing on empowering decision-makers within interconnected ecological and social frameworks to make difficult and urgent choices about various managerial and policy options. These are driven by a multitude of factors (Azzopardi et al. 2020). Various constraints and challenges related to human behaviour and governance, along with considerations regarding the limited effectiveness of increased scrutiny, vulnerability, and flexibility in reporting choices for adaptation, lead to concerns that adaptation plans often lack a connection to implementation (Azzopardi et al. 2020).

Moreover, decision-centred approaches are considered more adept at addressing the complexities of accounting for numerous confounding variables, such as cross-scale effects over time and space, various forms of unpredictability, and the strategic management of uncertain consequences arising from fluctuating and erratic values, preferences, and vulnerabilities among at-risk populations. The challenges have resulted in the loose coalition of methods and studies known as "adaptation science" exerting only a negligible impact on the effectiveness of adjustments in strategy, leadership, and policy (Egenhofer and Fujiwara 2003). Moreover, despite ongoing appeals for a focus on decision-making, the majority of adaptation actions implemented have been incremental and aimed at immediate causes; there have been limited reports of significant adjustments and big transformations. The UN System's Compact on Climate Change is the world's most significant and contentious global environmental treaty (Heller 2003).

Every year, negotiators from both developed and developing countries gather to further its advancement. This agreement addresses the primary challenges of

DOI: 10.1201/9781003492733-11

globalization, development, equity, and marginalization. It is expected to have substantial effects on matters ranging from immigration trends to global energy consumption. Global warming is likely the most significant long-term hazard to global safety in the contemporary period. The Intergovernmental Panel on Climate Change (IPCC), a significant international scientific initiative endorsed by states through the UN, has bolstered the legitimacy of the Convention by applying insights from past international environmental accords. During the Seventh Meeting of the Parties in November 2001 in Marrakech, Morocco, negotiators discussed mitigation options and adaptations to climate disruption, officially recognizing for the first time the issues confronting developing countries in this context (Ford and King 2015). This recognition manifested as financial resources to assist governments in adaptation. The importance of adaptability was reiterated in the Delhi Resolution during the eighth meeting of the Parties in November 2002. The Delhi Declaration has fundamentally connected the participation of poor nations in emissions reduction with their financial contributions and efforts to adapt to the effects of CC (Nieto et al. 2018).

This chapter emphasizes greenhouse gas (GHG) emissions, the status of CC, and the associated repercussions, hazards, and vulnerabilities. Furthermore, we have encapsulated the essential requirements for CC adaptation under the United Nations Framework Convention, clean development methods, the formulation of a climate vision, and strategies at the global level.

11.2 OVERVIEW OF GHG EMISSIONS

In 2019, the United Nations Environment Programme (UNEP) published a gap analysis indicating that of the 55.3 Gt of carbon dioxide emissions in 2018, 37.5 Gt CO_2 originated from fossil fuels utilized in energy generation and industrial operations. Globally, fossil fuel consumption and CO_2 emissions from GHGs have increased by 1.5% annually over the past decade, with a reported rise of 2% in 2018. In 2018, fossil fuel CO_2 emissions experienced a significant increase primarily attributed to heightened energy demand. Methane (CH_4), another significant GHG, has undergone a 1.7% increase in emission rate, compared to the 1.3% yearly increase observed during the previous decade. In 2018, nitrous oxide (N_2O) emissions increased by 0.8%, primarily due to manufacturing and agricultural practices, contrasting with the 1% yearly growth observed over the previous decade. Conversely, fluorinated emissions rose markedly in 2018 by 6.1%, in contrast to an annual increase of 4.6% over the previous decade. The UN Framework Convention on Climate Change (IPCC) recently published an article estimating that anthropogenic activities have caused a temperature increase of 1.0°C since the pre-industrial era, with a probable range of 0.8°C–1.0°C for context.

11.3 STATUS OF CC

The term "climate change" denotes a modification in atmospheric conditions mostly caused by the emission of GHGs. Global warming is chiefly attributed to the emission of GHGs, which facilitate the absorption of energy by the Earth's atmosphere

(Patil 2019). The natural environment and anthropogenic activities are the principal sources of these emissions. Forest fires, seismic activities, oceans, permafrost, wetlands, soil volcanoes, and eruptions exemplify natural systems. Conversely, human activities are chiefly linked to energy generation, industrial processes, land utilization, forestry, and changes in land use (Lo 2016). The Earth's ecological system is inherently self-regulating; nonetheless, anthropogenic emissions contribute to an excess of GHG pollution, according to a statistical analysis of global GHG emissions from human activities and biological systems.

11.3.1 Impacts of CC, Threats, and Vulnerabilities

Analysing the significant impacts of CC on individuals and ecosystems, along with the associated risks and vulnerabilities, is an essential first step in recognizing the gravity of our current predicament. The United Nations Climate Change Secretariat (UNCCS) has reported changes to climate indicators, including temperature, precipitation, sea level rise, ocean acidity, and extreme weather events. Documented climate risks included droughts, floods, hurricanes, severe storms, heat waves, landslides, wildfires, and extreme cold. In 2018, the worldwide community encountered 315 significant natural disasters (Nussbaumer 2009). These events constituted around 93% of the total expenses. Moreover, a significant contributor to CO_2 emissions is wildfires. Although forest fires are part of the organic structure, it is evident that human-induced emissions are directly disrupting and exacerbating the consequences of emissions from the natural system (Patil 2019). A key contributing component to most global natural disasters is anthropogenic CC (Patil 2019). In numerous nations, climate risks exacerbate vulnerability across various sectors due to factors such as temperature fluctuations, precipitation variability, modified seasonal patterns, altered disease distribution, desertification, oceanic effects, and degradation of soil and coastal regions (White 2011). The industry sectors most susceptible to CC impacts include nutrients, water, health, ecosystems, human housing, and infrastructure, with Africa identified as the region most vulnerable to climate variability, based on an empirical analysis of the risks and adaptation readiness of 192 United Nations member countries. It is essential to acknowledge the interrelationships among these industries and their resultant impacts. A range of climate realities was delineated comprehensively. The United Nations Economic Forum has published the latest edition of the 2020 Global Organization Risks Study, detailing the regions most affected. Risks encompass fatalities from diseases and natural disasters, as well as excessive strain on ecosystems, especially marine and aquatic systems (Patil and Bohara 2020). Moreover, the security of food and water is another domain that is significantly impacted. Severe weather, natural disasters, and rising sea levels are expected to result in increased migration. As governments attempt to exploit resources along terrestrial and aquatic borders, geopolitical tensions and disputes are likely to emerge. The research also addresses the adverse impact of increasing systematic risks on capital markets. Finally, WEF (2020) highlights the impact on supply chains and commerce (Bhatt 2018).

11.4 CLIMATE ACTION

The global acknowledgement of CC as a legitimate issue commenced with the inaugural global climate congress held in Geneva in 1979. The World Meteorological Organization (WMO) convened the worldwide climate conference in response to climatic records from the preceding decade (Canton 2006). The primary aim was to convene technological and professional scientists to evaluate recent discoveries on climate fluctuations and shifts induced by both anthropogenic and natural factors, as well as to examine potential hazards and future implications, with the goal of formulating recommendations for the future (Kumar 2022). This gathering may have been the inaugural event of its kind to discuss the adverse effects of CC. Consequently, the UNEP founded the IPCC in 1988 to furnish nations and regulatory bodies with scientific data and statistics for formulating CC-related policies (UNEP 2011). This was accomplished by the WMO. The Kyoto Protocol was ratified at the 3rd UNFCCC meeting of the parties (COP-3) in 1997 and came into effect in 2005. From 2008 through 2012, industrialized nations were mandated to decrease their emissions in accordance with the Kyoto Protocol. The Protocol provided three market-based strategies to achieve these objectives and delineated the pertinent mechanisms for policy, tracking, and reporting (Motisi 2022). The convention offered two project-orientated strategies: clean development and collaborative education (Wise et al. 2014). Affluent nations can finance and initiate emissions reduction initiatives in developing countries through the Clean Growth Process, promoting sustainable development (SD) in the recipient nation while mitigating the CO_2 emissions of the contributing party (Paudel et al. 2023). Nonetheless, such programmes may be established in other affluent nations that are signatories to the pact through collaborative implementation efforts, counterbalancing the excess emissions of the investing party. Furthermore, to facilitate the trade of annually allocated emissions, Protocol members have conserved those that exceed their limitations; hence, the Protocol incorporated a system for emissions exchange. The adoption of energy-efficient techniques, renewable energy sources, and afforestation/reforestation programmes has been instrumental in reducing emissions (Patil 2019). The concept of carbon pricing and a carbon emissions market was founded on the overarching framework and the components of the Kyoto Protocol. Numerous municipal and national administrations have established emissions trading programmes; some are mandatory, while others are discretionary. These programmes occasionally align with the Kyoto agreements and regulations. The European Carbon Trading Scheme is the largest carbon trading programme ever implemented. The effect of instituting a trading system for GHG emissions programmes on actual emission reductions across six primary emission sectors is under investigation (Sarkodie and Strezov 2019). The research produced multiple conclusions. The proposed technique can reduce GHG emissions by approximately 1.58% annually. Furthermore, in contrast to a scenario without implementation, a decrease of approximately 23.43% in emissions can be achieved 10 years post-launch (Villoria-Sáez et al. 2016). Carbon taxes are an additional mechanism for mitigating emissions that has garnered significant attention in academic discourse. Nonetheless, political opposition from local and business entities is the primary obstacle preventing numerous countries from implementing such a structure, despite increasing scientific evidence to the contrary.

At the 21st UNFCCC meeting, the Paris Protocol was ratified by each member. The COP-21 conference took place in Paris in 2015 and became effective in 2016. The Paris Agreement enhanced the existing framework for addressing CC by incorporating supplementary objectives, commitments, monitoring requirements, compliance demands, and advice systems (Warner 2018). The agreement incorporated new legally binding commitments, mandating each participant to execute officially designated roles and national activities to meet and potentially exceed these responsibilities. The contract promotes enhanced openness, regulation, and explicit monitoring and engagement. In addition to mandatory activity, the contract promotes voluntary collaboration between the two parties. Moreover, the agreement stipulates financial and technical assistance, as well as capacity-building programmes for poor nations (Winslow et al. 2009). Researchers assessed 188 countries represented by 161 planned contributions established by national governments to conduct a comprehensive, systematic analysis of the effectiveness of measures specified in the Paris Agreement (Mann 2009). The research assessed emissions in accordance with these INDCs and examined sectorial policies in each of these countries. The calculation indicates that the optimal scenario for global emissions in 2030 would entail an annual increase of around 19.3% compared to the baseline period of 2005–2015. An estimated 31.5% increase in global emissions would transpire in the absence of intervention. The conclusion indicates a minimum temperature increase of 3°C between 2030 and 2050, assuming emissions remain constant under the best-case scenario. Furthermore, if annual emissions continue to escalate, a 4°C increase would be assured (Vit 2018).

The current unconditional nationally determined contributions for 2030 are estimated to be between 52 and 58 Gt CO_2e per year. Human-induced GHG emissions must decrease by 45% by 2030 compared to a baseline level, with the goal of achieving net-zero emissions by 2050, as indicated by simulations related to a global warming scenario of 1.5°C. By 2030, we need to reduce emissions by approximately 25% from 2010 levels in order to maintain the current trajectory. By the end of the century, we expect the temperature increase to have reached 2°C. Net-zero emissions must be achieved by 2070 (Wesche and Armitage 2014). According to increasing statistics, the temperature objectives of the Paris Agreement cannot be achieved solely by current mitigation strategies or prospective emissions commitments. To achieve these objectives, further measures and innovative reduction strategies must be explored (Secretariat 2009).

11.5 UNFCCC ADAPTATION FUND

The CC adaptation is increasingly recognized by the international community as a serious issue, and official development aid (ODA) and multilateral financing are being used more frequently to support adaptation efforts in vulnerable nations. Founded as part of the UNFCCC's Kyoto Convention, the Adaptation Fund (AF) was created to supply funding for concrete adjustment schemes and efforts (referred to as "projects") in developing countries that are especially vulnerable to the consequences of the effects of CC (Siebenhüner 2003). The key issue still stands as to how to allocate funds among project supporters and the qualified nations where the projects are situated in a reasonable and equitable manner (Bedeke 2023).

A portion of the money received from authorized project operations must be used "to assist developing country parties that are particularly vulnerable to the adverse effects of CC in meeting the costs of adaptation." This establishes the legal foundation for the UNFCCC AIn 2001; the decision was made to establish an AF to provide funding for specific adaptation programmes and initiatives (Decision 10/CP.7). After that, a number of choices resulted in the AF's actual physical launch in 2007, and in April of 2010, the Air Force Base first published a call for ideas (Perdan and Azapagic 2011).

Partially overlapping eligibility and allocation criteria govern who is qualified to get benefits from the AF. In order to be considered eligible, a country must meet three requirements: (i) be classified as a developing nation; (ii) possess membership in the Kyoto Convention; and (iii) exhibit particular susceptibility. Currently, 149 nations meet the first two requirements, which are simple to confirm. In contrast, the AFB has not defined or applied "particularly vulnerable" clearly. The historical overview of the methods used to identify specific cases of susceptibility, such as tiny states on islands, nations with low-lying coastlines, nations with areas susceptible to catastrophic events, tiny island developed states (SIDS), least developed countries (LDC), and African countries, has been presented in a number of preambular writings, articles, and decisions.

The following factors will be considered when distributing funds from the Resilience Fund to eligible parties:

a. Vulnerability level;
b. Urgency level and delay risks;
c. Ensuring equitable and balanced access to the fund;
d. Gathering learning from the planning and execution of projects and programmes;
e. Securing regional co-benefits when practical;
f. Maximizing cross- or multi-sectorial advantages; and
g. Adaptive capability to the adverse impacts of CC.

Again, vulnerability is not a binary concept (that is, especially or not especially susceptible); rather, it is the comparative "level of" something. The AFB had multiple meetings to address vulnerability and has invited specialists to share research findings on measuring using indices and indicators. However, the AFB has decided against adopting a more operational definition if it can't be done more broadly within the UNFCCC. Therefore, it doesn't seem likely that a more restrictive definition of vulnerability will be implemented very soon. This is because projects should be driven by the nation and because it is in the self-interest of all AFB members not to want to potentially exclude their member nation from the top (Pirages and DeGeest 2003).

11.6 CLEAN DEVELOPMENT MECHANISMS

11.6.1 The Origins of the Clean Development Mechanism

The broader discourse regarding the relationship between "CC" and "SD," concepts that emerged in the late 1980s, constitutes the foundation of the Clean Development Mechanism (CDM)'s history. Historically, the concepts have largely remained segregated for a lengthy duration. A perspective centred on the social sciences and humanities has been employed to articulate the SD theme, whereas a naturalistic discipline has underpinned the CC argument (Matsuo 2003). The historical division is thoroughly examined and articulated in the academic literature. The divide continued until about 2001–2002, when the Third Assessment Report of the IPCC and attendees at the UN Conference on Sustainable Development formed forums focused on the integration and interrelations between CC and SD. Subsequently, an expanding body of research has examined the following subjects, highlighting trade-offs and opportunities between CC and SD: adaptation, poverty, equity, and the role of the CDM in advancing SD are four perspectives representing the Global South. The third section will analyse the final issue in relation to the article's subject matter. The Kyoto negotiations served as the impetus for the establishment of the CDM. Since the adoption of the Kyoto Protocol in 1997, various designations have been assigned to the "baby," including the CDM. These encompass "the Kyoto anomaly," "the mutually beneficial mechanism," "a pathway between the North and South," and "the principal challenger of the Kyoto Regime." The names signify the early enthusiasm and optimism that the CDM would resolve significant disparities regarding growth and CC between the North and the South (Gosnell et al. 2019). The latter stages of the negotiation allowed minimal time for deliberating terms and conditions; hence, the CDM agreement was established without any operational guidelines. The principal components of the "rule book" and the formation of the Executive Board for the CDM were established during COP-7 in Marrakesh, 2001 (Figure 11.1). Developing nations argued that a global standard for SD would infringe upon their sovereignty during the examination and determination of the institutional structure governing the CDM. Consequently, at the federal level, designated national authorities (DNAs) were assigned the responsibility of achieving SD. Since that time, deliberations on international policy have not explicitly considered the impact of the CDM on SD. It has been decontextualized and sensitively addressed in the research literature and in talks about "programmatic CDM" (Gupta and Grubb 2000).

11.6.2 CDM and SD

It remains challenging to ascertain whether the CDM contributes to SD overall and poverty alleviation specifically, as this evaluation is confined to the project's Designing Paperwork, which serves as its principal data source. Consequently, evaluations are confined to the specified objectives indicated in the project documentation, and the project developer's preliminary viability assurances are not subsequently validated. As a result, the government's DNA or a recognized operating entity, which validates Project Design Documents (PDDs) in violation of the CDM's regulations and protocols, does not provide ongoing oversight to ensure that the claims made in the PDD are implemented during project execution. In practical terms, this indicates that PDDs are evaluated by the DNA according to a predefined set of criteria that includes sustainability at both societal and ecological levels. Regrettably, the SD criteria of numerous host nations are confusing and imprecise, comprising predominantly qualitative characteristics that permit excessive interpretation. Furthermore, PDDs rarely contain quantitative data related to SD. Some argue that the "goodwill" of participants inside the carbon audit system influences the extent of the examination conducted.

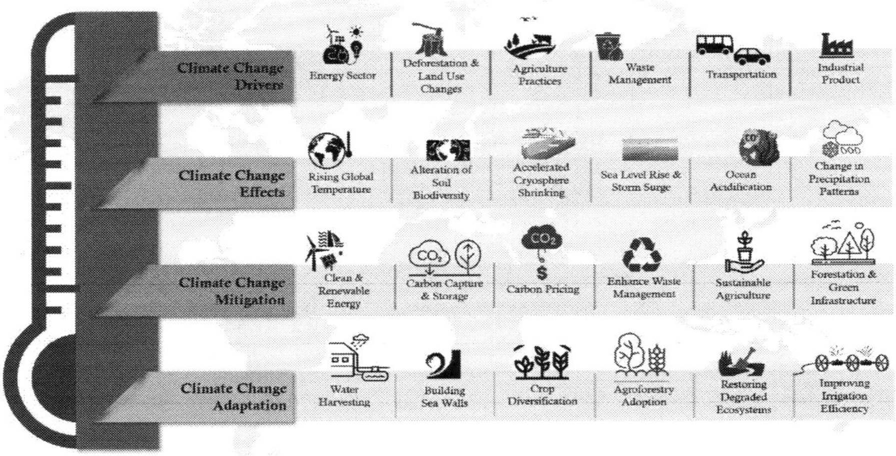

FIGURE 11.1 Current divers, effects of CC, mitigation, and strategies adaptation for CC.

Different host countries have very different views on SD, and some say that they don't look as closely at the expected benefits of SD as they do at reducing emissions, which is spelled out in the PDD. Although various nations prioritize different aspects of SD, the primary emphasis is often on financial considerations. Delegating the meaning of SD to the recipient nation likely resulted in its interpretation as "growth in GDP" in many instances. Furthermore, big CDM projects, especially when compared to countries with small CDM portfolios, show that host countries like China, India, and Brazil have a lower SD profile (Olsen and Fenhann 2006). Research demonstrates a significant association between host country influence and SD indicator values, particularly in the domain of social indicators. Moreover, the absence of universal norms for SD and the intense rivalry among developing countries to attract CDM initiatives result in a preference for cost-effective emission reduction programmes. This is intensified by the absence of substantial price incentives for investments in efforts that enhance SD (Patil 2019). As a result, the focus often shifts to prioritizing emission reduction over anticipated SD potential. This agreement may lead to a decline in SD requirements as host countries vie for carbon development mechanism investments. This competitive dynamic poses a substantial risk to the fundamental purpose of SD. Moreover, under the existing arrangement, both host nations and purchasing corporations do not receive direct incentives for maintaining elevated sustainability criteria. The ongoing competition aimed at securing "rapid and effortless certification" is likely to hinder the adoption of enhanced sustainability requirements (Patil 2019). A substantial body of research indicates that the CDM fails to advance the SD objectives of the host country, thereby not fulfilling its purpose of aiding poor nations in attaining SD. A continual difficulty is that SD advantages are generally undervalued in the CDM market framework. There is an urgent requirement for improvements to the CDM, especially regarding the establishment of standardized SD criteria and the execution of a comprehensive ex ante evaluation of the project's expected SD benefits (Pachauri et al. 2009). Furthermore, there is a demand for quality assurance procedures to guarantee that CDM projects authentically contribute to SD. Research emphasizes the necessity of a common understanding of what defines SD in order for programmes to facilitate effective participation. Initially, we will concentrate on the potential advantages offered by the CDM before discussing ideas for its enhancement (Lynn-Jones and Miller 1995).

11.6.3 Creating Climate Vision and Global Level Strategy

11.6.3.1 Protection of Natural Carbon Sink Resources

Using and safeguarding organic carbon sink resources is crucial for mitigating the effects of global warming. Forests, marshes, and oceans serve as natural carbon sinks by absorbing and sequestering carbon dioxide from the atmosphere. Mitigating anthropogenic land-use alterations, drainage, and deforestation can preserve these carbon sinks and prevent substantial quantities of CO_2 from being released into the atmosphere. Furthermore, the restoration and enhancement of natural carbon sinks might facilitate carbon sequestration, thereby mitigating the effects of global warming (Afionis 2017). Reforestation programmes can facilitate the restoration of forests lost due to land-use changes, while wetland rehabilitation can enhance carbon storage in aquatic environments (White 2011). Initiatives aimed at enhancing yields and carbon returns in agricultural systems, along with measures to mitigate carbon losses, substantially contribute to CC reduction. Ultimately, mitigating the impacts of CC necessitates the conservation and judicious utilization of carbon sink resources, and SD must prioritize initiatives aimed at enhancing and rehabilitating these ecosystems (Berrang-Ford et al. 2011). Organic soils contain around 5% of the global terrestrial organic carbon, despite constituting just 3% of the land area. Various types of organic soils exist, including fens, which obtain water from groundwater; bogs, which derive water from precipitation; and folic histosols, characterized by significant organic surface layers predominantly formed in cold climates (Boyd and Goodman 2011).

11.7 AFFORESTATION AND FOREST ECOSYSTEM RESTORATION

To keep global warming below the 2°C threshold set by the Paris Climate Agreement, emissions must be curtailed and GHGs must be extracted from the atmosphere. Consequently, in addition to the substantial decreases in fossil fuel consumption that are necessary, terrestrial ecosystems are crucial for alleviating the effects of CC. As previously noted, over the past decade, terrestrial ecosystems have absorbed around 30% of yearly human carbon emissions, with forests comprising the majority of this absorption. For many years, reforestation and afforestation have been crucial elements of natural climate solutions (NCSs) and are seen as key potential methods for alleviating the impacts of CC. Although terrestrial ecosystems are essential for carbon sequestration, their degradation has contributed around 25% of post-industrial GHG emissions. As a result, discussions about climate policy in numerous nations have significantly increased, with a focus on defining and quantifying the potential impact of ecosystem restoration on mitigating CC. Restoring 15% of changed land in critical areas globally may trap gigatonnes of CO_2 and save 60% of anticipated extinctions. This constitutes 14% of total emissions, or 30% of the complete rise in atmospheric CO_2 during the Industrial Revolution. Ecosystem restoration involves more than merely replanting or forest rehabilitation, despite these methods being among the most prevalent restoration strategies. Wetland and grassland ecosystems, along with various other non-forest ecosystems, have significant restoration potential for mitigating CC. The execution of restoration initiatives for carbon sequestration and biodiversity, however, may face considerable practical obstacles. It is crucial to recognize the risks of overstating the advantages of ecosystem restoration for the climate, as this may undermine mitigation initiatives and distract from the core objective of reducing carbon emissions in the energy and industrial sectors. Moreover, enduring ecosystem restoration relies on transitioning restoration methods from biologically centred approaches to a synergy that encompasses both climatic and socioeconomic dimensions. To realize the co-benefits of ecosystem restoration and climate mitigation, it is urgently necessary to develop long-term strategies, monitoring frameworks, and standardized techniques to comprehensively assess the benefits, costs, and risks associated with ecosystem restoration (Brown and Dworkin 2010).

11.8 DEVELOPING AND ADOPTING RENEWABLE AND CLEAN ENERGY TECHNOLOGIES

Fossil fuels are essential for facilitating globalization, rapid industrialization, and the expansion of the global population. Fossil fuels are hydrocarbon-containing chemicals that originated from the leftovers of plants and fauna in the Earth's crust. The three principal fossil fuels are natural gas, petroleum, and coal. Significant concealed expenses are linked to humanity's use of fossil fuels to sustain our modern and opulent lifestyles. By 2027, renewable energy is anticipated to prevail in the global electrical supply, transforming the worldwide power composition. More than 90% of grid-scale capacity is derived from hydropower; nevertheless, it is constrained by the substantial multibillion-dollar investment necessary and the availability of suitable locations (Fawzy et al. 2020). By 2027, solar photovoltaic (PV) systems will surpass coal in global installed electrical capacity. The development of large-scale energy storage technologies is crucial for the integration of variable and intermittent renewable energy into the electrical grid. The carbon capture, utilization, and storage (CCUS) approach involves extracting carbon dioxide from emissions generated by industrial facilities or power plants. We use the extracted CO_2 for diverse purposes and store the remaining CO_2 in long-term storage locations like geological formations. CCUS is considered a crucial technology for reducing GHG emissions and mitigating CC (Ford and King 2015). During this procedure, CO_2 is directly taken from the atmosphere or from emission sources, such as power plants. It is subsequently conveyed to geological locations for either long-term storage or conversion into goods. CCUS plays a crucial role in mitigating global warming by reducing CO_2 emissions from fossil fuels or directly extracting CO_2 from the atmosphere. In the absence of CCUS, maintaining the atmospheric CO_2 equivalent concentration at 450 parts per million would be impossible. CCUS has the potential to substantially diminish global GHG emissions. The IPCC 86 reports that CCUS has the potential to reduce CO_2

emissions by 3.0–6.8 billion metric tonnes per year by 2050 (Mariappan et al. 2023). Additionally, according to the International Energy Agency's SD scenario, CCUS might account for 15% of the overall CO_2 emission reductions necessary to achieve carbon neutrality by 2070. CCUS is the sole technology in the energy sector capable of attaining nearly zero emissions while utilizing fossil fuels, and coal power plants equipped with this technology can ensure the flexibility of the entire energy supply system (Hinkel and Bisaro 2016). Moreover, CCS can assist the steel and cement sectors in contributing approximately 34% and 48%, respectively, towards attaining net-zero CO_2 emissions. Furthermore, when combined with direct air capture (DAC) or bioenergy, CCUS can achieve zero CO_2 emissions, thereby freeing up emissions for other technologies. Despite being in the Research and Development (R&D) phase, DAC is expected to assume a pivotal role in the future (Koetz and others 2008). DAC may represent one of the ultimate technological approaches for the direct extraction of CO_2 from the atmosphere. Bioenergy with carbon capture and storage is attractive to energy systems as it both generates energy and sequesters carbon, often employed to manage emissions that go beyond allocated limits.

11.9 ADOPTING AND DEVELOPING ORGANIC FARMING

Approximately 30% of global GHG emissions are attributable to the food chain. Under a "business as usual" scenario, emissions from the food system are projected to reach 1356 Gt CO_2e from 2020 to 2100, surpassing the 1.5°C threshold between 2051 and 2063, hence obstructing the achievement of the 1.5°C–2°C targets, even if fossil fuel emissions were immediately curtailed. Five techniques exist for mitigating GHG emissions from the global food chain. Organic farming is a crucial approach to addressing the challenges posed by CC, providing a means to produce high-quality food using sustainable practices (Kohler 2019). Organic farming diminishes dependence on synthetic inputs, including chemical fertilizers and pesticides, by extending crop rotations, promoting herbaceous plant cover, and reducing tillage. The reintroduction of organic waste enhances soil health and fertility. This environmentally sustainable approach not only mitigates the effects of CC but also highlights the economic feasibility of resource use. Through the diversification of crop rotations, organic farming improves the resilience of agricultural systems to changing climate circumstances, especially in warmer and drier climates. Consequently, organic farming technologies become essential instruments in promoting climate-resilient agricultural methods. When cereals, small grains, and cover crops are planted in between maize and soybeans, they effectively reduce water stress in maize at both the canopy and leaf levels. This leads to a significant $17.1\% \pm 6.1\%$ drop in yield losses during drought conditions. The integration of herbaceous cover enhances species richness and aids insectivorous species experiencing population reductions due to CC (Mariappan et al. 2023). The capacity of organic farming systems to reduce global agricultural GHG emissions is rooted in their capability to improve soil organic carbon (SOC). Practices such as reducing tillage intensity, incorporating inter-row loosening and cutting, and shallow inversion tillage enhance soil carbon reserves. Despite a slight (~5.5%) decrease in production and negligible increases in weed prevalence, these techniques are crucial, as any type of tillage can result in carbon redistribution to deeper strata and the mineralization of labile carbon components. In integrated livestock systems (ICL), this makes the ratios of clover-grass leys higher than in shorter ley intervals without grazing, which is important for getting a net carbon benefit. The ability of ecosystems to function as carbon sinks is a contentious issue, shaped by numerous unique elements that impact carbon breakdown and sequestration. These parameters include soil composition, plant attributes, and the characteristics of organic additions. The complex interaction of various elements concerning carbon sequestration requires additional research for a thorough understanding (Berrang-Ford et al. 2011).

11.10 CONCLUSION

The rapid development of sustainable measures for adaptation and mitigation is imperative given the current climate crisis. CC can lead to numerous detrimental effects on people, including evident issues such as severe heat and flooding, as well as changed transmission of illnesses and increased malnutrition due to heightened competition for water and agricultural resources. As the Earth warms, it modifies the transmission of infections and elevates the prevalence of vector-borne diseases, thereby affecting summer mortality rates from illness, particularly among the elderly. One of the most highly recommended and effective mitigation strategies offered by the Kyoto Protocol is the Carbon Distortion Mechanism (CDM). Formulating new policies necessitates the evaluation of both current and revised legislation. It is essential to broaden the array of policy instruments that are being developed in their development of CC policies. Local perspectives need to be combined with national and international action scales for scaling up policy measures to combat CC in order to attain the levels of decrease in carbon dioxide required to prevent hazardous climatic consequences.

REFERENCES

Afionis, S. 2017. *The European Union in International Climate Change Negotiations*. Routledge, London.

Azzopardi, B, MV Balzan, S Cherif, E Doblas-Miranda, M dos Santos, P Dobrinski, and E Xoplaki. 2020. "Climate and Environmental Change in the Mediterranean Basin–Current Situation and Risks for the Future." First Mediterranean Assessment Report.

Bedeke, SB. 2023. "Climate Change Vulnerability and Adaptation of Crop Producers in Sub-Saharan Africa: A Review on Concepts, Approaches and Methods." *Environment, Development and Sustainability* 25 (2): 1017–1051.

Berrang-Ford, L, JD Ford, and J Paterson. 2011. "Are We Adapting to Climate Change?" *Global Environmental Change* 21 (1): 25–33.

Bhatt, RP. 2018. "Climate Change Assessment, Impacts of Global Warming, Projections and Mitigation of Ghg Emissions Endorsing Green Energy." *International Educational Scientific Research Journal* 4: 33–48.

Boyd, E, and MK Goodman. 2011. "The Clean Development Mechanism as Ethical Development? Reconciling Emissions Trading and Local Development." *Journal of International Development* 23 (6): 836–854.

Brown, MA, and M Dworkin. 2010. "The Environmental Dimension of Energy Security 1." In Marilyn A. Brown, Michael Dworkin (eds), *The Routledge Handbook of Energy Security*, 176–190. Taylor & Francis .

Canton, J. 2006. *The Extreme Future: The Top Trends That Will Reshape the World in the next 20 Years*. Penguin.

Du Monceau, T, AB CO_2logic, and N Fujiwara. 2011. "Briefing Paper 'Political Lock-in in the Context of the CDM.'" AEA A world leading energy and climate change consultancy, London.

Egenhofer, C, and N Fujiwara. 2003. "The Emerging Climate Change Regime: The UNFCCC, the Kyoto Protocol Forever, Kyoto Modified… or yet Something Else. Final Report of a Study Prepared for Environmental Studies Group Economic and Social Research Institute, Cabinet Office, Government of Japan." Centre for European Policy Studies (CEPS), Brussels.

Fawzy, S, AI Osman, J Doran, and DW Rooney. 2020. "Strategies for Mitigation of Climate Change: A Review." *Environmental Chemistry Letters* 18: 2069–2094.

Ford, JD, and D King. 2015. "A Framework for Examining Adaptation Readiness." *Mitigation and Adaptation Strategies for Global Change* 20: 505–526.

Gosnell, H, N Gill, and Michelle Voyer. 2019. "Transformational Adaptation on the Farm: Processes of Change and Persistence in Transitions to 'Climate-Smart' Regenerative Agriculture." *Global Environmental Change* 59: 101965.

Gupta, J, and MJ Grubb. 2000. *Climate Change and European Leadership: A Sustainable Role for Europe?* Vol. 27. Kluwer Academic Publisher, The Netherlands..

Heller, MPS. 2003. *Who Will Pay? Coping with Aging Societies, Climate Change, and Other Long-Term Fiscal Challenges*. International Monetary Fund, Washington, USA.

Hinkel, J, and A Bisaro. 2016. "Methodological Choices in Solution-Oriented Adaptation Research: A Diagnostic Framework." *Regional Environmental Change* 16: 7–20.

Koetz, T and others. 2008. "The Role of the Subsidiary Body on Scientific, Technical and Technological Advice to the Convention on Biological Diversity as Science–Policy Interface." *Environmental Science & Policy* 11 (6): 505–516.

Kohler, PM. 2019. *Science Advice and Global Environmental Governance: Expert Institutions and the Implementation of International Environmental Treaties*. Anthem Press, UK and USA.

Kumar, A. 2022. "Challenge of Climate Change." In *Ecosystem-Based Adaptation*, 39–104. Elsevier. doi:10.1016/B978-0-12-815025-2.00002-2.

Lo, V. 2016. "Synthesis Report on Experiences with Ecosystem-Based Approaches to Climate Change Adaptation and Disaster Risk Reduction." The Secretariat of the Convention on Biological Diversity, Quebec, Canada.

Lynn-Jones, SM, and SE Miller, eds. 1995. *Global Dangers: Changing Dimensions of International Security*. Mit Press, Massachusetts, London.

Mann, ME. 2009. "Do Global Warming and Climate Change Represent a Serious Threat to Our Welfare and Environment?" *Social Philosophy and Policy* 26 (2): 193–230.

Mariappan, S, AD Raj, S Kumar, and U Chatterjee. 2023. "Global Warming Impacts on the Environment in the Last Century." In Uday Chatterjee, Angela Oyilieze Akanwa, Suresh Kumar, Sudhir Kumar Singh, Abira Dutta Roy (eds), *Ecological Footprints of Climate Change: Adaptive Approaches and Sustainability*, 63–93. Springer International Publishing, Cham.

Matsuo, N. 2003. "CDM in the Kyoto Negotiations." *Mitigation and Adaptation Strategies for Global Change* 8 (3): 191–200.

Motisi, M. 2022. "*Treating Youth with Eco-Anxiety: An Acceptance and Commitment Therapy Model*." PhD Thesis, Widener University.

Nieto, MI, O Barrantes, L Privitello, and R Reiné. 2018. "Greenhouse Gas Emissions from Beef Grazing Systems in Semi-Arid Rangelands of Central Argentina." *Sustainability* 10 (11): 4228.

Nussbaumer, P. 2009. "On the Contribution of Labelled Certified Emission Reductions to Sustainable Development: A Multi-Criteria Evaluation of CDM Projects." *Energy Policy* 37 (1): 91–101.

Olsen, KH, and JV Fenhann. 2006. "Sustainable Development Benefits of Clean Development Projects."

Pachauri, RK, L Gnacadja, MZ Cutajar, A Steiner, S Briceno, J Ogwu, and S Dimas. 2009. *Facing Global Environmental Change: Environmental, Human, Energy, Food, Health and Water Security Concepts*. Vol. 4. UNEP Riso Centre, Energy Climate and Sustainable Development, Germany .

Patil, P. M., Mahamuni, P. P., Abdel-Daim, M. M., Aleya, L., Chougule, R. A., Shadija, P. G., & Bohara, R. A. (2019). Conversion of organic biomedical waste into potential fertilizer using isolated organisms from cow dung for a cleaner environment. *Environmental Science and Pollution Research*, 26(27): 27897–27904.

Patil, PM, and RA Bohara. 2020. "Nanoparticles Impact in Biomedical Waste Management." *Waste Management & Research* 38 (11): 1189–1203.

Paudel, D, RC Neupane, S Sigdel, P Poudel, and AR Khanal. 2023. "COVID-19 Pandemic, Climate Change, and Conflicts on Agriculture: A Trio of Challenges to Global Food Security." *Sustainability* 15 (10): 8280.

Perdan, S, and A Azapagic. 2011. "Carbon Trading: Current Schemes and Future Developments." *Energy Policy* 39 (10): 6040–6054.

Pirages, DC, and TM DeGeest. 2003. *Ecological Security: An Evolutionary Perspective on Globalization*. Rowman & Littlefield Publishers, Lanham, MD.

Sarkodie, SA, and V Strezov. 2019. "Economic, Social and Governance Adaptation Readiness for Mitigation of Climate Change Vulnerability: Evidence from 192 Countries." *Science of the Total Environment* 656: 150–164.

Secretariat, GCOS. 2009. "Implementation Plan for the Global Observing System for Climate in Support of the UNFCCC (2010 Update)." In *Proceedings of the Conference of the Parties (COP), Copenhagen, Denmark*, 7–18.

Siebenhüner, B. 2003. "The Changing Role of Nation States in International Environmental Assessments–the Case of the IPCC." *Global Environmental Change* 13 (2): 113–123.

Villoria-Sáez, P, VW Tam, M del Río Merino, CV Arrebola, and X Wang. 2016. "Effectiveness of Greenhouse-Gas Emission Trading Schemes Implementation: A Review on Legislations." *Journal of Cleaner Production* 127: 49–58.

Vit, M. 2018. "Desertification and Environmental Refugees: Concrete Problems in the Context of Climate Change." Università Ca' Foscari.

Warner, K. 2018. "Coordinated Approaches to Large-Scale Movements of People: Contributions of the Paris Agreement and the Global Compacts for Migration and on Refugees." *Population and Environment* 39 (4): 384–401.

Wesche, Sonia D, and Derek R Armitage. 2014. "Using Qualitative Scenarios to Understand Regional Environmental Change in the Canadian North." *Regional Environmental Change* 14: 1095–1108.

White, G. 2011. *Climate Change and Migration: Security and Borders in a Warming World.* OUP USA.

Winslow, M, M Akhtar-Schuster, M Cherlet, C Martius, S Sommer, R Thomas, and J Vogt. 2009. "Monitoring and Assessment Science to Support Decision-Making by the United Nations Convention to Combat Desertification (UNCCD)." In *American Geophysical Union*, Fall Meeting 2009, abstract id: GC21B-01, San Francisco, CA.

Wise, RM, I Fazey, MS Smith, SE Park, HC Eakin, EA Van Garderen, and B Campbell. 2014. "Reconceptualising Adaptation to Climate Change as Part of Pathways of Change and Response." *Global Environmental Change* 28: 325–336.

12 Low-Carbon and Climate-Resilient Cities

Amanullah Adeel, Pawar Pratiksha Shivaji, Sushama Tanaji Patil, and Abdulahad Ahadi

12.1 INTRODUCTION

Urbanization is considerably responsible for climate change. Several reasons, including cement, machinery, lifestyle, industrialization as well as heating and cooling, can be listed. The use of cement in urbanization has a high impact on climate change because of the significant greenhouse gas (GHG) emissions linked to cement manufacture (Chen et al. 2022). Since 2000, carbon dioxide (CO_2) emissions from the world's cement industry have increased significantly; in 2019, process emissions amounted to 1.57 Gt CO_2 (Guo et al. 2021). Moreover, lifestyle shapes the energy consumption, resulting in GHG emissions as the root of climate change (Hussain et al. 2019). In terms of urban industries, climate patterns are significantly shaped by urbanization (McDonald et al. 2011). This human-induced driver of climate change converts the greenery land to concrete land, increases transportation, and significantly increases GHG emission, namely carbon dioxide. Additionally, studies across the world show that there is a complex relationship between climate change and urban heating and cooling, emphasizing the need for energy-efficient building designs, precise climate projections, and sustainable building design to reduce the environmental impact of urban heating and cooling systems (Canales et al. 2020; Condori et al. 2022; Eshraghi et al. 2021; Verichev et al. 2020).

The term "Low Carbon" refers to various programmes and methods intended to lower GHG emissions, including carbon dioxide, to mitigate climate change and advance sustainability. This entails building low-carbon communities that eliminate or offset emissions caused by humans, switching from carbon-intensive sectors to cleaner, renewable ones, and creating low-carbon products like petrol with lower emissions (Boswell et al. 2019; Han 2023; Pan et al. 2022). On the other hand, "Low-Carbon" cities are defined as metropolitan areas that prioritize environmental friendliness and sustainability through the promotion of green technologies and the reduction of carbon emissions. Studies reveal that urban green technological innovation is greatly stimulated by policies such as China's low-carbon city pilot, especially through green funding mechanisms (Liu et al. 2024). In order to mitigate climate change, it is imperative that low-carbon cities be developed, with an emphasis on areas like energy structure, economic development, and technological innovation (Shen et al. 2023). A research on Chinese cities reported that low-carbon city plans, particularly in areas with robust government support and developed industry

DOI: 10.1201/9781003492733-12

structures, successfully led to low carbon; nevertheless, this achievement was possible by the implementation of environmental legislation and technological improvements (Zhao et al. 2023). Reducing carbon footprints and achieving low-carbon growth in urban areas improve energy security and pave the way for a more sustainable future.

Moreover, the ability of people, communities, and systems to foresee, plan for, respond to, and recover from the effects of climate change—such as extreme weather events and rising sea levels—is referred to as climate resilience (Adeel and Jadhav 2024). It entails adjusting to shifting circumstances, lowering susceptibility, and enhancing adaptive capability to resist and overcome climate-related difficulties (Lonsdale et al. 2024). Similar to low-carbon cities, in light of the growing effects of climate change, developing climate-resilient cities is imperative. An investigation highlighted the significance of urban adaptation policies in augmenting climate resilience via mechanisms such as innovation in technology, involvement of the public, and suitable fines for ecological harm (Huang and Yi 2024). Urban regions, which account for more than 70% of global emissions, must integrate resilience and enhance risk management into policies in order to mitigate the effects of climate change (Díaz et al. 2024). Recycling is essential for fostering resilience and the shift to a low-carbon economy, as well as for urban adaptation and GHG mitigation (Figueras and Pasqualetto 2024). A different study developed the Smart Resilient City Assessment Framework, which emphasizes integrated management and preparedness, to assess cities' resilience and smartness in managing climate change-induced difficulties (Khatibi et al. 2024). When it comes to strengthening urban resilience against climate change, foresight approaches in susceptible cities such as those in Bangladesh emphasize the role that governance plays in promoting sustainability and resilience (Hossain and Fernández-Güell 2024).

Globally, there is growing concern about the climate threats that urban regions confront, with an emphasis on African metropolises like Lagos and Port Harcourt in Nigeria. In a study, several kinds of floods can occur in urban settings, such as coastal flooding, major river flooding, small stream flooding, and inadequate drainage. In light of the effects of climate change, the study highlights the necessity for local, national, and international governments and organizations to assume accountability for lowering flood risks (Douglas et al. 2008). In this context, a pivotal effort used a social-ecological-technological systems (SETS) vulnerability index to create an urban flood vulnerability index (Chang et al. 2021). The study emphasizes how complicated urban floods may be and how important it is to have a spatially explicit risk framework that can be used in other areas with comparable problems. Along with flood risks, heatwaves and sea-level rise, which are climatic consequences, also threaten the cities. Climate change events, including heatwaves, are projected to disproportionately affect urban regions due to their diverse populations and vulnerabilities. This has led to discussions about how climate change may exacerbate urban inequality. A research reported the negative consequences of heatwaves on human health (Tapia et al. 2017). However, another effort has looked into how quickly expanding cities might use urban green technologies like urban agriculture and arboriculture as part of their natural solutions to adapt to climate change (Zhao et al. 2023). Sea-level rise is a pressing issue that poses significant threats to coastal cities around the world. In terms of this issue, many studies have focused on the consequences like economic effect of Hurricane Sandy in New York (Jevrejeva et al. 2016) and projected rapid sea-level rise (Yin et al. 2009) rather than a practical suggestion on how to prevent it.

Accordingly, this chapter will discuss strategies for building low-carbon cities, strategies for enhancing climate resilience, case studies and best practices, as well as challenges and barriers.

12.2 STRATEGIES FOR BUILDING LOW-CARBON CITIES

The ability of a system to "bounce back" or revert to a prior stable state following stressors brought on by any danger is known as resilience (Sarker et al. 2020). Cities are the major sources of GHG emissions; therefore, specific strategies are required to build a low-carbon city. Some examples of these strategies are energy efficiency and renewable energy, sustainable transportation, urban planning and green infrastructure, waste management, and circular economy.

The first strategy is building energy efficiency system via renewable sources. Green technology innovation is essential for encouraging the use of renewable energy sources and energy efficiency in the context of low-carbon emissions in cities. A study conducted on Chinese cities shows a strong correlation between low-carbon energy transition and green technological innovation, especially in non-resource towns with strict environmental restrictions (Zhao et al. 2024). Furthermore, by integrating external green power sources and encouraging ultra-low energy consumption buildings, electric cars, distributed photovoltaic systems, and heat pumps, urban energy planning that is centred on carbon reduction pathways can result in significant decreases in per capita emissions (Yu et al. 2023). Cities can make progress towards reaching carbon neutrality targets and promoting sustainable urban development by emphasizing energy efficiency and renewable energy projects. Achieving low-carbon strategies in urban areas needs some practical approaches. In this context, two essential tactics for cutting carbon emissions are retrofitting existing structures and putting smart grids in place. Accordingly, a study emphasized the importance of building retrofitting towards net zero carbon (NZC) in order to prevent climate change and meet sustainability targets (Ang et al. 2023). Another pivotal investigation suggested a system for scaling these assessments to all communities at a reasonable cost called urban building energy models (UBEMs), which are crucial for identifying technological possibilities for decreasing emissions (Berzolla et al. 2023).

Furthermore, lowering CO_2 emissions from existing buildings requires low-carbon retrofitting based on Green Building Rating Systems (GBRS), with an emphasis on energy and materials as crucial factors (Liu et al. 2022). Along with retrofitting, building smart grid plays a vital role in promoting a low-carbon strategy. Smart grids reduce carbon emissions by optimizing electricity management through the use of advanced technologies like demand response capabilities and digital communication (Tian 2024). Moreover, the deployment of smart grids increases the benefits associated with reducing carbon emissions by improving power use efficiency and minimizing power loss (Peng and Lin 2023). In addition, Market mechanisms and efficient dispatching further enhance the low-carbon growth of smart distribution grid systems, lowering overall operating costs and carbon emissions (Zhang et al. 2024). These tactics, when combined, can make a big difference in cities all over the world in meeting their low-carbon emission targets.

The second strategy that should be made for low-carbon cities is sustainable transportation. To attain sustainable transport in low-emission cities, it is imperative to tackle issues including urbanization, motorization, and the adverse effects on the environment, health, road safety, and mobility (Maheshwari and Gubreley 2023). Fighting global warming and lowering the hazards that climate change poses to people's health, way of life, and economic growth requires the development of low-carbon transportation networks (Kariuki and Kuria 2023). It has been demonstrated that switching from conventional fuels to low-carbon alternatives can effectively reduce emissions and improve performance in both on-road and marine transportation (Fattah et al. 2024). Adopting low-carbon city policies and plans, encouraging active transport such as cycling, and enhancing public transport infrastructure are essential measures for the development of sustainable urban transport systems with reduced carbon emissions. Cities can move towards sustainability and lower carbon emissions by concentrating on neighbourhood revitalization and creating integrated urban transportation systems that integrate all kinds of mobility (Ahmad 2023).

The third strategy for the purpose of low-carbon emission in cities is urban planning and green infrastructure. With the community serving as the fundamental unit for successful implementation, low-carbon urban community planning emphasizes resource conservation, low-waste emissions, high operating efficiency, and green and sustainable urban activities (Yu 2022). In order to achieve sustainable development, low-carbon city construction must lower carbon emissions, foster innovation, cooperation, greenness, openness, and sharing, and investigate eco-city planning concepts and techniques (Zhang 2021). In a research, 35 large Chinese cities had urban green infrastructure that can offset 0.33% of carbon emissions from using fossil fuels, but with the right laws and management techniques, this efficiency might rise significantly (Chen 2015).

The fourth strategy is waste management and promoting circular economy. Circular economy strategies are being adopted by cities more frequently in an effort to improve waste management and lower carbon emissions. Cities can attain economic growth, environmental sustainability, and social advantages by putting techniques like waste-to-energy, urban agriculture, industrial symbiosis, and sustainable transportation into practice (Salles et al. 2024). By valuing waste products, circular economy principles are converting waste into a useful resource and promoting urban and regional development (Savini 2023). The potential for low-carbon emission techniques in waste management is further demonstrated by the construction of closed-loop supply chain networks based on circular economy concepts, which can dramatically cut electronic waste carbon emissions in smart city developments (Gumte and Mitra 2023). These methods emphasize how crucial it is to involve multiple stakeholders (Figure 12.1), develop innovative business strategies, and work together with other cities to create an urban environment that is more low carbon and sustainable.

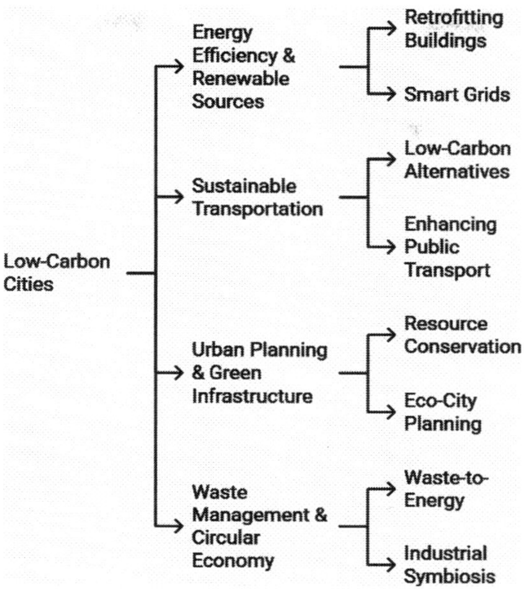

FIGURE 12.1 Strategies for building low-carbon cities.

12.3 STRATEGIES FOR ENHANCING CLIMATE RESILIENCE CITIES

Cities need to practice resilience strategies against the immediate consequences of climate change. The top suggested strategies are infrastructure resilience (e.g., flood defences, heat-resistant buildings), social resilience (community engagement and education), nature-based solutions (NBS) (urban forestry, wetlands, and green roofs), governance and policy frameworks (policies, regulations, and governance structures that support resilience). Firstly, by reducing the effects of climatic extremes on urban water management, infrastructure resilience is essential to the construction of climate-resilient cities (Yang 2024). This investigation also highlighted the need to comprehend resilience mechanisms to improve the efficiency of water-related infrastructures when confronted with climate-related difficulties. In smart cities, green infrastructure—which combines natural and artificial elements—is emphasized as a means of enhancing mental and physical well-being, lowering rates of illness and death, and encouraging social cohesiveness (Kapse and Jangale 2024). Furthermore, through public involvement, technological innovation, and appropriate environmental regulations, policy interventions that improve urban climate adaptation have been shown to significantly improve climate resilience, with varying effects across different regions and economic development levels (Huang and Yi 2024). To optimize infrastructure resilience and ensure the long-term sustainability of cities in the face of climate change problems, a comprehensive approach focusing on cognitive, functional, correlative, and organizational skills is necessary (Yang 2024).

Secondly, by addressing the varying degrees of susceptibility that urban inhabitants have to climate change, social resilience is essential to the construction of climate-resilient cities. Significant gains in climate resilience have been demonstrated by policies that support urban adaptation to climate extremes through mechanisms like technology innovation, public participation, and appropriate fines for environmental degradation (Huang and Yi 2024). With a focus on socio-ecological resilience and socio-spatial practices of resistance, resilient cities are understood as a collective practice of adaptation and mitigation to climatic changes (Gantus-Oliveira 2023). Cities are acknowledged as a component of the response to climate change, functioning as intricate techno-social systems that can spur change via peer-to-peer learning and networked governance, thereby supporting the creation of climate-resilient communities (Alexandra 2023). To minimize trade-offs and maximize co-benefits while constructing systemic climate resilience, cities can benefit from applying a systems approach to climate change that disentangles economic, social, and environmental systems (Matsumoto and Bohorquez 2023).

Thirdly, using natural resources and ecosystem services to lessen the effects of climate change, these NBS are essential for improving climate resilience in urban settings. These methods improve cities' liveability and resilience to harsh weather events by managing stormwater, regulating the climate, and implementing adaptation strategies (Aghaloo et al. 2024). Especially in areas vulnerable to climate oscillations like Colombia, implementing NBS in public spaces can help manage rainwater for reuse, lower urban temperatures, and get cities ready for climate change problems (Vazin et al. 2024). Furthermore, NBS provide co-benefits for health and well-being while also assisting in the reduction of carbon emissions, fostering adaptation, and enhancing resilience in both urban and wildland settings (Marino et al. 2024). NBS may effectively increase urban resilience to climate change by implementing a strategy process that involves identifying requirements, involving stakeholders, and evaluating results. This method offers a sustainable means of addressing environmental, economic, and social concerns in urban areas.

Lastly, governance and policy frameworks are crucial for climate-resilient cities. To better understand urban resistance to climate change, further research on the connections between technological, political, and ecological traditions is necessary for urban climate governance (Boyd and Juhola 2015). A governance framework was suggested by a study to improve the way the management of urban essential infrastructures is integrated into the city's climate resilience plan (Lomba-Fernández 2020). The framework was created using scholarly literature, and it has been enhanced by knowledge gained from expert and practitioner interviews and workshops. Thus, four facets of governance are addressed by the framework: (1) modes of governance; (2) roles of stakeholders; (3) processes; and (4) instruments. The vertical integration, cross-sectoral coordination, innovation and knowledge, and cooperation that are necessary to increase city resilience will be enhanced by these factors. Urban climate resilience governance frameworks encompass three key components: transformative capacity (community-oriented urban planning, communication technology, and collaboration) and adaptive capacity (education, health, food, and water), as well as absorptive capacities, protective infrastructure, and urban green space that offer community support (Zeng et al. 2022). An essential component of international efforts to mitigate the risk of disasters and the negative consequences of climate change is urban management (Figure 12.2). Resilience should be a key component of new urban policies (da Silva et al. 2019).

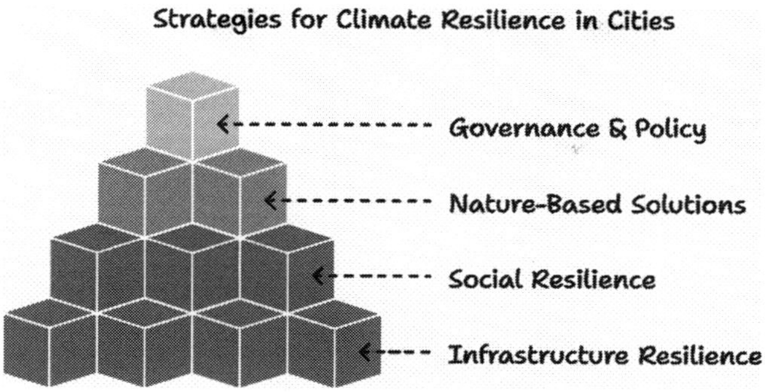

FIGURE 12.2 The diagram shows a stacked strategy pyramid for building climate resilience in cities, with infrastructure resilience as the cornerstone, followed by social resilience, nature-based solutions, and ending with governance and policy as the over-arching framework.

12.4 CASE STUDIES AND BEST PRACTICES FOR LOW-CARBON AND CLIMATE-RESILIENT CITIES

Since the growing climate change concerns, many cities have implemented excellent approaches towards climate resilience and low-carbon emissions. However, there is no city worldwide equipped with all aspects of resilience and low-carbon strategies, but some cities have implemented a few of the aspects successfully. According to the problem and solutions performed, the cities have been named as "Healthy City", "Garden City", "Resilient City", "Eco-city", "Sustainable City", "Green City", "Low-carbon City", "Ill City" and "Low-Carbon-Resilient City" (Figure 12.3). For further details about the above terms, readers are referred to He et al. (2020).

Vienna is a shining example of a city that has effectively adopted low-carbon and climate resilience initiatives. A research demonstrates Vienna's all-encompassing strategy, which combines technological advancements with NBS such as green infrastructure to dramatically lower ambient air temperatures (Bügelmayer-Blaschek et al. 2023). This method shows that urban greening projects have the ability to lower temperatures by up to 2°C (Bügelmayer-Blaschek et al. 2023). The study added that the public's strong support for sustainable urban development is demonstrated by the great readiness of Vienna's inhabitants to financially support the execution and upkeep of extensive greening initiatives. Another pivotal study analysed case studies from Copenhagen, Curitiba, Singapore, Stockholm, and Melbourne; the study demonstrates creative approaches to waste management, green space integration, renewable energy uptake, and sustainable transportation (Gürçam 2024). These cities serve as models, exhibiting effective solutions like well-run public transit networks, extensive waste recycling initiatives, support for renewable energy sources, and a focus on green areas. The main takeaway from this study is that in order to fully realize the potential of sustainable urbanization, coordinated, cross-sectoral initiatives are desperately needed. The biggest built network of cities for climate resilience is C40. This network included 100 mayors' agreement to decline their urban emission to 50% by 2030 and consequently keeping the temperature to 1.5°C (https://www.c40.org). In this great network, 5 Indian and 13 Chinese cities are

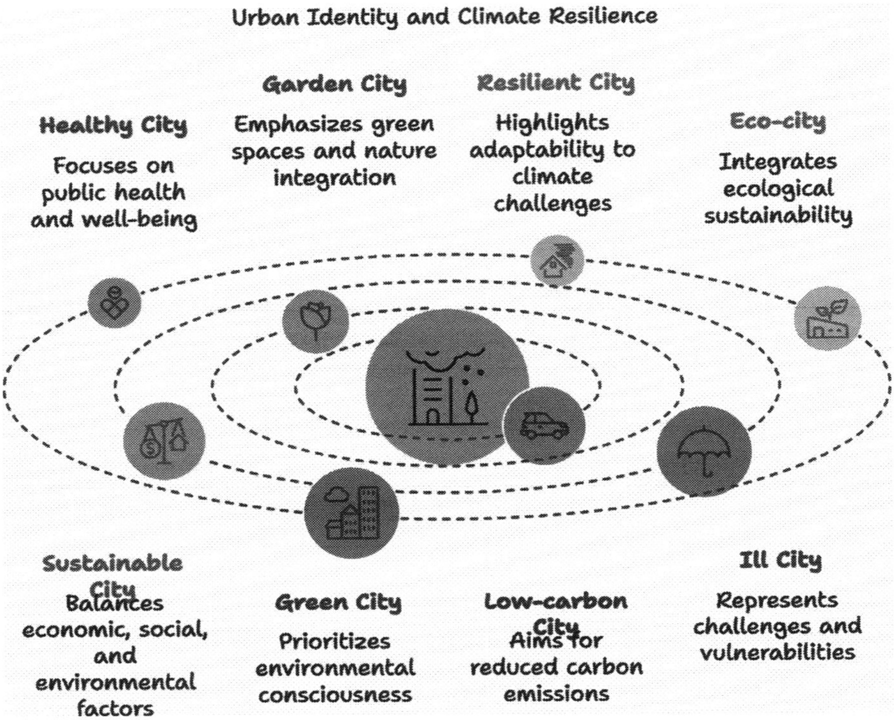

FIGURE 12.3 This chart plots different urban typologies, Healthy City, Garden City, Resilient City, Eco-city, showing how cities can create specific identities that enable climate resilience through health, sustainability, green infrastructure, carbon reduction, and social wellbeing as well as recognising vulnerabilities in the shape of the ill City.

included. C40 aimed to work together across borders; they take swift, cooperative action on climate change that is in line with science-backed aims. Their goal is to safeguard people and communities worldwide and to create a more resilient, sustainable, and equitable future.

12.5 CHALLENGES AND BARRIERS FOR LOW-CARBON AND CLIMATE-RESILIENT CITIES

12.5.1 Technical and Financial Barriers

One major factor impeding the development of low-carbon and climate-resilient cities is the presence of technical and financial constraints. The lack of professional interest, knowledge, and experience in sustainable building design, as well as uneven and insufficient regulatory frameworks, are examples of technical barriers (Wehbi and Messadi 2024). Cities confront financial difficulties because of their creditworthiness, lack of institutional capability, and lack of accountability. To overcome these obstacles and make projects financially feasible and attractive enough to draw in co-investments from the private sector, a move towards well-targeted concessional funding, blended finance approaches, and capacity-building programmes is necessary. Cities can advance towards social equity, energy efficiency, readiness for risks, efficient governance, and innovation in their pursuit of low-carbon resilience by tackling these financial and technical obstacles (He et al. 2020).

12.5.2 Social and Political Challenges

Building low-carbon and climate-resilient cities involves overcoming social and political obstacles. The social challenges include the exposure of vulnerable populations to climate disasters due to limited resources and capacity (Gaborit 2022), the systemic inequalities impeding just transitions (Shah 2024), and the necessity of understanding power relations between politics, stakeholders, and industry for relevant policies (Matsumoto and Bohorquez 2023). On the other hand, political challenges include the need for multi-stakeholder cooperation and dialogue to achieve transformative change and foster trust in climate disaster response processes (Gaborit 2022), the limited state capacity exacerbating other issues in urbanization management (Shah 2024), and the low-carbon transition research's neglected focus on institutional and political dimensions (Verma et al. 2023). Creating inclusive, resilient, and sustainable cities in the face of pressures from urbanization and climate change requires addressing the aforementioned issues.

12.5.3 Implementation and Monitoring

Inconsistent legislative frameworks, insufficient building rules, and professionals' lack of interest in, understanding of, and experience with sustainable building design are some of the difficulties for building low-carbon and climate-resilient cities (Wehbi and Messadi 2024). Additionally, it is challenging to choose indicators and benchmarks for efficient monitoring due to the complexity of adaptation measures like green space management and home rehabilitation (Díaz et al. 2024). In order to combat these problems, a theoretical framework that emphasizes the significance of institutional, regulatory, financial, technical, and communication viewpoints in overcoming implementation challenges has been proposed to analyse integrated strategies and key performance indicators for low-carbon resilient cities (He et al. 2020). Cities may advance social equality, energy efficiency, readiness for risks, effective governance, and innovation in their pursuit of low-carbon resilience by employing workable techniques and frameworks. Figure 12.4 highlights some major barriers and challenges that cities face in their efforts to become resilient and low carbon.

12.6 CONCLUSION

Low-carbon and climate-resilient cities are essential to address the two problems of growing urbanization and global climate change. These cities can be green, healthy, resilient, and eco-friendly when they are built with a lower carbon footprint and increased resistance to climate-related hazards like heat waves, flooding, and sea-level rise. Enhancing energy efficiency, implementing renewable energy,

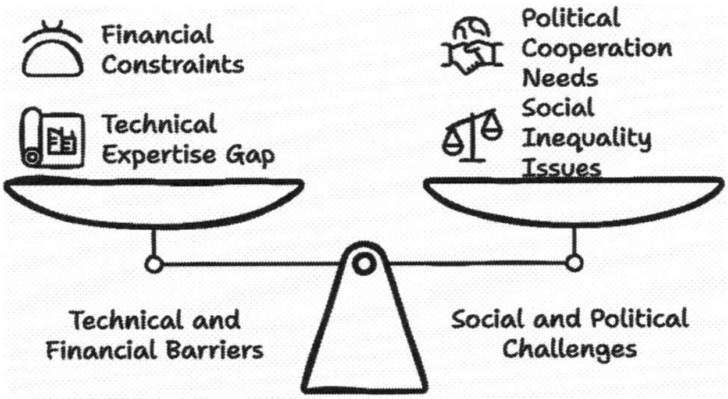

FIGURE 12.4 The scales reveal the delicate balance of challenges we face: on one side, technical and financial barriers; on the other, complex social and political issues. All must be weighed and addressed for true progress.

encouraging sustainable mobility, and incorporating green infrastructure into urban planning are among the strategies to accomplish these objectives. Reducing environmental effects also requires a circular economy approach and waste management techniques. Developing social resilience via community involvement and education, creating strong infrastructure, and putting NBS like urban forestry and green roofs into practice are all essential to building climate resilience. Strong policy frameworks and efficient governance are also necessary to support these activities. Global case studies offer insightful perspectives on effective tactics and impart lessons that might be implemented in other places. Nonetheless, some challenges and barriers need to be overcome, such as financial and technical limitations, social and political difficulties, and problems with implementation and monitoring. Notwithstanding these challenges, the significance of making the shift to low-carbon and climate-resilient urban systems is becoming increasingly apparent. To build a resilient and sustainable future, the chapter urges communities, urban planners, and legislators to take proactive measures. Cities can take the lead in reducing the effects of climate change and improving the quality of life for their citizens by adopting those strategies, which will ultimately help create more resilient and sustainable cities.

REFERENCES

Adeel, A., and A. S. Jadhav. 2024. "Identifying the Optimal Sugarcane Residue Percentage for Soil Enhancement: A Statistical and GIS Approach." *Circular Economy and Sustainability*,. doi:10.1007/s43615-024-00456-5

Aghaloo, Kamaleddin, Ayyoob Sharifi, Nader Habibzadeh, Tausif Ali, and Yie-Ru Chiu. 2024. "How Nature-Based Solutions Can Enhance Urban Resilience to Flooding and Climate Change and Provide Other Co-Benefits: A Systematic Review and Taxonomy." *Urban Forestry & Urban Greening* 95 (May): 128320. doi:10.1016/j.ufug.2024.128320

Ahmad, Tanveer. 2023. "Steering Cities towards Sustainability: Low-Carbon Transportation and Community Regeneration." doi:10.31219/osf.io/6agrc

Alexandra, Jason. 2023. "Cities and Climate-Resilient Development." In *Handbook on Climate Change and Technology*, edited by Frauke Urban and Johan Nordensvärd, 247–260. Edward Elgar Publishing. doi:10.4337/9781800882119.00027

Ang, Yu Qian, Zachary Michael Berzolla, Samuel Letellier-Duchesne, and Christoph F. Reinhart. 2023. "Carbon Reduction Technology Pathways for Existing Buildings in Eight Cities." *Nature Communications* 14 (1): 1689. doi:10.1038/s41467-023-37131-6

Berzolla, Zachary, Yu Qian Ang, Samuel Letellier-Duchesne, and Christoph Reinhart. 2023. "*An Eight-Step Simulation Based Framework to Help Cities Reach Building-Related Emissions Reduction Goals*." SSRN Scholarly Paper. Rochester, NY. doi:10.2139/ssrn.4528659

Boswell, Michael R., Adrienne I. Greve, and Tammy L. Seale. 2019. "Strategies for Creating Low-Carbon Communities." In *Climate Action Planning: A Guide to Creating Low-Carbon, Resilient Communities*, edited by Michael R. Boswell, Adrienne I. Greve, and Tammy L. Seale, 132–171. Washington, DC: Island Press/Center for Resource Economics. doi:10.5822/978-1-61091-964-7_5

Boyd, Emily, and Sirkku Juhola. 2015. "Adaptive Climate Change Governance for Urban Resilience." *Urban Studies* 52 (7): 1234–1264. doi:10.1177/0042098014527483

Bügelmayer-Blaschek, Marianne, Martin Schneider, Tanja Tötzer, Michael Friesenecker, Thomas Thaler, Antonia Schneider, Michael Getzner, Sebastian Seebauer, Claudia Hahn, and Maja Zuvela-Aloise. 2023. *Climate Resilience of the City of Vienna: Social Impact of Nature-Based Solutions*. EGU23–9178. Copernicus Meetings. doi:10.5194/egusphere-egu23-9178

Canales, Fausto A., Piotr Jadwiszczak, Jakub Jurasz, Marcin Wdowikowski, Bartłomiej Ciapała, and Bartosz Kaźmierczak. 2020. "The Impact of Long-Term Changes in Air Temperature on Renewable Energy in Poland." *The Science of the Total Environment* 729 (August): 138965. doi:10.1016/j.scitotenv.2020.138965

Chang, Heejun, Arun Pallathadka, Jason Sauer, Nancy B. Grimm, Rae Zimmerman, Chingwen Cheng, David M. Iwaniec, et al. 2021. "Assessment of Urban Flood Vulnerability Using the Social-Ecological-Technological Systems Framework in Six US Cities." *Sustainable Cities and Society* 68 (May): 102786. doi:10.1016/j.scs.2021.102786

Chen, Cuihong, Ruochong Xu, Dan Tong, Xinying Qin, Jing Cheng, Jun Liu, Bo Zheng, Liu Yan, and Qiang Zhang. 2022. "A Striking Growth of CO_2 Emissions from the Global Cement Industry Driven by New Facilities in Emerging Countries." Environmental Research Letters 17 (4): 044007. doi:10.1088/1748-9326/ac48b5

Chen, Wendy Y. 2015. "The Role of Urban Green Infrastructure in Offsetting Carbon Emissions in 35 Major Chinese Cities: A Nationwide Estimate." *Cities* 44 (April): 112–120. doi:10.1016/j.cities.2015.01.005

Condori, J., G. A. Lora, and S. D. Camargo. 2022. "The Impact of Climate Change on Electricity Demand in the City of Huancayo." *IOP Conference Series: Earth and Environmental Science* 1008 (1): 012001. doi:10.1088/1755-1315/1008/1/012001

da Silva, Cristiane Aparecida, Edicreia Andrade dos Santos, Stefania Maria Maier, and Fabricia Silva da Rosa. 2019. "Urban Resilience and Sustainable Development Policies: An Analysis of Smart Cities in the State of São Paulo." *Revista de Gestão* 27 (1): 61–78. doi:10.1108/REGE-12-2018-0117

Díaz, Carlota García, David Zambrana-Vasquez, and Carmen Bartolomé. 2024. "Building Resilient Cities: A Comprehensive Review of Climate Change Adaptation Indicators for Urban Design." Energies 17 (8): 1959. doi:10.3390/en17081959

Douglas, Ian, Kurshid Alam, Maryanne Maghenda, Yasmin Mcdonnell, Louise Mclean, and Jack Campbell. 2008. "Unjust Waters: Climate Change, Flooding and the Urban Poor in Africa." *Environment and Urbanization* 20 (1): 187–205. doi:10.1177/0956247808089156

Eshraghi, Hamed, Mehran Ansari, Shahab Moshari, and Javad Gholami. 2021. "Climatic Zoning and per Capita Demand Forecast of Iran Using Degree-Day Method." *Advances in Building Energy Research* 15 (6): 683–708. doi:10.1080/17512549.2019.1654918

Fattah, I. M. R., M. Mofijur, Arridina S. Silitonga, M. A. Kalam, T. M. I. Mahlia, and Jeffrey D. Ampah. 2024. "Blending Low Carbon Fuels for Sustainable Transportation."

In *Reference Module in Earth Systems and Environmental Sciences*. Elsevier. doi:10.1016/B978-0-323-93940-9.00221-8

Figueras, Luciana Vilardo de Freitas, and Antonio Pasqualetto. 2024. "Recycling in Integrated Urban Adaptation Actions and Greenhouse Gas Mitigation: A Sustainable Approach for Climate Resilient Cities." *International Journal of Business Administration* 15 (2): 66. doi:10.5430/ijba.v15n2p66

Gaborit, Pascaline. 2022. "Resilience and Climate Disaster Management in Cities: Transformative Change and Conflicts." *Journal of Peacebuilding & Development* 17 (3): 257–273. doi:10.1177/15423166221128793

Gantus-Oliveira, Talita. 2023. "Cidades resilientes e a disputa sobre o discurso da agenda de redução de riscos e desastres." GEOUSP Espaço e Tempo (Online) 27 (3). doi:10.11606/issn.2179-0892.geousp.2023.200724

Gumte, Kapil, and Kishalay Mitra. 2023. "A Circular Economy Approach Toward Managing E-Waste in Indian Smart City." In *Applications of Operational Research in Business and Industries*, edited by Angappa Gunasekaran, Jai Kishore Sharma, and Samarjit Kar, 273–294. Singapore: Springer Nature. doi:10.1007/978-981-19-8012-1_18

Guo, Rui, Jiaoyue Wang, Longfei Bing, Dan Tong, Philippe Ciais, Steven J. Davis, Robbie M. Andrew, Fengming Xi, and Zhu Liu. 2021. "Global CO_2 Uptake by Cement from 1930 to 2019." *Earth System Science Data* 13 (4): 1791–1805. doi:10.5194/essd-13-1791-2021

Gürçam, Selçuk. 2024. "Paving the Way for Climate Resilience through Sustainable Urbanization: A Comparative Study." Lectio Socialis 8 (1): 17–34. doi:10.47478/lectio.1396483

Han, Mengyao. 2023. "Low-Carbon Transition." In *International Encyclopedia of Geography*, 1–7. John Wiley & Sons, Ltd. doi:10.1002/9781118786352.wbieg2173

He, Xiaoni, Merrisa Lin, Tse-Lun Chen, Bowei Lue, Po-Chih Tseng, Wenzhi Cao, and Pen-Chi Chiang. 2020. "Implementation Plan for Low-Carbon Resilient City Towards Sustainable Development Goals: Challenges and Perspectives." *Aerosol and Air Quality Research* 20 (3): 444–464. doi:10.4209/aaqr.2019.11.0568

Hossain, Maruf, and José-Miguel Fernández-Güell. 2024. "Envisioning the Climate Change Resilient Cities in Bangladesh Using Foresight Approach." In . Vol. 15. Universitat Politècnica de Catalunya, Grup de Recerca en Urbanisme. doi:10.5821/siiu.12807

Huang, Qianqian, and Chengzhi Yi. 2024. "How Does the Construction of Climate Adaptive Cities Enhance Climate Resilience?" *Journal of Environmental Planning and Management*, June, 1–23. doi:10.1080/09640568.2024.2362920

Hussain, Mudassar, Abdul Rahman Butt, Faiza Uzma, Rafay Ahmed, Samina Irshad, Abdul Rehman, and Balal Yousaf. 2019. "A Comprehensive Review of Climate Change Impacts, Adaptation, and Mitigation on Environmental and Natural Calamities in Pakistan." *Environmental Monitoring and Assessment* 192 (1): 48. doi:10.1007/s10661-019-7956-4

Jevrejeva, Svetlana, Luke P. Jackson, Riccardo E. M. Riva, Aslak Grinsted, and John C. Moore. 2016. "Coastal Sea Level Rise with Warming above 2°C." *Proceedings of the National Academy of Sciences of the United States of America* 113 (47): 13342–13347. doi:10.1073/pnas.1605312113

Kapse, Neha, and Tejaswini Jangale. 2024. "Resilience and Sustainability through Green Infrastructure." *Journal of Recent Activities in Architectural Sciences*, May, 6–17.

Kariuki, Dorcas Wambui, and John Muna Kuria. 2023. "Potential for Transition to Low-Carbon Urban Road Transport in Developing Countries: Insights from Literature." *International Journal of Advanced Research* 6 (1): 23–37. doi:10.37284/ijar.6.1.1084

Khatibi, Hamed, Suzanne Wilkinson, Lukuba N. Sweya, Mostafa Baghersad, and Heiman Dianat. 2024. "Navigating Climate Change Challenges through Smart Resilient Cities: A Comprehensive Assessment Framework." Land 13 (3): 266. doi:10.3390/land13030266

Liu, Ke, Jianglan Tian, Jianping Chen, and Yueming Wen. 2022. "Low-Carbon Retrofitting Path of Existing Public Buildings: A Comparative Study Based on Green Building Rating Systems." Energies 15 (22): 8724. doi:10.3390/en15228724

Liu, Xiaojun, Hongyuan Bi, and Yu Teng. 2024. "An Empirical Study of the Policy of Low-Carbon City Pilot on the Level of Green Technological Innovation." *Climate Economics and Finance* 2 (1): 1–11. doi:10.58567/cef02010001

Lomba-Fernández, Cinta. 2020. "Urban Critical Infrastructure's Governance Framework for Climate Resilient Cities." *European Journal of Sustainable Development* 9 (3): 145–145. doi:10.14207/ejsd.2020.v9n3p145

Lonsdale, Kate, Nigel Arnell, Tim Coles, Kate Lock, Emer O'Connell, Paul O'Hare, and Emma Tompkins. 2024. "Climate Resilience: Interpretations of the Term and Implications for Practice." In *Quantifying Climate Risk and Building Resilience in the UK*, edited by Suraje Dessai, Kate Lonsdale, Jason Lowe, and Rachel Harcourt, 15–23. Cham: Springer International Publishing. doi:10.1007/978-3-031-39729-5_2

Maheshwari, Hariom, and Manvi Gubreley. 2024. "Strategies and Policy Frameworks for Low Carbon Mobility in Indian Cities." *IJFMR - International Journal For Multidisciplinary Research* 5 (5). Accessed July 27. doi:10.36948/ijfmr.2023.v05i05.5936

Marino, Raul, M. Kuller, B. Zurita, L. Carrasco, P. Gosciniak, K. Kustosik, and A. Riveros. 2024. "Assessment of Public Space Potential for Climate Change Adaptation in Three Colombian Cities: Nature-Based Solutions and Urban Design Guidelines." *Water Practice and Technology* 19 (5): 1696–1709. doi:10.2166/wpt.2024.089

Matsumoto, Tadashi, and Mateo Ledesma Bohorquez. 2023. *Building Systemic Climate Resilience in Cities*. Paris: OECD. doi:10.1787/f2f020b9-en

McDonald, Robert I., Pamela Green, Deborah Balk, Balazs M. Fekete, Carmen Revenga, Megan Todd, and Mark Montgomery. 2011. "Urban Growth, Climate Change, and Freshwater Availability." *Proceedings of the National Academy of Sciences of the United States of America* 108 (15): 6312–6317. doi:10.1073/pnas.1011615108

Pan, Jiahua, Guiyang Zhuang, Yan Zheng, Shouxian Zhu, and Qianyi Xie. 2022. "Clarification of the Concept of a Low-Carbon Economy and the Analysis of Its Core Elements." In *Political Economy of China's Climate Policy*, edited by Jiahua Pan, 179–198. Singapore: Springer Nature. doi:10.1007/978-981-16–8789-1_11

Peng, Fangyucheng, and Hongzhen Lin. 2023. "Research on Low-Carbon Benefits of Smart Grid Based on Multiple Linear Regression Model." In *2023 IEEE 3rd International Conference on Social Sciences and Intelligence Management (SSIM)*, 258–262. Taichung, Taiwan: IEEE. doi:10.1109/SSIM59263.2023.10469306

Salles, Adriana, Camila Cervantes, and Luís Bragança. 2024. "Assessing the Impacts of Urban Circular Economy Practices on Economic Growth, Environmental Sustainability, and Social Benefits: A Case Study Analysis." In *4th International Conference "Coordinating Engineering for Sustainability and Resilience" & Midterm Conference of CircularB "Implementation of Circular Economy in the Built Environment,"* edited by Viorel Ungureanu, Luís Bragança, Charalambos Baniotopoulos, and Khairedin M. Abdalla, 482–490. Cham: Springer Nature Switzerland. doi:10.1007/978-3-031-57800-7_44

Sarker, Md Nazirul Islam, Yang Peng, Cheng Yiran, and Roger C. Shouse. 2020. "Disaster Resilience through Big Data: Way to Environmental Sustainability." *International Journal of Disaster Risk Reduction* 51 (December): 101769. doi:10.1016/j.ijdrr.2020.101769

Savini, Federico. 2023. "The Circular Economy of Cities: The Good, the Bad, and the Ugly." In *Turning up the Heat*, edited by Maria Kaika, Roger Keil, Tait Mandler, and Yannis Tzaninis. Manchester University Press. doi:10.7765/9781526168016.00029

Shah, Jagan. 2024. "Getting a Grip on Green – Indian Cities and the Promise of Sustainability, Resilience, Inclusion." In *Indien Im 21. Jahrhundert – Auf Dem Weg Zur Postindustriellen Ökonomie: India in the 21st Century – On Its Way to a Post-Industrial Economy*, edited by Markus Hans-Peter Müller, 287–308. Wiesbaden: Springer Fachmedien. doi:10.1007/978-3-658-43014-6_17

Shen, Liyin, Haijun Bao, Yi Yang, Zhenchuan Yang, Xiangrui Xu, Lingyu Zhang, Shiju Liao, and Ziwei Chen. 2023. "DUAL PERSPECTIVE DIAGNOSIS ON LOW CARBON CITY PERFORMANCE." *Journal of Green Building* 18 (3): 167–184. doi:10.3992/jgb.18.3.167

Tapia, Carlos, Beñat Abajo, Efren Feliu, Maddalen Mendizabal, José Antonio Martinez, J. German Fernández, Txomin Laburu, and Adelaida Lejarazu. 2017. "Profiling Urban Vulnerabilities to Climate Change: An Indicator-Based Vulnerability Assessment for European Cities." *Ecological Indicators* 78 (July): 142–155. doi:10.1016/j.ecolind.2017.02.040

Tian, Xiaohe. 2024. "Impact of Smart Grid Applications on Improving Energy Efficiency and Reducing Carbon Emissions." *Highlights in Science, Engineering and Technology* 83 (February): 448–455. doi:10.54097/twwf6681

Vazin, Farnoosh, Daniel W. M. Chan, Toktam Hanaee, and Hadi Sarvari. 2024. "Nature-Based Solutions and Climate Resilience: A Bibliographic Perspective through Science Mapping Analysis." *Buildings* 14 (6): 1492. doi:10.3390/buildings14061492

Verichev, Konstantin, Montserrat Zamorano, and Manuel Carpio. 2020. "Effects of Climate Change on Variations in Climatic Zones and Heating Energy Consumption of Residential Buildings in the Southern Chile." *Energy and Buildings* 215 (May): 109874. doi:10.1016/j.enbuild.2020.109874

Verma, Pramit, Justyna Chodkowska-Miszczuk, Agata Lewandowska, and Łukasz Wiśniewski. 2023. "Local Resilience for Low-Carbon Transition in Poland: Frameworks, Conditions and Opportunities for Central European Countries." *Sustainable Development* 31 (3): 1278–1295. doi:10.1002/sd.2500

Wehbi, Hanan, and Tahar Messadi. 2024. "Challenges That Impact the Development of a Multi-Generational Low-Carbon Passive House in a Small City." Designs 8 (3): 52. doi:10.3390/designs8030052

Yang, Zhuyu. 2024. *Resilience Optimisation of Urban Water-Related Infrastructures under Multi-Risk Scenarios*. EGU24–9740. Copernicus Meetings. doi:10.5194/egusphere-egu24-9740

Yin, Jianjun, Michael E. Schlesinger, and Ronald J. Stouffer. 2009. "Model Projections of Rapid Sea-Level Rise on the Northeast Coast of the United States." *Nature Geoscience* 2 (4): 262–266. doi:10.1038/ngeo462

Yu, Dan, Xiaohan Zhou, Hui Qi, and Fanyue Qian. 2023. "Low-Carbon City Planning Based on Collaborative Analysis of Supply and Demand Scenarios." *City and Built Environment* 1 (1): 7. doi:10.1007/s44213-023-00008-x

Yu, Yan. 2022. "Explore the Theoretical Basis and Implementation Strategy of Low-Carbon Urban Community Planning." *Frontiers in Environmental Science* 10 (August). doi:10.3389/fenvs.2022.989318

Zeng, Xun, Yuanchun Yu, San Yang, Yang Lv, and Md Nazirul Islam Sarker. 2022. "Urban Resilience for Urban Sustainability: Concepts, Dimensions, and Perspectives." Sustainability 14 (5): 2481. doi:10.3390/su14052481

Zhang, Min. 2021. "Research on Strategies of Low-Carbon City Planning and Construction." *E3S Web of Conferences* 248. EDP Sciences: 02037. doi:10.1051/e3sconf/202124802037

Zhang, Qian, Daxin Wang, Chanjuan Zhao, Xunting Wang, Jinjin Ding, Haiwei Wang, and Xuemeng Zhang. 2024. "Low-Carbon Operation of Smart Distribution Grid Based on Life Cycle Assessment and Ladder-Type Carbon Trading." *Renewable Energy* 231 (September): 120816. doi:10.1016/j.renene.2024.120816

Zhao, Congyu, Kangyin Dong, Kun Wang, and Farhad Taghizadeh-Hesary. 2023. "How Can Chinese Cities Escape from Carbon Lock-in? The Role of Low-Carbon City Policy." *Urban Climate* 51 (September): 101629. doi:10.1016/j.uclim.2023.101629

Zhao, Jiacheng, Naika Meili, Xiang Zhao, and Simone Fatichi. 2023. "Urban Vegetation Cooling Potential during Heatwaves Depends on Background Climate." Environmental Research Letters 18 (1): 014035. doi:10.1088/1748-9326/acaf0f.

Zhao, Xin, Fangxia Pan, Hyoungsuk Lee, and Xiaowei Ma. 2024. "Towards a Low-Carbon Future: Driving Urban Energy Transformation through Green Technology Innovation." In *Reference Module in Social Sciences*. Elsevier. doi:10.1016/B978-0-44-313776-1.00262-2

13 Climate Change and Flood Disaster

Prakash T. Waghmare, Somanath D. Pawar, Sachin S. Panhalkar, and Tushar T. Waghmare

13.1 INTRODUCTION

According to Berz et al. (2001) and the International Strategy for Disaster Reduction (ISDR 2004), floods are recurrent and catastrophic natural disasters that occur worldwide, impacting various sectors such as agriculture, the environment, the economy, health, society, infrastructure, cultural heritage, and energy on a global scale (Schumann and Möller 2015). An estimated 104 billion US dollars are lost globally each year due to flood disasters (GAR 2015). Flood disasters have the greatest effect on the poorest populations in developing countries, as these regions are more vulnerable due to high population density, inadequate infrastructure, widespread poverty, and low levels of education. Consequently, they are less prepared and less able to cope with the aftermath of floods (Irasema 2002).

Flood disasters, characterized by the inundation of land due to excessive rainfall, river overflow, storm surges, or rapid snowmelt, have long been a natural hazard faced by communities worldwide. However, the fingerprints of climate change are increasingly evident in the frequency, intensity, and spatial distribution of these events. As global temperatures rise and weather patterns become more erratic, the hydrological cycle undergoes significant disturbances, leading to shifts in precipitation patterns and alterations in the magnitude and frequency of extreme rainfall events (IPCC 2007). Consequently, natural disasters such as floods, droughts, and cyclones are becoming more frequent and severe. The Intergovernmental Panel on Climate Change (IPCC 2007) states, "in many areas where mean precipitation is expected to increase, wet extremes are projected to become more severe, and in areas where mean precipitation is projected to decrease, dry extremes are projected to become more severe." This is expected to result in increased floods in the Asian monsoon region and other tropical locations.

In 2010, significant and destructive floods hit several countries, including Pakistan, India, China, Colombia, and Australia. China experienced the most severe damage, with an estimated total loss of $51 billion due to river floods, while Pakistan recorded nearly 2,000 immediate deaths from monsoon flooding (Syvitski and Brakenridge 2013). Similarly, in 2011, severe floods affected continents such as Africa, North America, South America, and Asia, impacting countries like Mozambique, Namibia, Brazil, the United States, and the Philippines. Fatalities exceeded 50 in each flood event, with over 1,000 deaths reported in the Philippines and Colombia. Significant material damage

DOI: 10.1201/9781003492733-13

was observed, particularly in developed countries (Kundzewicz et al. 2012). In 2012, several countries experienced "killer floods," resulting in over 50 fatalities in nations such as Niger, Nigeria, Bangladesh, and Russia.

Global databases like EM-DAT, NatCatSERVICE, and Sigma indicate a rising trend in the frequency of reported flood disasters and associated financial losses (EM-DAT 2023; NatCatSERVICE 2023; Sigma, 2023). India, as a highly vulnerable country, has over 40 million hectares out of its total geographical area of 329 million hectares prone to flooding. These recurring events result in substantial loss of lives and damage to livelihoods, property, infrastructure, and public utilities. The average annual flood damage in India rose from Rs. 1,805 crore during the period 1953–2005 to Rs. 4,745 crore in the last decade, driven by factors such as population growth, urbanization, economic development in flood-prone areas, and global warming (Kumar et al. 2005; Ministry of Home Affairs, Government of India 2008).

As climate change grows to be a major global concern in the 21st century, it causes serious problems for economies, cultures, and ecosystems worldwide. Understanding the connection between climate change and flood disasters is crucial for improving risk management, preparing for adaptation, and fortifying resilience measures (World Bank 2008; Ghosh and others 2012).

13.2 TYPES OF FLOOD

13.2.1 Riverine Floods

Riverine floods are triggered by factors such as heavy rainfall or snowmelt, resulting in a river surpassing its normal capacity. This overflow causes water to spill over its banks, submerging nearby regions. Unlike flash floods, riverine floods unfold over an extended period and often offer some advance notice. These floods impact extensive stretches along the river's path and can cause substantial harm to infrastructure, agriculture, and communities residing in flood-prone areas.

13.2.2 Flash Floods

Flash floods are commonly originated by intense rainfall, often occurring within a short timeframe, or sudden snowmelt, resulting in rapid runoff. They may also arise due to dam or levee failures, or debris flows in mountainous regions. These floods are distinguished by their abrupt onset and rapid, forceful currents. They can emerge with minimal warning, particularly in urban or steep terrain locations, posing significant hazards. Flash floods have the potential to inflict extensive devastation on property, infrastructure, and human lives.

13.2.3 Coastal Floods

Coastal floods primarily stem from storm surges, where powerful winds associated with tropical storms or hurricanes propel seawater onto the shore. Additional factors contributing to coastal flooding include elevated tides, intense rainfall, and tsunamis. These floods result in the submersion of low-lying coastal regions, erosion of beaches and dunes, and destruction of coastal infrastructure like buildings, roads, and ports. They pose substantial threats to both coastal communities and ecosystems.

TABLE 13.1
Flood Disaster Subtypes Occurring in India (2000–2023)

Disaster Subtype	Count	Percentage (%)
Coastal flood	1	0.52
Flash flood	27	13.99
Flood (general)	57	29.53
Riverine flood	108	55.96
Total	193	100.00

Source: https://www.emdat.be accessed at 09-May-2024.

13.2.4 Urban Flooding

Urban flooding is often caused by a lack of adequate drainage infrastructure, increased runoff from impermeable surfaces like buildings and roads, and the disappearance of wetlands and natural floodplains. Heavy rainfall overflows storm water drainage systems, causing water to collect on roadways and flood basements and low-lying areas. Urban flooding disrupts transportation networks, causes property damage, and poses health hazards due to water contamination. Its impact is particularly pronounced in densely populated urban areas with inadequate infrastructure.

Table 13.1 represents the distribution of flood disaster subtypes occurring in India. Data reveal that riverine floods are the most common subtype, accounting for 55.96% of all flood disasters in India. Flash floods, representing 13.99% of flood disasters, are characterized by their rapid onset and high-intensity rainfall, often occurring in urban or mountainous regions. Floods categorized as "General" account for 29.53% of flood disasters in India. This category likely includes various types of floods not specifically classified as riverine or flash floods, such as urban floods or floods caused by tropical cyclones. Coastal floods, representing only 0.52% of flood disasters, occur due to storm surges, high tides, or tsunamis, primarily affecting coastal regions. Figure 13.1 illustrates the various flood disaster subtypes that have occurred in India over the years 2000, 2005, 2010, 2015, 2020, and 2023. The subtypes include river floods, flash floods, urban floods, and coastal floods, with each subtype represented by a different color or pattern within the chart.

13.3 ORIGIN OF FLOOD IN INDIA (2000–2023)

Table 13.2 provides an overview of the origin of floods in India, with percentages indicating the proportion of each origin within the total occurrences. Monsoon rainfall is the primary cause of flooding in India, accounting for a significant portion of flood events at 48.19%. Heavy rainfall, excluding monsoon rain, contributes to 21.76% of flood events in India. These events may occur outside the monsoon season or in localized areas due to convective thunderstorms or cyclonic disturbances. A notable proportion of flood events (15.03%) have an unidentified origin, indicating challenges in

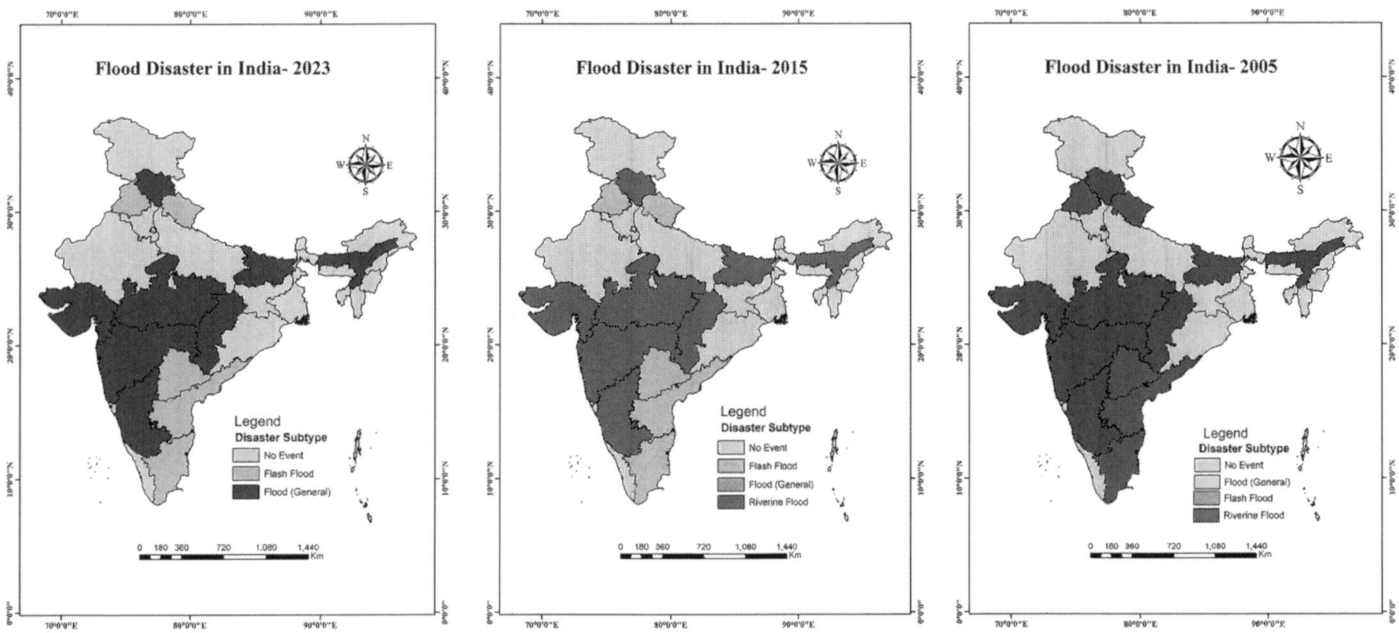

FIGURE 13.1 Flood disaster subtypes occurring in India (2000, 2005, 2010, 2015, 2020, & 2023). (*Source*: https://www.emdat.be accessed at 09-May-2024.)

(Continued)

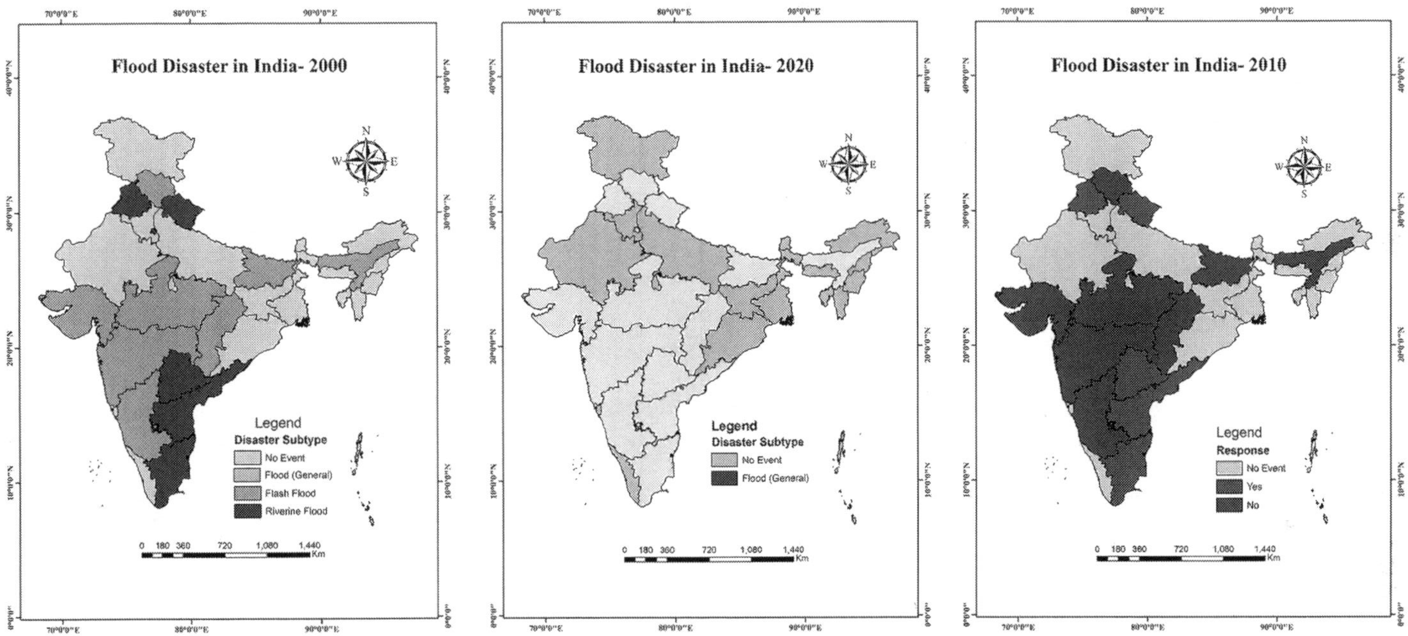

FIGURE 13.1 (*Continued*) Flood disaster subtypes occurring in India (2000, 2005, 2010, 2015, 2020, & 2023). (*Source*: https://www.emdat.be accessed at 09-May-2024.)

TABLE 13.2
Origin of Flood in India (2000–2023)

Origin of Flood	Count	Percentage (%)
Brief torrential rain	1	0.52
Early monsoonal rain	1	0.52
Heavy rain	42	21.76
Heavy rain and hailstorms	1	0.52
Heavy rain, overflowing rivers after dam opening in Nepal	1	0.52
Incessant rain	1	0.52
Long-term rainfall	1	0.52
Monsoon rain	93	48.19
Monsoon rain and cyclonic storm Ogni	1	0.52
Monsoon rain and tropical cyclone	1	0.52
Monsoon rain, dam releases	2	1.04
Monsoon rain, exacerbated by tail end of Cyclone Komen	1	0.52
Origin not identified	29	15.03
Pre-monsoon rains	2	1.04
Torrential monsoonal rain	2	1.04
Torrential monsoonal rains; Typhoon Kalmaegi	1	0.52
Torrential rain	10	5.18
Tropical Cyclone Nisha	1	0.52
Tropical depression in the Bay of Bengal	1	0.52
Tropical depression "Rovan"	1	0.52
Total	193	100.00

Source: https://www.emdat.be accessed at 09-May-2024.

accurately attributing the cause of floods. The table also lists various other origins of floods, including pre-monsoon rains, tropical cyclones, tropical depressions, hailstorms, and dam releases. These events contribute to smaller percentages of flood occurrences but highlight the diverse range of factors that can trigger flooding in India.

13.4 FREQUENCY OF FLOOD DISASTERS OF DIFFERENT RIVERS IN INDIA

The frequency of flood disasters in different states of India is visually represented using a word cloud generator in Figure 13.2. Each state's name is displayed with font size proportional to the frequency of flood disasters it has experienced. With the largest font size in the word cloud, Uttar Pradesh has experienced the highest frequency of flood disasters among the listed states (26 times). This highlights the significant vulnerability of Uttar Pradesh to flooding, primarily attributed to its location in the flood-prone Gangetic plains and its proximity to major rivers like the Ganges and Yamuna. Assam appears with a slightly smaller font size, indicating the state's

FIGURE 13.2 Frequency of flood disaster of different states in India. (*Source*: https://www.emdat.be accessed at 09-May-2024.)

second-highest frequency of flood disasters (25 times). Assam's susceptibility to flooding is well-documented, primarily due to its location in the Brahmaputra Valley, where heavy monsoon rainfall and the overflow of the Brahmaputra River lead to recurrent floods, affecting both rural and urban areas. Bihar is depicted with a font size reflecting its frequency of flood occurrences (20 times). Situated in the Gangetic plains, Bihar experiences severe flooding during the monsoon season, exacerbated by the overflow of rivers originating from the Himalayas. Floods in Bihar often result in widespread devastation, displacement, and loss of lives and livelihoods.

The states of West Bengal, Himachal Pradesh, Arunachal Pradesh, Orissa, Gujarat, Maharashtra, and Kerala appear with varying frequencies of flood disasters, reflecting the diverse geographical and climatic conditions across different regions of India. States like West Bengal, Orissa, and Kerala experience floods due to their coastal location and susceptibility to cyclonic storms and tidal surges, while Himachal Pradesh and Arunachal Pradesh face flood risks from rivers originating in the Himalayas. Gujarat and Maharashtra are prone to flooding due to heavy rainfall and riverine floods.

Figure 13.3 visually represents the frequency of flood disasters in different rivers of India using a word cloud generator. The river Ganga appears with the largest font size in the word cloud, indicating it has experienced the highest frequency of flood disasters among the listed rivers (13 times). The Ganga Basin is one of the most flood-prone regions in India, with floods occurring due to heavy rainfall, snowmelt from the Himalayas, and overflowing tributaries during the monsoon season. Following closely, the Brahmaputra River is depicted with a slightly smaller font size, representing its frequency of flood occurrences (10 times). The Brahmaputra Basin, particularly in Assam, experiences devastating floods, primarily driven by heavy monsoon rainfall and glacial melt from the Himalayas. Rivers like Bagmati, Kosi, Gandak, Godavari, Ghaghra, Rapti, and Vaitarani appear with varying font sizes corresponding to their respective frequencies of flood disasters. Each river basin faces unique flood risks influenced by factors such as geographical location, topography, rainfall patterns, and human activities.

FIGURE 13.3 Frequency of flood disaster of different rivers in India. (*Source*: https://www.emdat.be accessed at 09-May-2024.)

13.5 OFDA/BHA RESPONSE FOR FLOOD DISASTER

Table 13.3 presents the OFDA/BHA (Office of U.S. Foreign Disaster Assistance/Bureau for Humanitarian Assistance) response to flood disasters, indicating the percentage distribution of whether a response was provided or not. The majority of flood disasters, accounting for 95.85%, did not receive a response from the OFDA/BHA. This suggests that the affected areas did not receive direct assistance or support from this particular organization. Factors contributing to the lack of response may include the scale of the disaster, competing priorities, logistical challenges, or limitations in resources or capacity. In 4.15% of flood disasters, OFDA/BHA provided a response. This indicates instances where the organization intervened to provide humanitarian assistance, relief supplies, technical support, or funding to support disaster response and recovery efforts. The decision to provide a response may depend on various factors such as the severity of the disaster, the level of need, and the organization's mandate and available resources. While the percentage of flood disasters receiving a response from OFDA/BHA is relatively small, humanitarian assistance plays a crucial role in alleviating the impacts of disasters, including floods. Timely and effective response efforts can save lives, alleviate suffering, and support affected communities in their recovery and resilience-building endeavors. Figure 13.4 displays a map

TABLE 13.3
OFDA/BHA Response for Flood Disaster

Response	Count	Percentage (%)
No	185	95.85
Yes	8	4.15
Total	193	100.00

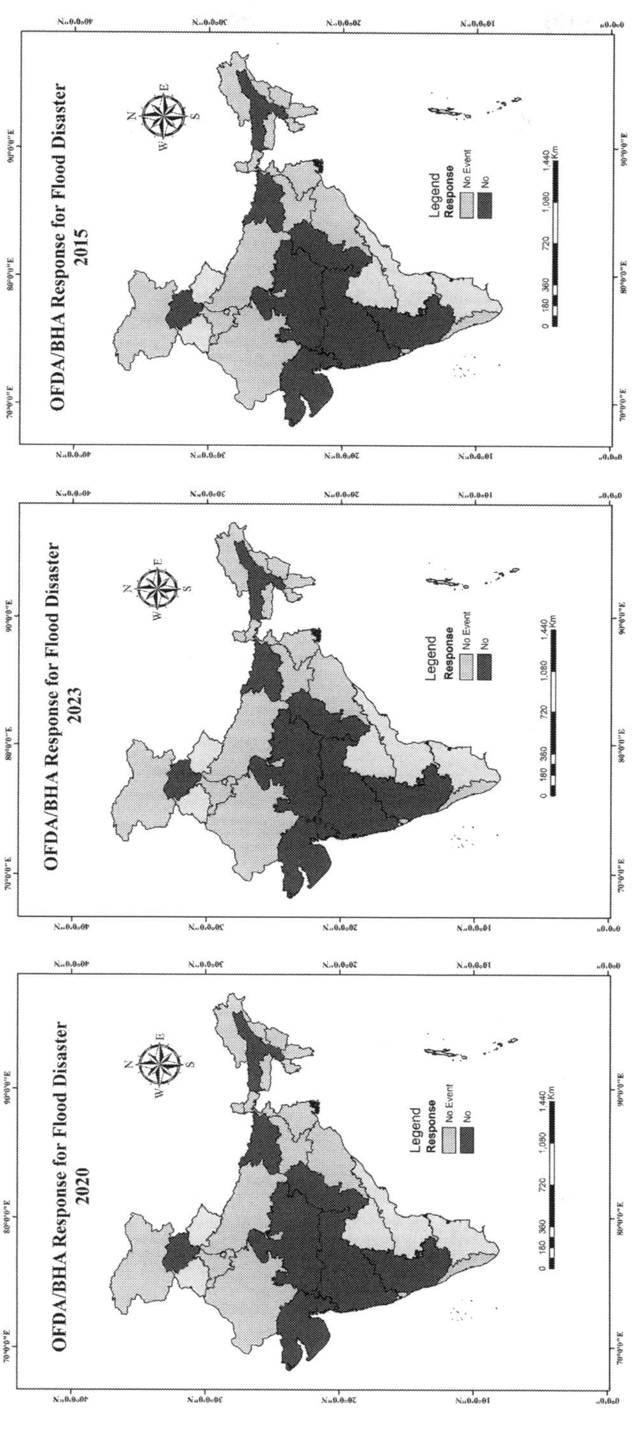

FIGURE 13.4 OFDA/BHA response for flood disaster (2000, 2005, 2010, 2015, 2020, & 2023). (*Source:* https://www.emdat.be accessed at 09-May-2024.)

(Continued)

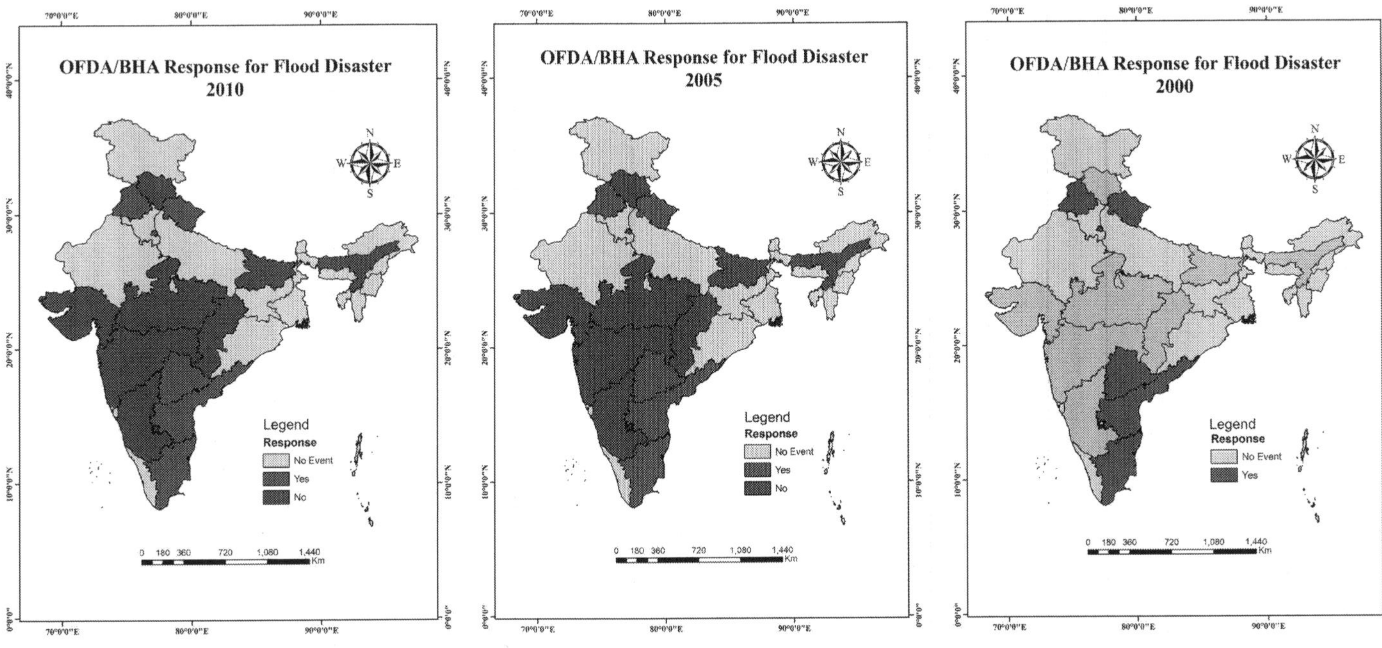

FIGURE 13.4 (*Continued*) OFDA/BHA response for flood disaster (2000, 2005, 2010, 2015, 2020, & 2023). (*Source*: https://www.emdat.be accessed at 09-May-2024.)

representing the response efforts of the OFDA and the BHA to flood disasters in India for the years 2000, 2005, 2010, 2015, 2020, and 2023. This map highlights the geographical areas affected by floods and the corresponding relief activities carried out by OFDA/BHA. Each year is marked to show the locations and scale of the response, providing a visual overview of how disaster assistance has been distributed across different regions in India over the selected years.

13.6 YEAR-WISE DISTRIBUTION OF FLOOD OCCURRENCES IN INDIA

Table 13.4 represents the distribution of flood occurrences in India over the years, with percentages indicating the proportion of floods in each respective year. The data show fluctuations in the occurrence of floods in India over the years. While some

TABLE 13.4
Year Wise Distribution of Flood Occurrences in India (2000–2023)

Year	Count	Percentage (%)
2000	6	3.11
2001	9	4.66
2002	6	3.11
2003	6	3.11
2004	6	3.11
2005	17	8.81
2006	17	8.81
2007	16	8.29
2008	8	4.15
2009	6	3.11
2010	8	4.15
2011	7	3.63
2012	6	3.11
2013	5	2.59
2014	7	3.63
2015	10	5.18
2016	8	4.15
2017	9	4.66
2018	9	4.66
2019	5	2.59
2020	5	2.59
2021	9	4.66
2022	3	1.55
2023	5	2.59
Total	193	100.00

Source: https://www.emdat.be accessed at 09-May-2024.

years witness higher percentages of flood events, others have relatively lower percentages. For example, the years 2005, 2006, and 2007 had notably higher percentages of flood occurrences compared to other years.

13.6.1 Month-wise Distribution of Flood Occurrences in India

Table 13.5 provides the distribution of flood occurrences in India on a monthly basis from 2000 to 2023, with percentages indicating the proportion of floods in each respective month. The data highlight the significant influence of the monsoon season on flood occurrences in India. Months from June to September, coinciding with the peak of the monsoon season, collectively account for a substantial majority (78.72%) of flood events. This underscores the critical role of monsoon rainfall in driving floods, particularly riverine floods and inundation of low-lying areas. July emerges as the month with the highest percentage of flood occurrences at 23.32%, followed closely by August at 21.76%. These months typically experience the heaviest and most widespread rainfall across the country, leading to elevated river levels, flash floods, and waterlogging in urban areas. The concentration of flood events in these peak monsoon months emphasizes the need for heightened preparedness and response measures during this period.

Flood occurrences are also observed in the months preceding and following the peak monsoon season. May sees a notable percentage of floods at 5.70%, marking the onset of pre-monsoon rainfall and potential localized flooding. Similarly, September retains a significant proportion of flood occurrences at 14.51%, reflecting the lingering effects of monsoon rainfall and continued flood risk during the post-monsoon

TABLE 13.5
Month Wise Distribution of Flood Occurrences in India (2000–2023)

Months	Count	Percentage (%)
February	1	0.52
March	3	1.55
April	3	1.55
May	11	5.70
June	35	18.13
July	45	23.32
August	42	21.76
September	28	14.51
October	16	8.29
November	7	3.63
December	2	1.04
Total	193	100.00

Source: https://www.emdat.be accessed at 09-May-2024.

period. Flood events are relatively less frequent during the remaining months of the year. October, November, December, February, and March collectively contribute to only 15.83% of flood occurrences. However, it's essential to note that floods can still occur during these months due to factors such as cyclonic disturbances, heavy localized rainfall, or snowmelt in northern regions.

13.7 NUMBER OF DEATHS DUE TO FLOOD DISASTER IN INDIA

Table 13.6 provides summary statistics of the number of deaths due to flood disasters in India from 2000 to 2023, sourced from the EM-DAT international disaster database. The total number of deaths recorded due to flood disasters during the period is 34,985. This staggering figure underscores the significant human toll of floods in India over the past two decades. Floods can result in loss of lives through drowning, injuries, waterborne diseases, and other direct and indirect causes, particularly in vulnerable communities with limited access to resources and infrastructure. The average number of deaths per flood event is approximately 194.36. This statistic provides an insight into the typical magnitude of fatalities associated with flood disasters in India. While some flood events may result in minimal casualties, others can lead to large-scale loss of lives, highlighting the variable nature and severity of flood impacts. The maximum number of deaths recorded in a single flood event is 6,054. This statistic highlights the potential for catastrophic loss of lives during extreme flood events, particularly when coupled with factors such as population density, inadequate infrastructure, and limited emergency response capacity.

13.8 NUMBER OF INJURED DUE TO FLOOD DISASTER IN INDIA

Table 13.7 represents summary statistics of the number of individuals injured due to flood disasters in India from 2000 to 2023. The total number of injuries recorded due to flood disasters during the period is 7,576. Flood-related injuries can result from various causes, including physical trauma, waterborne diseases, and accidents during evacuation or rescue operations. The average number of injuries per flood event is

TABLE 13.6
Number of Deaths due to Flood Disaster in India (2000–2023)

Summary Statistics	Deaths
Total	34,985
Average	194.36
Min	1
Max	6,054
Median	47

Source: https://www.emdat.be accessed at 09-May-2024.

TABLE 13.7
Number of Injured due to Flood Disaster in India (2000–2023)

Summary Statistics	Injured
Total	7,576
Average	303.04
Min	2
Max	4,473
Median	25

Source: https://www.emdat.be accessed at 09-May-2024.

approximately 303.04. The minimum number of injuries recorded in a single flood event is 2, while the maximum is 4,473. These statistics demonstrate the wide range of injury counts observed across different flood events. While some events may have relatively few injuries, others can result in a large number of individuals requiring medical attention, particularly in densely populated areas or where evacuation and rescue efforts are challenging.

13.9 AFFECTED TOTAL PEOPLE DUE TO FLOOD DISASTER IN INDIA

Table 13.8 illustrates summary statistics on the total number of people affected by flood disasters in India from 2000 to 2023. The data reveal the significant and widespread impact of flooding on the population over this period. The total number of people affected by floods in India during the 23-year period is a staggering 360,605,367. This figure highlights the immense scale of flood disasters and their far-reaching consequences on the country's population. Floods disrupt lives, displace communities, and lead to significant economic and social challenges. On average,

TABLE 13.8
Affected of Total People due to Flood Disaster in India (2000–2023)

Summary Statistics	Total Affected
Total	360,605,367
Average	2225959.056
Min	4
Max	42,000,000
Median	100,000

Source: https://www.emdat.be accessed at 09-May-2024.

each flood disaster affects approximately 2,225,959 people. This average reflects the typical extent of disruption caused by floods in India, illustrating how even an average flood event can impact millions of individuals, disrupting their daily lives, livelihoods, and access to essential services.

The minimum number of people affected in a single flood event is 4, indicating that some floods have a much localized impact, potentially affecting only a small community or area. In contrast, the maximum number of people affected in a single flood event is an astonishing 42,000,000. This maximum value underscores the potential for catastrophic impacts during severe flood events, particularly in densely populated or vulnerable regions. The median number of people affected by flood disasters is 100,000. This median value provides a sense of the central tendency of the data, indicating that half of the flood events affected fewer than 100,000 people and half affected more. The median is a useful measure for understanding the typical flood event's impact, particularly in the context of highly variable data.

13.10 TOTAL DAMAGE ('000 US$) DUE TO FLOOD DISASTER IN INDIA

Table 13.9 shows summary statistics on the total economic damage caused by flood disasters in India from 2000 to 2023, reported in thousands of US dollars. The data, sourced from the EM-DAT international disaster database, underscores the significant financial impact of flooding over this period. The total damage incurred from flood disasters in India over the 23-year period amounts to $76,951,327,000. This substantial figure highlights the severe economic burden that flooding places on the country. The damage encompasses destruction to infrastructure, homes, agriculture, and other economic activities, leading to extensive financial losses. On average, each flood disaster results in damage amounting to $1,115,236,623. This average reflects the typical financial impact of flood events in India, indicating that even moderate flood events can cause considerable economic disruption and require significant recovery efforts. The minimum recorded damage for a single flood event is $2,000, highlighting that some flood events have a relatively minor financial impact, possibly due to their limited scope or effective mitigation measures. In stark contrast, the maximum

TABLE 13.9
Total Damage ('000 US$) due to Flood Disaster in India (2000–2023)

Summary Statistics	Total Damage (000 US$)
Total	76,951,327
Average	1115236.623
Min	2,000
Max	16,000,000
Median	163,000

Source: https://www.emdat.be accessed at 09-May-2024.

TABLE 13.10
Consumer Price Index during Flood Disaster in India (2000–2023)

Summary Statistics	CPI
Average	72.62
Min	56.51
Max	100
Median	70.66

Source: https://www.emdat.be accessed at 09-May-2024.

damage recorded for a single flood event is $16,000,000,000. This maximum value underscores the potential for catastrophic economic losses during severe flood events, particularly those affecting densely populated or economically critical regions.

13.11 CONSUMER PRICE INDEX DURING FLOOD DISASTER IN INDIA

Table 13.10 presents summary statistics on the Consumer Price Index (CPI) during flood disasters in India from 2000 to 2023. The CPI is an economic indicator that measures changes in the price level of a basket of consumer goods and services purchased by households. Analyzing the CPI during flood events provides insight into how these disasters impact the cost of living and inflation. The average CPI over the 23-year period is 72.61763766. This average indicates the general price level during the years of flood disasters. It shows that floods likely contribute to an overall increase in consumer prices, possibly due to supply chain disruptions, damage to agricultural production, and increased costs of goods and services. The minimum recorded CPI during flood years is 56.51429056, while the maximum is 100. The minimum value suggests that in some years, the impact of floods on consumer prices was relatively moderate, possibly due to effective management and mitigation measures. The maximum value, which is 100, reflects years with significant inflation, likely exacerbated by severe flood events that caused substantial disruptions in economic activities and supply chains. The median CPI is 70.66008903. This value indicates that in half of the years studied, the CPI was below this level, and in the other half, it was above. The median provides a central tendency measure, suggesting that in most years, floods had a notable but varied impact on consumer prices.

13.12 CONCLUSION

Flood disasters have increasingly become a significant concern in India due to their frequent occurrence and devastating impacts on various sectors, including agriculture, economy, infrastructure, and society. The interplay between climate change and flood disasters is particularly noteworthy, as changes in global weather patterns, rising temperatures, and altered hydrological cycles exacerbate the intensity and frequency of these natural hazards. The data from 2000 to 2023 underscore the profound human and economic toll of floods in India, with significant loss of life, injuries, displacement, and economic damage.

13.13 KEY FINDINGS

13.13.1 Frequency and Impact of Floods

Floods have become more frequent and severe, impacting millions of people annually. Over 360 million individuals were affected between 2000 and 2023, highlighting the widespread disruption caused by these events. Uttar Pradesh, Assam, and Bihar are among the states most frequently affected by floods, primarily due to their geographical locations in flood-prone regions like the Gangetic plains and the Brahmaputra Valley.

13.13.2 Human and Economic Costs

Floods resulted in 34,985 deaths and 7,576 injuries over the 23-year period, emphasizing the significant human cost of these disasters. The economic damage from floods totaled approximately $76.95 billion, with an average of over $1 billion per flood event, indicating substantial financial losses and the extensive impact on infrastructure and livelihoods.

13.13.3 Role of Climate Change

Climate change has contributed to more erratic and extreme weather patterns, intensifying monsoon rains and increasing the frequency of severe flooding. The IPCC reports predict further increases in both the frequency and intensity of flood events due to rising global temperatures, which will likely exacerbate the vulnerability of flood-prone regions.

13.13.4 Geographical Distribution and Causes

Monsoon rainfall is the predominant cause of floods in India, accounting for nearly half of the flood events. Heavy rainfall outside the monsoon season also contributes significantly to flood occurrences. River basins, particularly those of the Ganga and Brahmaputra, are highly susceptible to flooding due to heavy monsoon rains and snowmelt from the Himalayas.

13.13.5 Response and Preparedness

The response from international organizations like the OFDA/BHA has been limited, with only a small percentage of flood disasters receiving direct assistance. This underscores the need for enhanced local and national preparedness and response mechanisms. Effective risk management, adaptation strategies, and resilience-building measures are crucial to mitigating the impacts of floods. This includes improving infrastructure, early warning systems, and community-based disaster preparedness programs.

13.13.6 Economic Indicators

Floods have a notable impact on the CPI, reflecting increased costs of living and inflation during flood years. This further underscores the economic strain caused by these disasters on affected populations. To address the growing threat of flood disasters in India, there is an urgent need for comprehensive and integrated approaches that encompass climate change mitigation, improved infrastructure, and effective disaster management strategies. Strengthening early warning systems, investing in resilient infrastructure, and enhancing community awareness and preparedness are essential steps to reduce the vulnerability of populations and mitigate the adverse impacts of floods. Additionally, addressing the underlying factors contributing to climate change and its effects on weather patterns will be critical in curbing the increasing frequency and severity of flood disasters. By adopting a multi-faceted and proactive approach, India can better safeguard its communities and infrastructure against the devastating effects of floods.

REFERENCES

Berz, Gernot, Wolfgang Kron, Thomas Loster, Eberhard Rauch, Jorg Schimetschek, Jörg Schmieder, and others. 2001. "World Map of Natural Hazards—a Global View of the Distribution and Intensity of Significant Exposures." *Natural Hazards* 23 (2). Springer: 443–465.

Census of India. (2011). *Population Dynamics and Urbanization Trends in India*. Office of the Registrar General, India.

EM- DAT: The International Disaster Database. (2023). *Flood Disaster Reports: 2000–2023*. Centre for Research on the Epidemiology of Disasters (CRED).

Ghosh, S and others. 2012. "Flood Management in India: Emerging Issues and Solutions." *Hydrology Journal* 3 (4): 23–36.

Irasema, Alcántara-Ayala. 2002. "Impacts of Natural Disasters on Developing Economies: Implications for the International Development and Disaster Community." *Disaster Prevention and Management* 11 (1): 14–23.

Intergovernmental Panel on Climate Change (IPCC). (2007). *Climate change 2007: synthesis report*. Geneva: IPCC.

International Strategy for Disaster Reduction (ISDR). (2004). *Living with Risk: A Global Review of Disaster Reduction Initiatives*. Geneva: United Nations.

NatCat SERVICE. (2023). *Global Trends in Natural Catastrophes*. Munich Re.

National Institute of Disaster Management (NIDM). (2014). *India's Vulnerability to Natural Disasters*. New Delhi: NIDM.

Kumar, R, RD Singh, and KD Sharma. 2005. "Water Resources of India." *Current Science* 89 (5): 794–811.

Kundzewicz, Zbigniew W, Shinjiro Kanae, Srikantha I Seneviratne, John Handmer, Neville Nicholls, Pascal Peduzzi, and others. 2012. "Flood Risk and Climate Change: Global and Regional Perspectives." *Hydrological Sciences Journal* 57 (1): 1–27.

Ministry of Home Affairs, Government of India. 2008. *Disaster Management in India*. New Delhi: Government of India.

Schumann, Gerrit, and Dirk Möller. 2015. "Microwave Remote Sensing in Flood Monitoring: Recent Developments and Perspectives." *Physics and Chemistry of the Earth, Parts A/B/C* 83–84: 84–95.

Sigma. (2023). *Natural Catastrophes and Man-Made Disasters in 2023: Economic Losses and Insured Losses*. Swiss Re Institute, Zurich, Switzerland.

Syvitski, James PM, and Gregory R Brakenridge. 2013. "Causation and Avoidance of Catastrophic Flooding along the Indus River, Pakistan." *GSA Today* 23 (1): 4–10.

World Bank. 2008. *Climate Resilient Cities: A Primer on Reducing Vulnerabilities to Disasters*. Washington, DC: The World Bank.

14 Entrepreneurship Ideas and Climate Change Mitigation Mechanism

Amanullah Adeel, Pooja Ramchandra Dharmoji, and Abdulahad Ahadi

14.1 INTRODUCTION

The concept of entrepreneurship is intricate and has evolved throughout time. In 1993, Kao defined it as the process of creating something innovative and unique that enhances individual prosperity and benefits society as a whole. In 1998, this process involved identifying opportunities, gathering resources, and creating enterprises that provide products or services. In the corporate realm, entrepreneurship is regarded as an inherent occurrence encompassing diverse skills, innovations, and managerial strategies. Certain researchers classify entrepreneurship based on specific characteristics, such as technology or procedure, while others focus on its operational elements. An entrepreneur is an individual or collective responsible for identifying opportunities, establishing enterprises, and overseeing their operations. An individual who demonstrates initiative and original thought, can organize social and economic frameworks to optimize resources and conditions, and is prepared to embrace risk and potential failure is characterized as an entrepreneur. Although often associated with entrepreneurship, traits such as growth, creativity, and resilience are regarded as consequences rather than catalysts for entrepreneurial endeavors.

Climate change is a complex concept that has evolved throughout time, encompassing both natural and anthropogenic environmental alterations. It presents a significant challenge for contemporary worldwide scientific, economic, political, and cultural spheres (Tóth 2023). The concept includes anthropogenic factors and their impact on the Earth's climate system, comprising the atmosphere, biosphere, cryosphere, hydrosphere, and lithosphere, along with climatic variability (Kumar 2022). Prolonged international endeavors to address climate change have entailed stringent negotiations regarding state obligations and responsibilities. The discourse on climate change has become paramount in social communication networks, altering conceptions of the relationship between nature and society and impacting broader meaning systems (Tóth 2023). The complexity of climate change research encompasses numerous approaches, including Earth system science, which provides essential concepts and strategies for investigating global environmental changes.

Climate refers to the anticipated weather conditions in a specific place during a particular season. Climate change transpires across time, and the extent of change

DOI: 10.1201/9781003492733-14

is contingent upon the selected timeline for analysis. Alterations in the past decade appear trivial in comparison to those occurring during the preceding million years. Climate change denotes enduring alterations in temperature and meteorological phenomena. Such alterations may occur naturally, resulting from variations in solar activity or significant volcanic eruptions. Climate change refers to a long-term alteration in the average circumstances, including temperature and precipitation, within a certain region. Global climate change denotes the average long-term alterations across the entire Earth. The consequences of Earth's warming encompass increasing temperatures and alterations in precipitation patterns. For instance, rising ocean levels, the reduction of mountain glaciers, and the accelerated ice loss in Greenland, Antarctica, and the Arctic are accompanied by fluctuations in flowering and planting seasons. Prior to the advent of humanity, Earth's climate had undergone continual transformation.

The Paris Agreement and the 2030 Agenda for Sustainable Development foresee a low-carbon, climate-resilient, and sustainable future for all. It is essential that we follow that aim. The global average temperature has increased by around 1°C since the Industrial Revolution, with 2017 ranking as the third hottest year recorded. The prospects for sustainable development are increasingly threatened by the intensifying impacts of global climate change. Countries have developed national adaptation plans, mid-century strategies, and nationally determined commitments to address climate change in accordance with the Paris Agreement. Currently, states and the international community are focused on implementing them.

Climate change mitigation refers to efforts aimed at reducing or preventing greenhouse gas (GHG) emissions, with the goal of limiting the rise in global temperatures (Panepinto et al. 2021). Mitigation measures encompass the adoption of novel technology, enhancement of energy efficiency, and modification of management procedures or customer behavior. To achieve the Paris Agreement's objective of maintaining a temperature increase well below 2°C relative to pre-industrial levels, both the reduction of emissions and the removal of carbon dioxide are essential. Negative emission methods, such as microalgae cultivation for CO_2 bio-fixation, are emerging as viable options (Panepinto et al. 2021). Since 2005, the notion of transformation has become significant in climate change mitigation research, highlighting the necessity for profound, extensive, and swift societal alterations (Moore et al. 2021).

The Asia-Pacific area is experiencing a rise in environmental and social entrepreneurship, targeting climate change, poverty, and hunger. Climate-smart activities and green technologies exemplify novel solutions that are beginning to surface. Women and youth are prominently spearheading these initiatives, promoting a more sustainable and equitable future throughout the region. The increasing trend of female and youth entrepreneurship in the Asia-Pacific area presents significant opportunities for innovative solutions to socioeconomic and climate change challenges. The AGREA Agricultural Communities International Foundation and ESCAP's Youth Climathon illustrate how education and engagement can foster practical and innovative solutions in climate-smart agriculture, just transition, and ecosystem restoration. These projects seek to eradicate inequalities that heighten the vulnerability of women and youth while also mitigating the impacts of climate change. Through the promotion of inclusive green employment and sustainable development, firms headed by women and youth are establishing a more resilient and equitable future, exemplifying the transformative potential of innovation in effecting enduring positive change.

This chapter examines many entrepreneurship-related subjects concerning the mitigation of climate change. The discussion begins by examining mitigation strategies and emphasizing the crucial role of company owners in addressing climate-related issues. The chapter introduces innovative business models that could significantly contribute to climate change mitigation. It examines the opportunities and challenges encountered by business proprietors in this sector, providing an objective evaluation of the circumstances. The chapter presents practical examples of successful entrepreneurship via case studies of effective climate mitigation enterprises. Finally, it offers guidance and prospective pathways, instructing potential entrepreneurs on how to engage in and enhance this significant domain.

14.2 MITIGATION STRATEGIES

Innovation is crucial for efforts to alleviate and adjust to climate change. Establishing resilient and resource-efficient climate-smart agricultural systems necessitates technological, managerial, and institutional transformations. To foster environmentally sustainable innovation and its dissemination, it is imperative to implement a price on GHG emissions with public research and development support and international technology-focused accords. Innovation can lead to the substitution of fossil fuels with non-carbon alternatives, and the long-term development of energy systems is highly dependent on the initial robustness and duration of policy (Anderson and Winne 2007). Nevertheless, owing to challenges such as insufficient infrastructure, limited financing sources, and a deficiency in research competence, numerous developing countries allocate minimal resources to the advancement of technologies pertaining to climate change. To fully actualize the potential of innovation in addressing climate change issues, political actions to eliminate these obstacles are essential (Adenle et al. 2015).

14.3 THE ROLE OF ENTREPRENEURSHIP IN CLIMATE CHANGE MITIGATION

14.3.1 Entrepreneurial Ecosystem

The entrepreneurial ecosystem is essential for climate change mitigation as it fosters sustainable practices and innovation. Research demonstrates that corporations not only substantially contribute to GHG emissions but also play a pivotal role in developing solutions to alleviate the impacts of climate change (Sundaram 2023). Governments can enhance human ties with the environment and promote intergenerational fairness by advocating for education on the entrepreneurship ecosystem at higher education institutions and integrating traditional knowledge related to biodiversity and ecology (Agarwal et al. 2022). Furthermore, many business sectors are impacted by climate change policy in distinct manners. Some encounter difficulties such as elevated operational expenses and regulatory compliance challenges, while others gain from heightened demand for eco-friendly products and enhanced energy efficiency (Alqarni 2024). Companies can enhance their market value by emphasizing climate risk mitigation strategies such as investing in renewable energy and producing environmentally sustainable products. This underscores the importance of integrating sustainable practices into company plans to successfully address climate change (Nguyen et al. 2023).

14.3.2 Innovation and Technology

The integration of advanced technologies, sustainable regulations, and collaborative stakeholder engagement fosters resource-efficient solutions that drive economic growth and environmental sustainability (Deng 2024). Entrepreneurs leverage innovation to disrupt industries and cultivate a more equitable future, thus advancing sustainable development. Smart farming technologies utilize sensors, artificial intelligence, and the Internet of Things (IoT) to enhance the sustainability of agricultural systems and tackle global food concerns (Judijanto et al. 2024). Innovative technology predominantly propels effective solutions for minimizing climate change. Disruptive technologies provide a temporary solution by promoting greater decarburization and improved energy efficiency, both crucial for emission reduction (Ogunmodede 2024). Technological advances are essential to attain substantial reductions in GHG emissions at a socially acceptable cost. The GHG emissions of the chemical industry are highly influenced by the costs associated with research and development (R&D) and technological equipment and machinery (TEM), highlighting the need for innovation in mitigating detrimental emissions (Apostu et al. 2023).

14.4 INNOVATIVE ENTREPRENEURIAL IDEAS FOR CLIMATE CHANGE MITIGATION

14.4.1 Renewable Energy Solutions

In 2008, more than 80% of global human GHG emissions originated from energy use, significantly influencing climate change (Akpan and Akpan 2012). Significant emitters encompass the transportation, heating, and electrical sectors; since 1850, fossil fuels have predominated global energy consumption. Besides socioeconomic factors, climate change is anticipated to increase future energy consumption, with projections estimating a 25%–58% increase by 2050 under severe warming scenarios (Hou et al. 2022). Consequently, an urgent replacement for the current energy paradigm is essential for mitigating climate change. The potential for sustainable development and climate change mitigation through renewable energy sources is substantial (Arvizu et al. 2011). Solar energy, the most plentiful resource globally, can provide numerous energy services (Arvizu et al. 2011). Other notable alternatives to fossil fuels are geothermal, hydropower, wind, and biomass energy (Dutta 2020). Renewable energy sources provide opportunities for socioeconomic development, energy security, accessibility, and reduced negative environmental impacts (Owusu and Asumadu-Sarkodie 2016). The viability of renewable energy is hindered by challenges such as market failures, communication deficiencies, and raw material availability, notwithstanding their potential (Owusu and Asumadu-Sarkodie 2016).

The adoption of renewable technology depends on continuous innovation, cost management strategies, and regulatory support (Arvizu et al. 2011). The advancement and advocacy of these clean energy alternatives are crucial for meeting climate change mitigation objectives since they can lead to sustainable development and reduced GHG emissions (Dutta 2020).

Notable case studies worldwide, including Brazil's ethanol initiative and Denmark's wind energy policy, demonstrate the feasibility of adopting renewable energy (Brown and Sovacool 2012). The residential sector in South Africa significantly contributes to GHG emissions, with fossil fuels serving as the predominant energy source.

14.4.2 Sustainable Agriculture and Food Systems

A crucial element in addressing and adapting to climate change is sustainable agricultural and food systems. Despite providing opportunities for mitigation across the entire food system, from production to consumption, agriculture significantly contributes to GHG emissions (Niles et al. 2018). Climate-smart agriculture can be advanced through synergies among production, adaptation, and mitigation (Loboguerrero et al. 2019). Mitigation involves reducing emissions and enhancing carbon sinks, whereas adaptation strategies encompass altering crop varieties, adjusting the timing of agricultural practices, and improving pest management (Adeel and Jadhav 2024). To meet our climate goals, we need a food system strategy that includes all sectors and stakeholders, as well as clear policies that reduce trade-offs and increase synergies (Niles et al. 2018).

Case studies highlight efficient corporate farming methodologies, particularly those centered on adaptation and mitigation, rooted in systems thinking and corporate social responsibility (Reaves 2019). Startups such as Green Agr-evolution Pvt Ltd are revolutionizing the agricultural sector by providing comprehensive services from seed to market, thereby reaching millions of farmers (Sinha et al. 2019). Agricultural landscapes possess significant potential for mitigating GHG emissions through methods such as carbon storage in soils and plant tissues (Braimoh et al. 2016). Cost-effective agricultural mitigation measures, including zero-tillage, improved rice-water management, and optimized fertilizer use, have been discovered for India, a major GHG emitter. By 2030, these initiatives might reduce emissions by 85.5 $MtCO_2e$ per year, with 80% potentially achieved only through cost-reduction strategies (Sapkota et al. 2019).

14.4.3 Waste Management and Circular Economy

Strategies for the circular economy and waste management are crucial for mitigating climate change in urban regions. Implementing circular economy principles, such as the 9R framework, can effectively diminish resource consumption and GHG emissions (Muljaningsih et al. 2023). In developing countries, innovative strategies for managing municipal solid waste have led to significant reductions in GHG emissions and waste management costs (Paes et al. 2024; Adeel and Jadhav 2024). Cities pursuing climate neutrality are focusing on circular economy activities to optimize waste streams and reduce their environmental impact (Möslinger et al. 2023). Collaborations, intergovernmental collaboration, environmental education, and the enhancement of local expertise are crucial enablers of effective implementation (Paes et al. 2024). Digital tools can motivate individuals to minimize waste, such as through the development of websites featuring life cycle assessment calculators and circular economy principles (Nair et al. 2023). The circular economy represents a novel strategy for waste management. The principle of a circular economy is to transform contemporary garbage into future raw materials.

Business waste management techniques have experienced significant transformations over the past five years. Businesses are increasingly prioritizing sustainable solutions, including product reuse and waste-to-energy conversion. In 2020, 97.5% of enterprises worldwide used impact management for generated garbage, while 66% employed non-hazardous waste disposal methods, according to a study of 780 firms across 28 countries that highlighted the growing focus on waste management (Aibar-Guzmán et al. 2023). Tereshchenko et al. (2023) conducted a comprehensive analysis identifying eight key sustainability measures for small and medium-sized enterprises (SMEs). SMEs have emerged as significant contributors to eco-innovation and sustainable waste management. Despite the continuous advancement of waste management technology, achieving comprehensive sustainability objectives necessitates a heightened focus on user knowledge, education, and empowerment (Farooq et al. 2022).

14.4.4 Green Building and Urban Planning

Recent research emphasizes innovative approaches to energy-efficient building design and sustainable construction as strategies to alleviate the impacts of climate change. Research emphasizes the application of renewable energy, sustainable materials, and innovative designs (Rogalev et al. 2022) in construction practices. The primary options for constructing green buildings include solar energy conversion devices, energy-efficient control systems, and optimal lighting. Integrated design techniques and model building designs are crucial for enhancing energy efficiency (Aslan and Bulut 2022). In conjunction with advanced technologies like 3D-printed structures and smart glass facades, innovative materials are being explored, including bio-based insulation, carbon-storing concrete, and recycled composites, which can dramatically reduce GHG emissions (Najmi 2024).

The implementation of green building techniques is increasingly recognized as an effective strategy for mitigating urban warming. Case studies demonstrate diverse urban solutions, with the most potential for diminishing GHG emissions arising from technology-based interventions in the waste, transportation, and energy sectors (Sethi et al. 2020). Green building grading systems, such as GRIHA in India, promotes the application of mitigation strategies and alignment with national climate targets (Kochhar et al. 2022). However, challenges like law enforcement and pre-existing non-compliant infrastructure persist in hindering widespread implementation. Generally, mitigation techniques are more advanced than adaptation strategies. The capacity of green construction initiatives to diminish emissions is significant, contingent upon their effective implementation and integration across

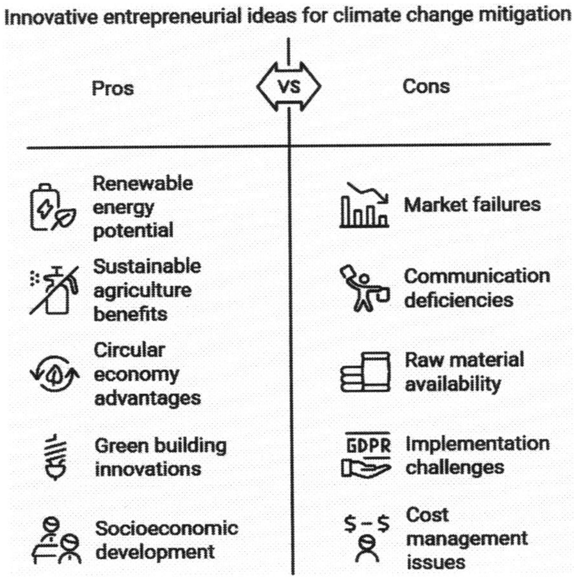

FIGURE 14.1 Pros and cons of new investment in climate change mitigation.

urban sectors, as indicated by Sethi et al. (2020). In Figure 14.1, we highlight the advantages and disadvantages of investment in climate change mitigation.

14.5 CHALLENGES AND OPPORTUNITIES FOR ENTREPRENEURS

14.5.1 Market and Financial Barriers

Entrepreneurial concepts aimed at alleviating climate change are occasionally hindered by financial and commercial limitations. Corporate climate risk is undervalued because the market isn't working well because there isn't enough detailed asset-level data, old risk assessment methods are still used, incentives aren't fair, and regulators don't have enough power to keep an eye on things (Condon 2021). Ambiguous policy frameworks hinder the private sector in South Africa from translating policy into action, hence impacting the kind and scope of climate change mitigation measures implemented (Jongh and Möllmann 2014). Financial impediments, including perceived constraints, technological dependencies, and investment ambiguities, hinder entrepreneurs from adopting environmental advances, highlighting the pivotal role of finance in shaping the trajectory of the green economy. Substantial financial prospects for climate initiatives can be realized by overcoming these barriers via regulatory reforms, including climate risks in investment decisions, and offering risk subsidies for sustainable projects (Purkayastha and Sarkar 2021).

Global investment in creative climate change mitigation faces financial challenges. For instance, owing to the vagueness of governmental structures, the corporate sector in South Africa employs a "take position, wait and see" strategy, demonstrating an unwillingness to actively participate in climate change mitigation

(Jongh and Möllmann 2014). Globally, financial institutions encounter issues like the mispricing of climate-related risks, data shortages, reliance on outdated risk assessment methods, mismatched incentives, and regulatory capture (Condon 2021). Moreover, the potential devaluation of traditional assets complicates banks' ability to fund sustainable enterprises and deters investment in green technologies. The failure of climate-oriented initiatives to generate profit and the inadequate alignment of the existing financial system with climate finance goals restrict the influx of climate funding (Purkayastha and Sarkar 2021).

14.5.2 Policy and Regulatory Environment

Policies and regulations play a crucial role in shaping entrepreneurial activities, particularly in addressing climate change mitigation. Policy entrepreneurs utilize many strategies to influence perceptions and attitudes on climate issues, emphasizing the importance of authenticity, accountability, and equity in their actions. Research indicates that the impact of climate policy on businesses is heterogeneous; certain enterprises gain advantages, such as heightened demand for sustainable products, while others encounter challenges, like elevated operational costs and regulatory compliance (Alqarni 2024). Furthermore, research indicates that to achieve net-zero carbon emissions, environmental rules must effectively promote clean technology and deter detrimental practices (Balcilar and Agan 2023). Moreover, there is an increasing acknowledgment of the significance of supportive policies in fostering sustainable entrepreneurship, emphasizing the necessity for enhanced policy tools and their alignment with objectives to direct businesses toward sustainable practices (Cai et al. 2024).

14.6 CASE STUDIES OF SUCCESSFUL CLIMATE MITIGATION ENTERPRISES

Progressive enterprises are employing innovative strategies to address the pressing issue of climate change. These enterprises are at the vanguard of entrepreneurial concepts and climate change mitigation strategies. These enterprises are actively involved in the advancement of eco-efficiency initiatives, renewable energy solutions, and sustainable technologies (Ayers et al. 2017). To enhance their market value, they are focusing on reducing carbon emissions, developing sustainable products, and investing in environmental initiatives (Nguyen et al. 2023). Startups are essential in advancing innovation in climate technology by contributing to AI solutions, big data analysis, and the creation of sustainable solutions for a greener future (Hakovirta 2023). Moreover, these enterprises recognize that effective execution of climate change mitigation and adaptation strategies necessitates multilateral cooperation, stable conditions, and international accords (Hillman and Baydoun 2020). A technology review website has identified the top 15 organizations engaged in climate change mitigation within their production processes (https://www.technologyreview.com). Examples of these companies are Fervo Energy and Orsted for renewable energy generation, H_2 Green Steel for hydrogen, GEM for batteries, and NotCo for food,

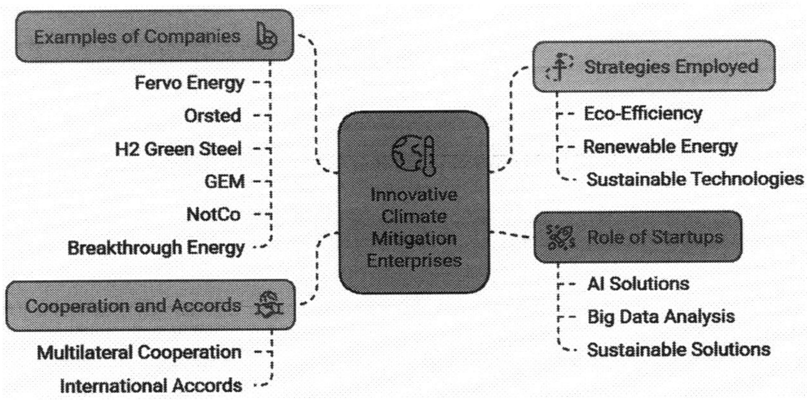

FIGURE 14.2 This mind map shows how innovative climate mitigation businesses are addressing environmental issues through various strategies such as eco-efficiency and renewable energy, where startups are utilizing AI and big data, all facilitated through international collaboration.

among others. In Figure 14.2, we classified the successful case studies worldwide; however, many others may be missed in this figure.

One of the priciest investments in climate change prevention is the firm "Breakthrough Energy," owned by Bill Gates. This company captures CO_2 from the atmosphere and stores it in tanks, subsequently converting it into other energy forms, such as electro-fuels utilized for jet propulsion, according to CSB News (https://www.youtube.com/watch?v=N7HWL1g1sek). The report indicated that a total of 9,000 tons of CO_2 per year is sequestered underground and repurposed by two facilities. This technique captures CO_2 from the atmosphere and generates renewable energy. However, there is insufficient evidence to determine whether this most expensive company is also the most effective one.

14.7 FUTURE DIRECTIONS AND RECOMMENDATIONS

Future advancements in entrepreneurship-focused climate change mitigation are expected to concentrate on the development of eco-inclusive technology, such as cleantech solutions, driven by entrepreneurs' engagement in sustainable innovation (Alexoaei and Robu 2024). Startups will remain crucial in advancing climate technology innovation toward a carbon-neutral society by leveraging AI, big data analytics, and strategic partnerships (Hakovirta 2023). However, challenges hinder the global dissemination of low-carbon technology; following a fall in patenting rates post-2013, increased technological transfers to developing countries are necessary (Probst et al. 2021). Policymakers must possess a comprehensive understanding of the motivations driving entrepreneurs to engage in climate change mitigation—whether financial, socio-ecological, or a combination of both—to effectively address the issue and enhance the impact of those already involved (Kaesehage et al. 2019).

To foster a promising future for climate change mitigation, the primary responsibility of concerned teams and individuals is to inform consumers today. This recommendation facilitates individuals' comprehension of potential risks they may encounter in the near future. To capture their attention, we ought to employ compelling phrases such as "rescue your family," "a clean environment is a universal right," and "your financial situation has impacted my health." If consumers favor the acquisition of more environmentally sustainable products, the industrial sectors will respond by producing to meet that demand. Nonetheless, transparent and ethical regulations, together with a monitoring procedure, must be established and enforced to identify industries that engage in eco-friendly practices.

14.8 CONCLUSION

This chapter elucidates the substantial connection between entrepreneurship and climate change mitigation, emphasizing the essential significance of innovative entrepreneurial initiatives in addressing a paramount global challenge. By defining essential concepts such as entrepreneurship and climate change mitigation, we established a foundation for understanding how innovative economic initiatives may drive effective environmental solutions. Previous achievements in environmental conservation demonstrate the efficacy of innovative problem-solving, establishing innovation as a crucial element in combating climate change. This chapter illustrated how the struggle against climate change presents numerous commercial opportunities that facilitate the advancement of sustainable firms and technologies. Entrepreneurs possess a distinctive ability to capitalize on these opportunities and convert environmental issues into lucrative commercial ventures due to their capacity for innovation and adaptability.

Despite the promise, business entrepreneurs in this sector face certain hurdles. Financial and commercial impediments, including scalability issues, funding difficulties, and entry barriers, might hinder growth. The legal and policy framework is essential, as legislation can either facilitate or hinder entrepreneurial efforts to address climate change. The case studies of successful climate mitigation enterprises provided demonstrable examples of pioneering companies exerting considerable impact. Furthermore, future projections indicate that this sector will likely continue to evolve as new trends in entrepreneurship-driven climate change mitigation emerge. Prospective entrepreneurs are encouraged to monitor these advancements and utilize them as a source of inspiration for innovative solutions. Expert advice and strategies were provided, emphasizing the importance of adaptability, resilience, and a comprehensive understanding of the legal and commercial landscapes.

REFERENCES

Adeel, A., and A. S. Jadhav. 2024. "Identifying the Optimal Sugarcane Residue Percentage for Soil Enhancement: A Statistical and GIS Approach." *Circular Economy and Sustainability*, November. doi:10.1007/s43615-024-00456-5

Adenle, A. A., H. Azadi, and J. Arbiol. 2015. "Global Assessment of Technological Innovation for Climate Change Adaptation and Mitigation in Developing World." *Journal of Environmental Management* 161: 261–275. doi:10.1016/j.jenvman.2015.05.040

Agarwal, R., S. Pundir, and A. K. Pokhriyal. 2022. "The Role of Indian Entrepreneurship Ecosystem in the Era of Climate Change." *AMC Indian Journal of Entrepreneurship* 5 (2). doi:10.17010/amcije/2022/v5i2/171470

Aibar-Guzmán, B., S. Monteiro, F. David, and F. M. Somohano-Rodríguez. 2023. "The Waste Hierarchy at the Business Level: An International Outlook." Mathematics 11 (22): 4574. doi:10.3390/math11224574

Akpan, U. F., and G. E. Akpan. 2012. "The Contribution of Energy Consumption to Climate Change: A Feasible Policy Direction." *International Journal of Energy Economics and Policy* 2 (1): 21–33.

Alexoaei, A. P., and R. G. Robu. 2024. "Eco-Inclusive Entrepreneurship: Addressing Climate Change through Technological Innovation. The Case of Cleantech Industry." In *Proceedings of the 14th Economics & Finance Conference, Lisbon.*

Alqarni, A. 2024. "Assessing the Economic Implications of Climate Change Mitigation Policies on Local Business Ecosystems." Preprints. doi:10.20944/preprints202404.0011.v1

Anderson, D., and S. Winne. 2007. "Energy System Change and External Effects in Climate Change Mitigation." *Environment and Development Economics* 12 (3): 359–378. doi:10.1017/S1355770X07003580

Apostu, S. A., E. M. Nichita, C. L. Manea, A. M. Irimescu, and M. Vulpoi. 2023. "Exploring the Influence of Innovation and Technology on Climate Change." *Energies* 16 (17): 6408. doi:10.3390/en16176408

Arvizu, D., P. Balaya, L. F. Cabeza, K. G. T. Hollands, A. Jäger-Waldau, M. Kondo, C. Konseibo, et al. 2011. "Direct Solar Energy." In *Renewable Energy Sources and Climate Change Mitigation*, edited by O. Edenhofer, R. Pichs-Madruga, Y. Sokona, K. Seyboth, S. Kadner, T. Zwickel, P. Eickemeier, et al., 333–400. Cambridge University Press. doi:10.1017/CBO9781139151153.007

Aslan, M. M., and Y. Bulut. 2022. "İklim Değişikliğine Dirençli Kentlerin Oluşumunda Yenilikçi Yaklaşımlar: İklim Değişikliği Yönetişimi Ve Enerji Verimli Binalar." *Kent Akademisi* 15 (3): 1023–1036. doi:10.35674/kent.1126928

Ayers, J., S. Huq, H. Wright, A. M. Faisal, and S. T. Hussain. 2017. "Mainstreaming Climate Change Adaptation into Development in Bangladesh." In *Community-Based Adaptation*, 15–27. Routledge. https://www.taylorfrancis.com/chapters/edit/10.4324/9781315658735-8/mainstreaming-climate-change-adaptation-development-bangladesh-jessica-ayers-saleemul-huq-helena-wright-arif-faisal-syed-tanveer-hussain

Balcilar, M., and B. Agan. 2023. "On the Impact of Climate Change Policies on the Clean Technology Innovation: Evidence from Patent Data Analysis." *Environmental and Climate Technologies* 27 (1): 850–863. doi:10.2478/rtuect-2023-0062.

Braimoh, A. K., X. Hou, C. Heumesser, and Y. Zhao. 2016. "*Greenhouse Gas Mitigation Opportunities in Agricultural Landscapes*." Washington, DC: World Bank. doi: 10.1596/25171

Brown, Marilyn A., and B. Sovacool. 2012. "Climate Change and Global Energy Security: Technology and Policy Options." *Choice Reviews Online* 49 (7): 49–3825. doi:10.5860/CHOICE.49-3825.

Cai, P., S. Zhou, and P. Huang. 2024. "Seasonally Varying SST Changes in the Joining Area of Asia and Indian-Pacific Ocean from Boreal Spring to Summer." *Climate Dynamics* 62: 1–14. doi:10.1007/s00382-023-07037-6

Condon, M. 2021. "Market Myopia's Climate Bubble." SSRN Scholarly Paper. doi:10.2139/ssrn.3782675

Deng, B. 2024. "Innovation and Sustainable Development: Management Strategies for Driving Future Economic Growth." In *Advances in Economics, Management and Political Sciences*, 136–140. Atlantis Press. doi:10.54254/2754-1169/93/20241107

Dutta, R. 2020. "Use of Clean, Renewable and Alternative Energies in Mitigation of Greenhouse Gases." In *Encyclopedia of Renewable and Sustainable Materials*, 821–834. Elsevier. doi:10.1016/B978-0-12-803581-8.11048-3

Farooq, M., J. Cheng, N. U. Khan, R. A. Saufi, N. Kanwal, and H. A. Bazkiaei. 2022. "Sustainable Waste Management Companies with Innovative Smart Solutions: A Systematic Review and Conceptual Model." Sustainability 14 (20): 13146. doi:10.3390/su142013146

Hakovirta, M. 2023. "Carbon Neutrality and Entrepreneurship." In *Carbon Neutrality: Follow the Money*, edited by M. Hakovirta, 127–142. Springer Nature Switzerland. doi:10.1007/978-3-031-45202-4_7

Hillman, J. R., and E. Baydoun. 2020. "An Overview of Innovation and Entrepreneurship to Address Climate Change." In *Higher Education in the Arab World: Building a Culture of Innovation and Entrepreneurship*, edited by A. Badran, E. Baydoun, and J. R. Hillman, 141–181. Springer International Publishing. doi:10.1007/978-3-030-37834-9_6

Hou, Rui, Chantel Fong, Benny D. Freeman, Matthew R. Hill, and Zhiyong Xie. 2022. "Current Status and Advances in Membrane Technology for Carbon Capture." *Separation and Purification Technology* 300: 121863. doi:10.1016/j.seppur.2022.121863

Jongh, D. D., and C. Möllmann. 2014. "Market Barriers for Voluntary Climate Change Mitigation in the South African Private Sector." *South African Journal of Economic and Management Sciences* 17 (5). doi:10.4102/sajems.v17i5.532

Judijanto, L., M. M. Lubis, S. R. S. Riyadi, D. A. Krisprimandoyo, and Y. Koentjoro. 2024. "Economic Innovation for Global Food Crisis: Technology-Based Sustainable Solutions Review." *Journal of Business Management and Economic Development* 2 (1). doi:10.59653/jbmed.v2i01.574

Kaesehage, K., M. Leyshon, G. Ferns, and C. Leyshon. 2019. "Seriously Personal: The Reasons That Motivate Entrepreneurs to Address Climate Change." *Journal of Business Ethics* 157 (4): 1091–1109. doi:10.1007/s10551-017-3624-1.

Kochhar, P., N. Mahal, S. Seth, and M. Singh. 2022. "Green Rating for Integrated Habitat Assessment—A Green-Building Rating System for Catalysing Climate-Change Mitigation/Adaptation in India." F1000Research 11. doi:10.12688/f1000research.108826.1

Kumar, A. 2022. "Challenge of Climate Change." In *Ecosystem-Based Adaptation*, 39–104. Elsevier. doi:10.1016/B978-0-12-815025-2.00002-2

Loboguerrero, A., B. Campbell, P. Cooper, J. Hansen, T. Rosenstock, and E. Wollenberg. 2019. "Food and Earth Systems: Priorities for Climate Change Adaptation and Mitigation for Agriculture and Food Systems." Sustainability 11 (5): 1372. doi:10.3390/su11051372

Moore, B., C. Verfuerth, A. M. Minas, C. Tipping, S. Mander, I. Lorenzoni, C. Hoolohan, A. J. Jordan, and L. Whitmarsh. 2021. "Transformations for Climate Change Mitigation: A Systematic Review of Terminology, Concepts, and Characteristics." WIREs Climate Change 12 (6): e738. doi:10.1002/wcc.738

Möslinger, M., G. Ulpiani, and N. Vetters. 2023. "Circular Economy and Waste Management to Empower a Climate-Neutral Urban Future." *Journal of Cleaner Production* 421: 138454. doi:10.1016/j.jclepro.2023.138454

Muljaningsih, S., W. Andayani, M. Ekawaty, and D. A. Nur Asrofi. 2023. "Scenario for Mitigating Climate Change in Indonesia: Circular Economy-Based Waste Management (9r)." *IOP Conference Series: Earth and Environmental Science* 1268 (1): 012001. doi :10.1088/1755-1315/1268/1/012001

Nair, P. A., A. F. Jelince, and P. Aiswarya. 2023. "Circulomania: Rethinking Waste Through a Circular Economy Framework." In *2023 International Conference on Intelligent Technologies for Sustainable Electric and Communications Systems (iTech SECOM)*, 334–339. IEEE. doi:10.1109/iTechSECOM59882.2023.10435128

Najmi, A. 2024. "Innovative Materials and Techniques for Sustainable Building Structures." *International Journal of Scientific Research in Engineering and Management* 8 (4): 1–5. doi:10.55041/IJSREM30133

Nguyen, M.-H., Q. L. Nguyen, V.-P. La, M. I. Bhatti, and V.-Q. Hoang. 2023. "Enterprise's Strategies to Improve Financial Capital under Climate Change Scenario – Evidence of the Leading Country." doi:10.21203/rs.3.rs-2800372/v1

Niles, M. T., R. Ahuja, T. Barker, J. Esquivel, S. Gutterman, M. C. Heller, N. Mango, et al. 2018. "Climate Change Mitigation beyond Agriculture: A Review of Food System Opportunities and Implications." *Renewable Agriculture and Food Systems* 33 (3): 297–308. doi:10.1017/S1742170518000029

Ogunmodede, P. 2024. "The Potential for Disruptive Technologies to Mitigate and Adapt to Climate Change." ScienceOpen Preprints. doi:10.14293/PR2199.000657.v1

Owusu, P. A., and S. Asumadu-Sarkodie. 2016. "A Review of Renewable Energy Sources, Sustainability Issues and Climate Change Mitigation." *Cogent Engineering* 3 (1): 1167990. doi:10.1080/23311916.2016.1167990

Paes, M. X., J. A. Puppim De Oliveira, S. D. Mancini, and J. Rieradevall. 2024. "Waste Management Intervention to Boost Circular Economy and Mitigate Climate Change in Cities of Developing Countries: The Case of Brazil." *Habitat International* 143: 102990. doi:10.1016/j.habitatint.2023.102990

Panepinto, D., V. A. Riggio, and M. Zanetti. 2021. "Analysis of the Emergent Climate Change Mitigation Technologies." *International Journal of Environmental Research and Public Health* 18 (13): 6767. doi:10.3390/ijerph18136767

Probst, B., S. Touboul, M. Glachant, and A. Dechezleprêtre. 2021. "Global Trends in the Innovation and Diffusion of Climate Change Mitigation Technologies." doi:10.21203/rs.3.rs-266803/v1

Purkayastha, D., and R. Sarkar. 2021. "Getting Financial Markets to Work for Climate Finance." *The Journal of Structured Finance* 27 (2): 27–41. doi:10.3905/jsf.2021.1.122

Reaves, D. R. 2019. "Successful Climate Change Strategies in Corporate Farming." *Journal of Social Change* 11 (1). doi:10.5590/JOSC.2019.11.1.05

Rogalev, Nikolay, Anton Rogalev, Vikas Kindra, Oleksii Zlyvko, and Pavel Bryzgunov. 2022. "Review of Closed SCO2 and Semi-Closed Oxy–Fuel Combustion Power Cycles for Multi-Scale Power Generation in Terms of Energy, Ecology and Economic Efficiency." *Energies* 15 (23): 9226. doi:10.3390/en15239226

Sapkota, T. B., S. H. Vetter, M. L. Jat, S. Sirohi, P. B. Shirsath, R. Singh, H. S. Jat, P. Smith, J. Hillier, and C. M. Stirling. 2019. "Cost-Effective Opportunities for Climate Change Mitigation in Indian Agriculture." *Science of the Total Environment* 655: 1342–1354. doi:10.1016/j.scitotenv.2018.11.225

Sethi, M., W. Lamb, J. Minx, and F. Creutzig. 2020. "Climate Change Mitigation in Cities: A Systematic Scoping of Case Studies." Environmental Research Letters 15 (9): 093008. doi:10.1088/1748-9326/ab99ff.

Sinha, A., S. Jha, and A. Amritesh. 2019. "Green Agrevolution Pvt Ltd: Delivering 360° 'Seed-to-Market' Solution." *Emerald Emerging Markets Case Studies* 9 (1): 1–23. doi:10.1108/EEMCS-06-2018–0148

Sundaram, A. K. 2023. Chapter 2: Business and Climate Change. In: A. K. Sundaram & R. G. Hansen (Eds.), *Forthcoming in the Handbook of Business and Climate Change*. Edward Elgar Publishing. https://www.elgaronline.com/edcollchap/book/9781839103001/book-part-9781839103001-8.xml.

Tereshchenko, E., A. Happonen, J. Porras, and C. A. Vaithilingam. 2023. "Green Growth, Waste Management, and Environmental Impact Reduction Success Cases from Small and Medium Enterprises Context: A Systematic Mapping Study." *IEEE Access* 11: 56900–56920. doi:10.1109/ACCESS.2023.3271972

Tóth, B. 2023. "On Social Scientific Approaches to the Concept of Climate Change." *Belvedere Meridionale* 35 (1): 5–13. doi:10.14232/belv.2023.1.1

15 Environmental Education as a Key Solution for Climate Change

V. S. Khandagale and Amanullah Adeel

15.1 INTRODUCTION TO CLIMATE EDUCATION

The role of education is vital in fostering long-lasting and effective climate action. However, climate change is negatively affecting educational outcomes. As the primary source of funding for education and the leading multilateral provider of climate action support in developing nations, the World Bank aims to leverage educational resources to address climate change mitigation and adaptation strategies (Tripathy et al. 2024). Environmental education raises awareness about climate change, its causes, and effects, which is crucial for informed decision-making (Tripathy et al. 2024). Studies show that individuals exposed to environmental education demonstrate greater consciousness regarding their environmental choices (Dahiya and Mishra 2023). Education encourages the adoption of sustainable practices, fostering innovation and solutions to mitigate climate impacts (Priatna and Khan 2024). Integrating climate change education across various disciplines can enhance understanding and promote action-oriented approaches (Damoah 2023). Educational institutions play a significant role in policy advocacy and community engagement, contributing to climate action initiatives (Priatna and Khan 2024). The chapter also states that climate change education can support the achievement of Sustainable Development Goals (SDGs), particularly SDG 13, by promoting ethical and sustainable values. While environmental education is a powerful tool for climate action, challenges remain, such as climate change scepticism and the need for global cooperation. Addressing these obstacles is essential for maximizing the impact of educational initiatives on climate change mitigation (Ikporukpo 2023).

Integrating climate education into mainstream curricula will transform climate-related thinking and actions, affecting not only students but also their families, local communities, and governing bodies. This approach will additionally enhance overall readiness and adaptability to climate-related challenges across diverse populations. In this context, several studies have attempted to raise information on climate crisis. For example, Velasco-Martínez et al. (2020) demonstrated how learners are persuaded that embracing small eco-friendly practices (such as switching off lights, disconnecting electronic devices, opting for public transportation, and utilizing solar and wind energy for electricity generation) can contribute

DOI: 10.1201/9781003492733-15

to addressing climate change. The Environmental Education component of the EU Project "I-CHANGE" (https://ichange-project.eu/) focuses on fostering global citizenship, with the aim of creating new knowledge and more engaged informed methods of interacting with the environment. Munerol et al. (2024) examined the effects of this study and reported that, even after 15–30 days, pro-environmental behaviours had increased and remained stable. Ephraim (2019) conducted interviews with individuals he claimed to have "Environmental Literacy" in a seminal study. The process of demonstrating a set of knowledge, abilities, attitudes, and mental habits enable people to relate to their surroundings in a way that preserves or restores sustainable relationships with others and the environment as a whole; the capacity and willingness to make environmental decisions are compatible with the quality of the environment and the quality of human life. This is known as environmental literacy education. In this regard, educational institutions play a pivotal role in promoting community engagement and policy advocacy related to climate change by fostering awareness, facilitating partnerships, and integrating climate education into their curricula. These institutions not only equip students with the necessary skills to address climate challenges but also serve as hubs for research and community collaboration. Universities engage in partnerships with local communities, providing hands-on experiences that bridge theoretical knowledge and practical application (Phiri 2024). Institutions advocate for sustainable policies by conducting research that informs local and national climate strategies (Kaplan 2023). They also promote student leadership and community engagement in policy discussions, ensuring diverse voices are included in climate action initiatives (Khalo and Damoah 2023).

On the other hand, the institutions face many challenges for sitting their targets for climate change combating. Effective partnerships between educational institutions and communities often face coordination issues, as stakeholders may have differing priorities and expectations (Phiri 2024). Institutions struggle to foster genuine community engagement, as many initiatives fail to connect with local identities and social networks (Ruiz-Mallén et al. 2022). Empowering community members to take active roles in climate action is challenging, particularly in rural areas where resources and planning capabilities are limited (Pattison et al. 2021). Inconsistent communication among community members and institutional representatives can hinder effective participation in climate initiatives. Despite these challenges, some institutions have successfully implemented community climate commons, which can enhance collective action by fostering social cohesion and providing necessary support structures (Nässén et al. 2024).

The primary objective of this chapter is to investigate the indispensable function of environmental education in confronting the myriad challenges posed by climate change through the enhancement of awareness, the cultivation of actionable knowledge, and the promotion of sustainable behaviours among individuals and communities. This chapter aspires to elucidate how the incorporation of climate change paradigms into educational curricula can empower citizens, policymakers, and stakeholders to make enlightened decisions and undertake proactive initiatives aimed at climate mitigation and adaptation.

15.2 CLIMATE LITERACY

Climate literacy frameworks are essential for equipping individuals with the knowledge and skills to understand and respond to climate change. Various organizations, including the United Nations Educational, Scientific and Cultural Organization (UNESCO) and the Intergovernmental Panel on Climate Change (IPCC), have developed guidelines that emphasize the integration of scientific knowledge and social understanding. These frameworks aim to foster a comprehensive approach to climate education, ensuring that learners can critically engage with climate issues. Frameworks emphasize the importance of grasping the physical science behind climate change, including natural climate variability and anthropogenic impacts (Harker-Schuch and Watson 2019; Harker-Schuch 2022). Effective climate education should promote not only knowledge but also the ability to take informed actions against climate change. A comprehensive framework for environmental literacy includes concepts, contexts, and competencies. Consensus among experts highlights the importance of defining sub-dimensions and competencies, as well as identifying responsible institutions and groups for developing environmentally literate individuals (Kaya and Elster 2019).

15.3 LINKING CLIMATE EDUCATION WITH SDGs

Their significance lies in several key aspects. Formally referred to as transforming our world agenda, the SDGs signal a significant advancement in global collaboration (Priatna and Khan 2024). These 17 interrelated goals, which were accepted by all UN members in 2015, provide a thorough framework for tackling the most important issues confronting people and the environment by 2030 (Kim 2023). They are significant in a number of important ways.

- **Holistic approach**: The SDGs transcend traditional siloed development strategies. They recognize the complex interdependencies between issues like poverty, hunger, and climate change. Progress in one area necessitates progress in others, demanding a holistic approach to achieve lasting change.
- **Universality**: Unlike previous development agendas, the SDGs are not targeted solely at developing nations. They represent a global commitment, ensuring that all countries, regardless of their current development status, work collaboratively towards a more sustainable future.
- **Ambitious realism**: The goals strike a critical balance between ambition and achievability. They are undeniably ambitious, demanding significant transformation across various sectors. However, they are also grounded in a realistic assessment of what can be accomplished through collective effort from governments, businesses, civil society organizations, and individuals.
- **Action orientation**: The SDGs are not merely aspirational statements. Each goal is accompanied by specific targets and measurable indicators to track progress. This action-oriented framework fosters accountability and enables continuous improvement efforts.

- **Catalyst for change**: The SDGs act as a powerful driver for innovation and investment. They incentivize countries and organizations to develop and implement sustainable solutions across various sectors, accelerating the pace of progress towards a more sustainable future.

In essence, the SDGs represent a shared vision for a more just, equitable, and sustainable world. By providing a comprehensive framework for collective action, they offer a roadmap for tackling our planet's most pressing challenges and creating a brighter future for all.

15.3.1 Preparing the Next Generations for Climate Change

The shadow of climate change looms large, casting its uncertainty over the world our children will inherit. The burden of addressing this crisis falls not just on our shoulders today, but on the shoulders of those who will carry the torch forward. Equipping them with the knowledge, skills, and, most importantly, the hope to navigate this complex challenge is paramount.

The first step to prepare the next generation is weaving climate literacy into the fabric of learning. Climate change education shouldn't be relegated to a single subject. Instead, let's weave it into the very fabric of learning. In science classes, explore the dynamics of our planet's climate system and the human impact on it (Chew Hung 2022). In literature, analyse dystopian narratives to spark discussions about potential futures and the importance of responsible choices. In social studies, examine historical climate shifts and their societal consequences. This holistic approach fosters a deeper understanding of the interconnectedness of environmental issues and empowers students to make informed decisions.

The second way is to spark curiosity through exploration. Facts and figures are crucial, but knowledge needs fertile ground to grow. Ignite a love for the natural world by fostering a sense of wonder. Organize nature hikes, where students can experience first-hand the intricate dance of life in a healthy ecosystem. Plant a school garden together, nurturing a connection to the source of our food and the delicate balance required for its growth. Volunteer for local environmental projects, allowing young people to witness the power of collective action in protecting their surroundings. By fostering a sense of wonder and connection with nature, we cultivate a desire to protect it.

Third is empowering through action. Knowledge is power, but action is transformative. Let's move beyond rote memorization by providing opportunities for young people to engage in solutions. Organize school clean-up drives, where students can take ownership of their environment and witness the tangible impact of their efforts. Partner with local organizations to plant trees, a symbolic act of renewal and a testament to their role in shaping the future of our planet. Encourage participation in climate marches or letter-writing campaigns, empowering them to use their voices to advocate for change (Syropoulos et al. 2023). By giving them a platform for action, we nurture a sense of agency and equip them with the tools to make a difference.

The fourth step is hope as the fuel for change. Climate change can be a daunting topic, a seemingly insurmountable obstacle. It's crucial to acknowledge the challenges but also to showcase the incredible work already happening. Highlight stories of innovation, resilience, and successful environmental initiatives. Share the tale of Greta Thunberg, a young woman whose unwavering voice ignited a global movement (Nedevska 2023). Showcase communities transitioning to renewable energy sources or pioneering sustainable agricultural practices. Let them see the power of human ingenuity and the positive ripple effects of collective action. By nurturing hope, we fuel their determination to create a better future.

The last step is leading by example: actions speak louder than words. Our children are keen observers, constantly absorbing the world around them. Make sustainable choices in your own life, demonstrating a commitment to environmental responsibility. Conserve energy at home, reduce waste by embracing reusable alternatives, and support eco-friendly businesses. Let your actions speak louder than words, inspiring them to follow suit. By living a sustainable lifestyle, we set a powerful example and encourage them to become responsible stewards of the planet.

Preparing the next generation for climate change isn't just about equipping them with knowledge; it's about cultivating hope, fostering a sense of ownership over the environment, and empowering them to take action. By weaving climate literacy into education, sparking curiosity through exploration, and providing opportunities for action, we can empower them to become the architects of a sustainable future (Razzaq et al. 2021). Let's work together to ensure they inherit a planet not just teeming with life, but one where they have the tools and the will to create a lasting legacy for generations to come.

15.4 NET ZERO AS A PATH TO A 1.5°C WORLD

Net-zero emissions represent a state of carbon neutrality within the global carbon cycle. Essentially, it signifies reaching equilibrium where anthropogenic greenhouse gas (GHG) emissions are balanced by an equivalent amount of removals through natural or technological solutions (Schumer 2024). This doesn't necessarily mean complete elimination of emissions; residual emissions can be offset by an equivalent amount of carbon dioxide removal. The goal is to stop the net accumulation of GHGs in the atmosphere, effectively halting the further acceleration of climate change. Figure 15.1 shows a few steps toward net zero strategy at 1.5°C in the future.

- **Multi-pronged approach to emission reduction**: The primary focus is on drastically reducing GHG emissions across all sectors – energy, transportation, industry, agriculture, and land-use change. This necessitates a rapid transition towards cleaner energy sources like renewables, significant improvements in energy efficiency, and widespread adoption of sustainable practices. Imagine a transformation of our cities into hubs of green transportation with electric vehicles and improved public transit systems. Envision industries embracing circular economy principles, minimizing waste, and maximizing resource efficiency (Teske et al. 2023). Picture agricultural practices that prioritize soil health, reduce reliance on chemical inputs, and promote carbon sequestration.

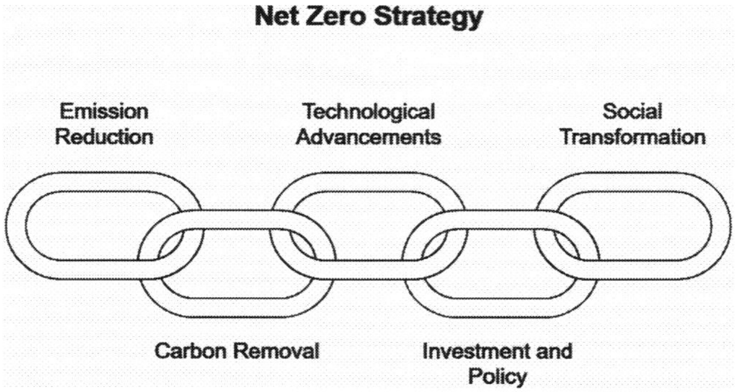

FIGURE 15.1 Net-zero strategies.

- **Complementary role of carbon removal strategies**: Even with significant emission reductions, some emissions will inevitably remain. This is where carbon removal strategies become crucial. These encompass both natural solutions like afforestation and ecosystem restoration, alongside technological options such as direct air capture and storage. Think of vast swathes of degraded land being revitalized through scientific reforestation programmes, creating vital carbon sinks (Dafnomilis et al. 2024). Imagine innovative technologies capturing carbon dioxide directly from the atmosphere, either for storage underground or for potential future utilization in sustainable applications.
- **Challenges and opportunities**: **a global transformation**: The transition to a net-zero economy is a global undertaking, presenting both challenges and opportunities.
- **Technological advancements**: **the engine of innovation**: Developing and scaling up carbon capture technologies on a large scale is essential for achieving complete net-negative emissions. Imagine a future where captured carbon dioxide is utilized in the production of sustainable fuels or building materials, creating a truly circular carbon economy.
- **Investment and policy frameworks**: **enabling change**: Significant investments in clean energy infrastructure, research, and development are required. Additionally, supportive policies and regulations can incentivize the shift towards a low-carbon economy and create a level playing field for sustainable practices. Governments can play a pivotal role by setting ambitious yet achievable net-zero targets, providing financial incentives for clean energy adoption, and streamlining regulations to expedite the transition.
- **Social transformation a collective effort**: Transitioning to net zero necessitates a collective effort and social transformation. Individuals and communities can contribute by adopting sustainable lifestyles, reducing consumption, and supporting businesses that are committed to net zero goals. Imagine a future where energy conservation becomes a societal norm, where consumers actively seek out eco-friendly products and services, and where communities collaborate on local renewable energy projects.

- **A Shared responsibility for a sustainable future**: Net zero represents a global challenge, but achieving it is no longer optional. By working collaboratively as a global community – scientists, policymakers, businesses, and individuals alike – we can chart a course towards a net-zero future that secures a 1.5°C world for generations to come. This future will be powered by clean energy, driven by innovation, and built on a shared responsibility for safeguarding our planet's well-being.

15.5 TEACHING CLIMATE LITERACY FOR BEHAVIOUR CHANGE AND COLLECTIVE ACTION

Climate change, a defining challenge of our era, necessitates a response that transcends mere knowledge acquisition. Higher education institutions are uniquely positioned to equip future generations with the skills and understanding to navigate this complex crisis. This requires a shift from traditional climate science education towards a curriculum that fosters climate literacy for behaviour change and collective action (Figure 15.2).

15.5.1 Reframing Climate Education for Transformation

- **From passive learning to active engagement**: Climate literacy education needs to move beyond information transmission. Course design should encourage active learning through real-world case studies, scenario planning exercises, and simulations. These activities can challenge students to

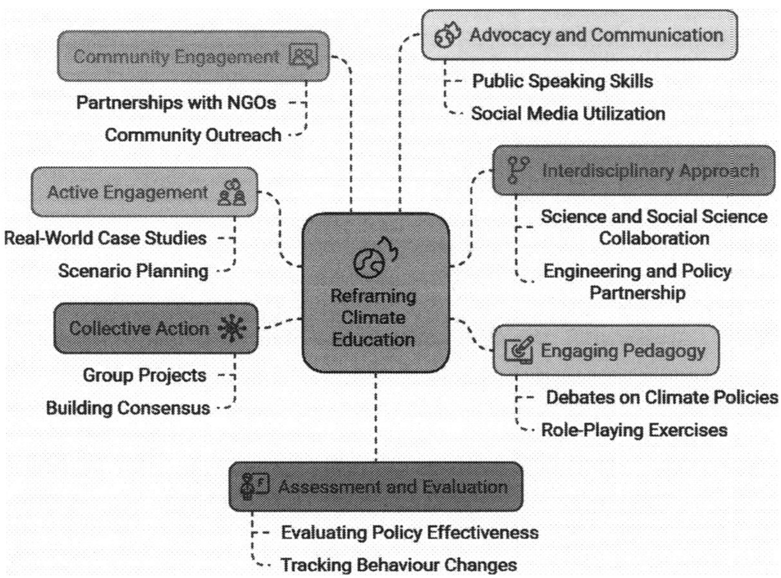

FIGURE 15.2 Reframing climate education.

critically analyse solutions, adaptation strategies, and the ethical considerations surrounding climate action (Papagiannaki et al. 2024). For instance, a business school course could incorporate case studies of corporations implementing sustainable practices, allowing students to analyse the financial and environmental benefits of such initiatives.

- **Interdisciplinary: a catalyst for holistic understanding:** Climate change is a multifaceted issue with profound social, economic, and political dimensions. To foster a holistic comprehension, higher education must encourage collaboration across disciplines. A science course could partner with a social science course to develop communication strategies for climate action campaigns. Engineering students could collaborate with policy students to create and advocate for sustainable infrastructure development plans.
- **Engaging pedagogy igniting curiosity and critical thinking**: Moving beyond traditional lectures, the curriculum should embrace interactive and engaging pedagogy. Debates on climate policies, role-playing exercises simulating international negotiations, or service-learning projects can be powerful tools. Imagine a course where students work with local NGOs to develop educational materials promoting sustainable waste management in schools and communities (Tang 2022). These engaging approaches can ignite curiosity, cultivate critical thinking skills, and encourage a deeper understanding of climate issues.
- **Empowering collective action for a sustainable future: building synergy and consensus:** Encouraging group projects where students develop and present solutions to specific climate challenges can be highly effective. This fosters teamwork, negotiation skills, and the ability to build consensus – essential for successful collective action.
- **Community engagement: bridging the gap between Theory and Practice:** Breaking down the walls between academia and the outside world is crucial. Partner with local NGOs or environmental organizations to create opportunities for students to engage in community outreach and co-create solutions with local stakeholders. Imagine a project where environmental engineering students collaborate with a coastal community to develop climate-resilient infrastructure solutions.
- **Advocacy and communication**: Integrate public speaking and communication skills development into climate-related courses. Workshops on crafting compelling narratives and utilizing social media platforms can empower students to amplify climate awareness and mobilize action.
- **Assessment and evaluation**: **measuring impact beyond the classroom**: Evaluation should shift beyond traditional exams focused on memorization and incorporate assessments that gauge students' ability to apply their knowledge and skills to real-world climate solutions. This could involve evaluating the effectiveness of a proposed environmental policy or assessing the sustainability of a business model.
- **Behaviour change as a key outcome**: Consider incorporating metrics to track behaviour changes beyond the classroom. Pre- and post-course surveys can assess shifts in attitudes and self-reported behaviour changes

regarding sustainable practices. This feedback loop allows educators to refine their teaching methods and measure the long-term impact of climate literacy education.

- **Cultivating a generation of change maker**: Integrating climate literacy for behaviour change and collective action is about more than simply adding content to existing curricula; it's about reframing the purpose of higher education. By embracing these strategies, we can cultivate a generation of informed, engaged, and action-oriented citizens, equipped to tackle the climate challenge together. Let us not merely graduate students; let us cultivate a generation of change makers.

15.6 CULTIVATING SKILLS FOR A "JUST TRANSITION" TO A GREEN ECONOMY

The transition to a green economy, while essential for environmental sustainability, presents potential challenges like job displacement and economic disruption in certain sectors. As educators and curriculum framers, we have a critical responsibility to ensure a just transition – one that prioritizes equity, social justice, and a smooth adaptation for the workforce. This framework outlines a comprehensive approach to integrate "just transition" skills development within both K-12 and higher education systems, equipping students with the knowledge, skillset, and values needed to navigate this transition effectively and contribute positively to a sustainable future (Figure 15.3).

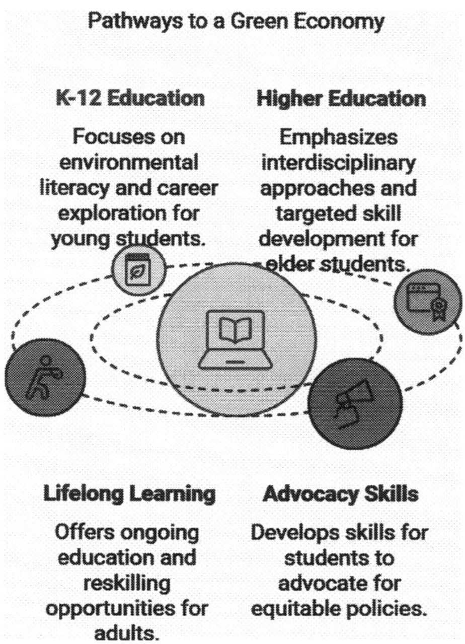

FIGURE 15.3 Ways to green economy.

15.6.1 K-12 Education (Ages 5–14)

- **Fostering environmental literacy and empathy**: Environmental education should be seamlessly woven throughout the curriculum. Interactive lessons and engaging activities highlight the interconnectedness of nature and human well-being. Discussions on climate change, resource scarcity, and the need for a sustainable future cultivate environmental awareness and a sense of responsibility. Additionally, promoting empathy for all living things and the communities most vulnerable to climate impacts fosters a sense of global citizenship and social justice. Integrate children's literature that explores themes of environmental stewardship and social responsibility. Encourage students to connect local environmental issues with global challenges, sparking a sense of agency and an understanding of their role in creating positive change.
- **Project-based learning for practical sustainability solutions**: Hands-on learning experiences are crucial for developing critical thinking and problem-solving skills. Project-based learning allows students to tackle local environmental challenges directly. This could involve initiatives such as planting trees in school grounds, designing energy-efficient buildings through model building, or conducting community waste audits. These practical projects empower students to become active agents of change while nurturing a sense of ownership for environmental solutions. Consider partnering with local environmental organizations to identify relevant projects and ensure real-world applicability.
- **Career exploration with a sustainability lens**: Introduce students to diverse career paths within the green economy (renewable energy, sustainable agriculture, clean technology) from a young age. Highlighting the skills and knowledge required for these jobs helps students explore future possibilities aligned with environmental consciousness. Consider inviting guest speakers from various green sectors to share their career journeys and inspire students with real-world examples. This early exposure allows students to make informed decisions about their academic pursuits and career paths, aligning their education with the demands of a green economy.

15.6.2 Higher Education (Ages 15+)

- **Interdisciplinary approach to "just transition"**: Create dedicated courses or modules that explore the socioeconomic dimensions of climate change alongside the science. Integrate topics like social justice, worker rights, and community resilience into discussions about emission reduction strategies and green technologies. This interdisciplinary approach fosters a holistic understanding of the complexities of climate change and the transition to a green economy. Encouraging collaboration between social sciences, economics, and environmental studies departments further strengthens this comprehensive learning experience. Partner with social justice organizations to ensure these courses address the concerns of marginalized communities most impacted by environmental and economic change.

- **Targeted green skills development**: Offer specialized programmes and workshops that equip students with the specific skills needed for green jobs. These skills might encompass renewable energy installation and maintenance, sustainable agricultural practices, eco-design principles, green building technologies, and environmental data analysis. Collaboration with relevant industries can ensure programmes provide the most up-to-date and practical skillsets for the current green job market. Consider offering industry certifications alongside academic degrees, enhancing employability and ensuring graduates are well-prepared for the specific demands of green jobs.
- **Lifelong learning opportunities for upskilling and reskilling**: The workforce requires ongoing adaptation. Developing continuing education programmes and certification courses caters to upskilling and reskilling existing professionals for the green economy. Offering these programmes in flexible formats, such as online or evening courses, ensures accessibility for working adults. Partner with local community colleges or technical training institutes to expand reach and cater to the needs of diverse learners.
- **Case studies and industry collaborations**: Integrate real-world case studies into curriculum to showcase successful examples of just transition initiatives. Partnering with green businesses and organizations can provide internship opportunities and guest lectures. Exposing students to the practical application of sustainability principles and the realities of the green workforce fosters a deeper understanding of the path forward. Encourage student participation in industry-sponsored design competitions or hackathons, allowing them to propose solutions to real-world sustainability challenges faced by businesses.
- **Advocacy and activism skills development**: Empower students to become advocates for a just transition. Train them in communication and negotiation skills, enabling them to advocate for policies that promote green jobs, protect worker rights, and ensure a fair and equitable transition away from fossil fuel dependence. Engaging in mock policy debates, simulations of community engagement exercises, or volunteer work with environmental justice organizations can provide valuable practical experience.

Climate change presents an existential threat, demanding ongoing adaptation from communities across the globe. Environmental education plays a critical role in fostering resilience. By equipping individuals and communities with the knowledge, skills, and resources necessary to understand and respond to these evolving challenges, environmental education empowers them to navigate the complexities of climate change and adapt effectively.

15.7 BUILDING THE CAPACITY TO ADAPT

Effective adaptation hinges on a deep understanding of local climate risks. Environmental education programmes must move beyond broad climate science overviews and localize the message. Utilizing data visualization tools, interactive exercises, and geospatial mapping can highlight local climate projections, potential impacts like rising sea levels or extreme weather events, and the specific vulnerabilities faced by different communities.

For instance, consider interactive web applications that allow users to explore projected changes in temperature and precipitation patterns specific to their postal code. This approach fosters a sense of ownership and empowers communities to identify areas where adaptation is most needed, such as flood-prone neighbourhoods or communities heavily reliant on rain-fed agriculture. Environmental education should not be a passive exercise; it should equip individuals with actionable knowledge. Training programmes can focus on flood preparedness, heat stress management, drought-resistant agricultural practices, and sustainable water management techniques. Consider incorporating workshops on building climate-resilient infrastructure tailored to specific needs. In coastal communities, workshops could focus on constructing seawalls using locally available materials (Jack et al. 2024). In drought-prone regions, rainwater harvesting and soil conservation techniques could take centre stage.

15.7.1 Fostering Collaborative Innovation and Resource Management

Climate change disrupts established systems, demanding innovative solutions. Environmental education programmes can bridge the gap between traditional ecological knowledge (TEK) held by indigenous communities and modern scientific understanding. Facilitating knowledge exchange workshops can promote the integration of TEK into adaptation strategies. This fosters a more holistic approach, leveraging the wisdom of past generations alongside the advancements of modern science. Imagine workshops where elders from indigenous communities share their knowledge of drought-resistant crops alongside agricultural scientists exploring the development of new climate-resilient crop varieties.

Resource scarcity challenges are often exacerbated by climate change. Environmental education programmes can equip communities with the knowledge and skills to manage their resources sustainably. This could involve training on water conservation techniques like drip irrigation, greywater reuse, and rainwater harvesting systems. Additionally, programmes can promote soil health improvement strategies to reduce erosion and increase water retention capabilities. Consider incorporating hands-on training sessions on building and maintaining community gardens that demonstrate efficient water usage and sustainable food production methods, promoting local food security in the face of potential disruptions to traditional agricultural practices.

15.8 THE CHALLENGES AT HAND, AND HOW THE EDUCATION SECTOR CAN HELP EMPOWERING PARTICIPATION AND DECISION-MAKING: BUILDING COLLECTIVE ACTION

Successful adaptation requires not just knowledge and skills, but also active participation from local communities. Environmental education programmes can become a catalyst for community engagement. By facilitating workshops, roundtable discussions, and citizen science initiatives, programmes can encourage community members to identify local vulnerabilities, brainstorm solutions, and develop adaptation plans tailored to their specific needs. This fosters ownership and empowers communities to take charge of their future and advocate for their needs in regional and national adaptation strategies. Imagine community forums where residents discuss potential solutions to rising sea levels, such as constructing elevated walkways or reinforcing coastal defences with locally sourced materials.

Environmental education should equip individuals with the skills to communicate effectively and collaborate with diverse stakeholders. Training workshops can focus on negotiation skills, conflict resolution strategies, and public speaking techniques. This empowers communities to engage with policymakers, advocate for their adaptation needs, and foster collaboration with other stakeholders like NGOs and scientists. Imagine workshops on effective communication strategies, equipping community leaders with the skills to present their adaptation plans to local government officials or potential donors, securing the resources necessary to implement their plans.

15.8.1 Leveraging Technology for Accessible Capacity-Building: Knowledge at Your Fingertips

The digital revolution presents exciting possibilities for environmental education. Online learning platforms can offer accessible educational resources and training modules on climate adaptation strategies in various languages. Develop mobile applications with localized climate data, early warning systems for extreme weather events, and information on available resources and support programmes. Imagine a mobile app that not only provides real-time weather updates but also offers localized agricultural advice based on current and projected climate conditions, empowering farmers to make informed decisions about crop selection and planting schedules.

Digital storytelling techniques, like documentaries or interactive maps, can showcase successful adaptation strategies implemented in other regions. Organize online forums and virtual conferences to foster knowledge exchange and collaboration between communities across geographical boundaries. Imagine an interactive map showcasing successful adaptation projects around the world, allowing communities to learn from each other's experiences and adapt existing solutions to their own unique contexts.

15.8.2 Building a Culture of Lifelong Learning: Continuous Adaptation

Climate change is a dynamic challenge, demanding continuous learning and adaptation. Environmental education programmes should foster a culture of lifelong learning. This can be achieved by integrating climate literacy into adult education programmes, vocational training courses, and community outreach initiatives. Additionally, establishing partnerships with local universities and research institutions can provide access to ongoing research and arrive at a contextual solution.

The environmental challenges of our time – climate change, biodiversity loss, resource depletion – demand a collective response of unprecedented scale. We need to believe the education sector holds the key to unlocking a more sustainable future. However, to fully harness this potential, we must acknowledge the existing hurdles and chart a course that fosters critical thinking, empowers action, and cultivates a sense of shared responsibility for our planet.

15.9 CONCLUSION

By integrating climate change education into the curriculum, engaging students in hands-on projects, providing continuous teacher training, involving the community, and using innovative teaching methods, schools can play a pivotal role in enhancing awareness and technical knowledge about climate change. This comprehensive approach not only educates students about the science and impacts of climate change but also empowers them to act and become advocates for a sustainable future.

REFERENCES

Chew Hung, Chang. 2022. *Climate Change Education: Knowing, Doing and Being*. 2nd ed. London: Routledge. doi:10.4324/9781003093800

Dafnomilis, Ioannis, Michel Den Elzen, and Detlef Van Vuuren. 2024. "Paris Targets within Reach by Aligning, Broadening and Strengthening Net-Zero Pledges." *Communications Earth & Environment* 5 (1): 48. doi:10.1038/s43247-023-01184-8

Dahiya, Manju, and Naman Mishra. 2023. "A Study Regarding the Effectiveness of Environmental Education on Climate Change." In *The Impact of Climate Change and Sustainability Standards on the Insurance Market*, edited by Kiran Sood, Simon Grima, Peter Young, Ercan Ozen, and Balamurugan Balusamy, 1st ed., 165–177. Wiley. doi:10.1002/9781394167944.ch10

Damoah, Benjamin. 2023. "Reimagining Climate Change Education As a Panacea to Climate Emergencies." *International Journal of Environmental, Sustainability, and Social Science* 4 (4): 977–987. doi:10.38142/ijesss.v4i4.590

Ephraim, Kpeno Amon. 2019. "Environmental Literacy Education: A Means for Curbing the Effect of Climate Change." *African Research Review* 13 (4): 144. doi:10.4314/afrrev.v13i4.13

Harker-Schuch, Inez. 2022. "Defining Climate Literacy: Developing a Working Definition on What It Means to Be Climate Literate." doi:10.5194/egusphere-egu22-2302

Harker-Schuch, Inez, and Michel Watson. 2019. "Developing a Climate Literacy Framework for Upper Secondary Students." In *Climate Change and the Role of Education*, edited by Walter Leal Filho and Sarah L. Hemstock, 291–318. Climate Change Management. Cham: Springer International Publishing. doi:10.1007/978-3-030-32898-6_17.

Ikporukpo, Chris O. 2023. "Environmental Deterioration and Environmental Education: The Challenge of Climate Change." *Advances in Social Sciences Research Journal* 10 (9): 363–381. doi:10.14738/assrj.109.15538.

Jack, Helen E., Ali Giusto, Alexandra L. Rose, Rukudzo Mwamuka, Imani Brown, Tarisai Bere, Ruth Verhey, et al. 2024. "Mutual Capacity Building Model for Adaptation (MCB-MA): A Seven-Step Procedure for Bidirectional Learning and Support during Intervention Adaptation." *Global Health Research and Policy* 9 (1): 25. doi:10.1186/s41256-024-00369-8.

Kaplan, Mitchell Alan. 2023. "The Role of Institutions of Higher Education in Promoting the Development of Sustainable Green Energy Policies That Facilitate Climate Justice and Environmental Equality:" In *Advances in Educational Marketing, Administration, and Leadership*, edited by Gamze Sart, 26–47. IGI Global. doi:10.4018/978-1-6684-8356-5.ch002

Kaya, Volkan Hasan, and Doris Elster. 2019. "A Critical Consideration of Environmental Literacy: Concepts, Contexts, and Competencies." *Sustainability* 11 (6): 1581. doi:10.3390/su11061581

Khalo, Xolani, and Benjamin Damoah. 2023. "Reinvigorating Climate Change Education in Universities a Social Transformative Agenda." *Environmental Science & Sustainable Development*, September, 19–26. doi:10.21625/essd.v8i4.1013

Kim, Yunji. 2023. "A Case Study of University Education for Sustainable Development Linked to SDGs." *Research Institute of Education Science, Jeju National University* 25 (3): 119–139. doi:10.15564/jeju.2023.9.25.3.119

Munerol, Francesca, Antonio Parodi, Lara Polo, Massimo Milelli, Nicola Loglisci, Nadia Rania, Fabrizio Bracco, and Ilaria Coppola. 2024. "Can Climate Change Training Promote Pro-Environmental Behavior in the Long Term? A Pilot Study with Teenagers." doi:10.5194/egusphere-egu24-6110

Nässén, Nessica, Maja Lilja, Stefan Sjöberg, and Johan Colding. 2024. "Community Climate Commons for Collective Climate Action." *Sustainable Development*, October, sd.3227. doi:10.1002/sd.3227

Nedevska, Jasmina. 2023. "Climate Change and Institutions for Future Generations: The Litigation Option." In *Handbook of the Philosophy of Climate Change*, edited by Gianfranco Pellegrino and Marcello Di Paola, 1229–1245. Handbooks in Philosophy. Cham: Springer International Publishing. doi:10.1007/978-3-031-07002-0_132

Papagiannaki, Katerina, Vassiliki Kotroni, and Konstantinos Lagouvardos. 2024. "Assessing the Effects of Citizen Climate Literacy and Attitudes on Their 'Greening' Behaviour in a Climate Change Hotspot Region of the Eastern Mediterranean." *Climate* 12 (9): 146. doi:10.3390/cli12090146

Pattison, Andrew, Christopher R. Henke, and John Pumilio. 2021. "Community-Based Climate Action Planning as an Act of Advocacy: A Case Study of Liberal Arts Education in a Rural Community." *Journal of Environmental Studies and Sciences* 11 (2): 183–193. doi:10.1007/s13412-020-00655-0

Phiri, Chimwemwe. 2024. "University Community Partnerships for Climate Change Adaptation in Malawi: Exploring Contributions, Potential Opportunities and Challenges." *Scholarship of Teaching and Learning in the South* 8 (2): 6–32. doi:10.36615/sotls.v8i2.415

Priatna, Dolly, and Shujaul Mulk Khan. 2024. "The Importance of Education and Role of Educational Institutions in Climate Change Mitigation and Achieving UN SDG 13 Climate Action." *Indonesian Journal of Applied Environmental Studies* 5 (1): 1–5. doi:10.33751/injast.v5i1.10559

Razzaq, Ali, Parwinder Kaur, Naheed Akhter, Shabir Hussain Wani, and Fozia Saleem. 2021. "Next-Generation Breeding Strategies for Climate-Ready Crops." Frontiers in Plant Science 12 (July): 620420. doi:10.3389/fpls.2021.620420

Ruiz-Mallén, Isabel, Mar Satorras, Hug March, and Francesc Baró. 2022. "Community Climate Resilience and Environmental Education: Opportunities and Challenges for Transformative Learning." *Environmental Education Research* 28 (7): 1088–1107. doi:10.1080/13504622.2022.2070602

Schumer, Clea. 2024. "Net-Zero Emissions." In *Elgar Encyclopedia of Climate Policy*, edited by Daniel J. Fiorino, Todd A. Eisenstadt, and Manjyot Kaur Ahluwalia, 178–183. Edward Elgar Publishing. doi:10.4337/9781802209204.ch33

Syropoulos, Stylianos, Gordon Kraft-Todd, Andrea Mah, Ezra Markowitz, and Liane Young. 2023. "Responsibility to Future Generations: A Moral Reframing Strategy for Combatting Climate Change." doi:10.31219/osf.io/wknuq

Tang, Kuok Ho Daniel. 2022. "A Model of Behavioral Climate Change Education for Higher Educational Institutions." Environmental Advances 9 (October): 100305. doi:10.1016/j.envadv.2022.100305

Teske, Sven, Jonathan Rispler, Sarah Niklas, Maartje Feenstra, Soheil Mohseni, Simran Talwar, and Saori Miyake. 2023. "Net-Zero 1.5 °C Sectorial Pathways for G20 Countries: Energy and Emissions Data to Inform Science-Based Decarbonization Targets." *SN Applied Sciences* 5 (9): 252. doi:10.1007/s42452-023-05481-x

Tripathy, Atal Bihari, Bimal Charan Swain, and Subhashree Mishra. 2024. "Environmental Sustainability for a Sustainable Future And Role Of Education (In Climate Change Perspectives)." *Educational Administration Theory and Practices*, May. doi:10.53555/kuey.v30i5.5952

Velasco-Martínez, Leticia-Concepción, Juan-Jesús Martín-Jaime, Ligia-Isabel Estrada-Vidal, and Juan-Carlos Tójar-Hurtado. 2020. "Environmental Education to Change the Consumption Model and Curb Climate Change." *Sustainability* 12 (18): 7475. doi:10.3390/su12187475

Index

P

R

S

T

U

V

W

Made in the USA
Las Vegas, NV
19 September 2026

9c38479c-831d-4871-b70e-0b10f376b1e4R01